Cognitive Technologies

Titles in this series now included in the Thomson Reuters Book Citation Index and Scopus!

The Cognitive Technologies (CT) series is committed to the timely publishing of high-quality manuscripts that promote the development of cognitive technologies and systems on the basis of artificial intelligence, image processing and understanding, natural language processing, machine learning and human-computer interaction.

It brings together the latest developments in all areas of this multidisciplinary topic, ranging from theories and algorithms to various important applications. The intended readership includes research students and researchers in computer science, computer engineering, cognitive science, electrical engineering, data science and related fields seeking a convenient way to track the latest findings on the foundations, methodologies and key applications of cognitive technologies.

The series provides a publishing and communication platform for all cognitive technologies topics, including but not limited to these most recent examples:

- Interactive machine learning, interactive deep learning, machine teaching
- Explainability (XAI), transparency, robustness of AI and trustworthy AI
- Knowledge representation, automated reasoning, multiagent systems
- Common sense modelling, context-based interpretation, hybrid cognitive technologies
- Human-centered design, socio-technical systems, human-robot interaction, cognitive robotics
- Learning with small datasets, never-ending learning, metacognition and introspection
- Intelligent decision support systems, prediction systems and warning systems
- Special transfer topics such as CT for computational sustainability, CT in business applications and CT in mobile robotic systems

The series includes monographs, introductory and advanced textbooks, state-of-the-art collections, and handbooks. In addition, it supports publishing in Open Access mode.

Gregor Thurmair

Knowledge-Driven Multilingual Text Analysis and Transparent Information Retrieval

Language Technology for Industrial Applications

Springer

Gregor Thurmair
Munich, Bayern, Germany

ISSN 1611-2482 ISSN 2197-6635 (electronic)
Cognitive Technologies
ISBN 978-3-031-91740-0 ISBN 978-3-031-91741-7 (eBook)
https://doi.org/10.1007/978-3-031-91741-7

This Springer imprint is published by the registered company Springer Nature Switzerland AG
The registered company address is: Gewerbestrasse 11, 6330 Cham, Switzerland

Preface

Today's success of AI systems is based on large language models, computed by analysing huge amounts of textual data. Some of the results are impressive and have generated the current hype, in particular if the queries match the prepared answers; others are less impressive and sometimes even wrong. As these results are not always reliable it would be necessary to correct them. This however is not easy as they are results of computations which are not transparent to end users. This fact creates additional effort, for instance in cases where machine-translated output needs to be post-edited for a given customer or domain.

It may be an option, in order to improve and stabilise the computation results, to consider the specific properties of their input data, i.e. the texts and their language. It makes a difference whether the input of computation is just any string the machine can find, or the input is pre-processed, by providing linguistically processed data, like creating lemmata instead of textforms, removing spelling errors, collapsing multiword terms, splitting clitics and compounds, considering specification relations, and the like. If not considered, even in cases of huge data volumes the quality of the results deteriorates, and noise is introduced into the systems, which sometimes disturbs the overall results. The techniques required for such linguistic (pre)processing are presented in this book.

A focus of the book is multilinguality. All linguistic components must be able to support Transparent Information Retrieval (TIR): This goes beyond cross-language retrieval (CLIR) as it requires to return document hits in the same language as the query was: For end users it should be transparent in which language the document which answers their search request was written. Obviously such a requirement must include two translation steps. These translation steps, query translation and document re-translation, must satisfy specific conditions and should include all existing linguistic knowledge, collected in a multilingual network of concepts.

The following book presents all components and knowledge sources needed for Transparent Information Retrieval. It deals with all language components required, what their challenges are, and how they can be implemented. It starts from the analysis of the (text) data and collects the results of the text analysis in a conceptual network, which is an index extended by both monolingual and multilingual relations, this way enabling Transparent Information Retrieval.

The task requires to analyse the data on which the computational analysis is supposed to work in much more detail, identify rules and exceptions, and special cases, with the intention to have a higher-quality output for each analysis component, leading to a higher overall quality. Many sections of the book have evaluation results proving this claim.

This book deals with Language Engineering and also supports readers interested in developing software components for text analysis.

In a bottom-up approach, it investigates all text analysis components required for multilingual indexing. It starts from preparatory work like language and topic identification, continues with sentence splitting and tokenisation (including Chinese), and describes lexical analysis, including multiword entries and Named Entities. Entries are then disambiguated both on syntactic (by a tagger) and semantic level (by multilingual word sense disambiguation).

The interoperability between the components is ensured by a common knowledge representation in form of a document model, and a common linguistic description inventory, in form of both a 3-tier multilingual tagset and a corresponding feature-value representation.

The analysis results are collected in a dynamic multilingual ConceptNet, an index structure extended by monolingual relations (like synonyms or head-modifier links) as well as multilingual ones (translations). This ConceptNet is used in retrieval for query expansion and translation and for document retranslation and ranking. This way a Transparent Information Retrieval can be built: Search and result presentation in the users' native language, while the documents may be written in a different language.

The book concludes with a description of components needed to build the required resources, like crawlers, bilingual term extraction, and tools for defaulting linguistic annotations.

For each component, readers will find a technology overview, a discussion of its main challenges in computational treatment, a description of the technical solution selected, and evaluation information. Some 'non-standard' languages like Turkish, Arabic, Persian, and Chinese are also included, and innovative solutions, e.g. for tokenisation, lexical analysis, tagging etc. are presented.

The different chapters of this book have been developed in the context of a project called TINA. TINA (German 'TextINhaltsAnalyse', text content analysis) was a long term project for text analysis in German and other languages. The components of the TINA technology have been developed in an industrial context, as part of R&D departments of Language Technology companies (Siemens, Lernout&Hauspie, Linguatec, Lucy Software). This fact has influenced their design.

All components presented in the following chapters have been implemented and tested, albeit not in all languages. They were used in several industrial applications, be it in products or in customer-related projects, often in contexts of translation or multilingual information extraction, but also in stand-alone applications. So there is a certain amount of experience behind the descriptions given in this book.

The descriptions of the system components obviously follow the knowledge-based paradigm. They result from specifications how such components should be designed and implemented, to produce the best possible results in the different user setups. Corpus-based methods are often required to achieve this goal; however the driving force has always been the human expertise. This is the main difference to data-driven methods.

Of course such projects could not be carried out by a single person. Special thanks goes to Christoph Schwarz, Alois Baumer, Gerda Ruge, Ute Ziegenhain, Juan Alberto Alonso, Vera Aleksić, Christoph Goller, Thilo Will, Mehr Newid. Thanks also to Roberto Cencioni, Bernard Olthues, Stephan Bodenkamp, Daniel Grasmick, Nicoletta Calzolari, Nuria Bel, Bente Maegaard, and Stelios Piperidis for their support and consultancy.

Munich, Germany Gregor Thurmair

Competing Interests The author has no competing interests to declare that are relevant to the content of this manuscript.

Contents

1 System Design 1
1.1 Outline 1
1.2 TINA System Architecture 3
1.2.1 System Requirements 3
1.2.2 TINA System Architecture 8
1.2.3 TINA Component Architecture 11
References 12

2 TINA Analysis Strategy 15
2.1 Rationale 15
2.2 Deep Versus Shallow Analysis 16
2.2.1 Deep Analysis (LtDeep) 16
2.2.2 LtShallow 24
2.3 Interoperability 26
2.3.1 The LtDoc Document Model 27
2.3.2 The LtTagset 37
References 67

3 Text Analysis Preprocessing 71
3.1 Text Handling (LtText) 71
3.2 Language Identification (LtLanguageIdentifier) 74
3.2.1 Setup 74
3.2.2 LtLanguageIdentifier 76
3.3 Topic Identification (LtTopicIdentifier) 81
3.3.1 Rationale 81
3.3.2 LtTopicIdentifier 83
References 90

4 Text Segmentation 93
4.1 Sentence Splitting (LtSSplit) 93
4.1.1 Rationale 93
4.1.2 LtSsplit 96

4.2 Tokenisation (LtTokeniser) 101
4.2.1 Rationale 101
4.2.2 LtTokeniser 107
4.3 Normalisation (LtNormaliser) 113
4.3.1 String Normalisation 114
4.3.2 Spelling Correction (LtSpeller) 116
References 118

5 Lexical Analysis 121
5.1 Design 122
5.2 Shallow Analysis (LtLemmatiser) 125
5.2.1 Dictionary Lookup (Chinese) 125
5.2.2 Lemmatisation (LtLemmatiser): English, German, Italian, Spanish, French, Portuguese, Russian, Arabic 128
5.2.3 Clitic Elements (LtClixer): English, French, Italian, Spanish, Arabic 134
5.2.4 Decomposition (LtDecomposer): German 137
5.2.5 Local Defaulting (LtDefaulter) 143
5.2.6 Large Scale Test: English, German 147
5.3 Lexical Analysis with Segmentation: Turkish 150
5.3.1 Existing Approaches 152
5.3.2 Lexical Analyser Turkish 153
5.3.3 Evaluation 157
5.4 Deep Lexical Analysis: Persian, Pashto 161
5.4.1 Context 162
5.4.2 Lexical Analysis Persian and Pashto 163
5.4.3 Evaluation of Persian and Pashto 172
5.5 Stemming (LtStemmer) 173
5.5.1 Rationale 173
5.5.2 Definition 173
5.5.3 Approaches 174
5.5.4 LtStemmer 176
5.5.5 Stemming Results 177
5.5.6 Evaluation 178
5.5.7 Clustering 179
References 181

6 Special Entries 187
6.1 Multiword Recognition 187
6.1.1 Rationale 188
6.1.2 Description 189
6.1.3 Analysis of Multiwords 199
6.2 Named Entity Recognition (LtNamer) 207
6.2.1 Design 208

6.2.2 Analysis . . . 214
6.2.3 Test and Evaluation, Domain Adaptation. . . . 225
References. . . . 233

7 **Disambiguation** 239
7.1 Syntactic Disambiguation (LtTagger) 239
7.1.1 Design 240
7.1.2 LtTagger Rules. . . . 242
7.1.3 LtTagger Component 248
7.1.4 Tests 249
7.2 Semantic Disambiguation (LtSensExtract) 252
7.2.1 Sense Definition. . . . 252
7.2.2 Features for WSD Analysis 255
7.2.3 LtSensExtract. . . . 258
7.3 Syntactic Analysis and Knowledge-Based Term Association 268
7.3.1 Comparisons. . . . 269
7.3.2 TINA Term Association 270
7.4 Output of the Document Analysis: LtDoc FOOTER. . . . 274
7.4.1 Terms 275
7.4.2 Names. . . . 276
References. . . . 277

8 **Transparent Information Retrieval (TIR) and the LtConceptNet** . . . 281
8.1 Transparent Information Retrieval (TIR). . . . 281
8.1.1 Definition 281
8.1.2 System Design 282
8.1.3 Translation Integration in CLIR. . . . 287
8.1.4 Transparent Information Retrieval (TIR). . . . 293
8.1.5 The Retrieval Resource: LtConceptNet 294
8.2 The Dynamic Multilingual LtConceptNet. . . . 296
8.2.1 LtConceptNet: Design and Implementation 297
8.2.2 LtConceptNet Layout. . . . 308
8.2.3 Updating the LtConceptNet 313
8.3 Transparent Information Retrieval Using the LtConceptNet. . . . 317
8.3.1 Support for Query Processing 317
8.3.2 Two Examples 321
References. . . . 332

9 **Resources** 339
9.1 Resource Design 339
9.1.1 Resource Requirements 339
9.1.2 Design 342
9.2 The TINA Lexical Model and Its Derivations 343
9.2.1 Basic Lemma Dictionaries (BLD) 345
9.2.2 Full Word Dictionaries (FWD) 353

9.2.3 Named Entity Gazetteer (NEG) . . . 355
9.2.4 Word Sense Repository (WSR) . . . 359
9.2.5 Bilingual Dictionary (XFR) . . . 360
9.3 Tools for Resource Production . . . 363
9.3.1 Focused Crawler (ILSP-FC) . . . 364
9.3.2 Term Extraction (LtTermExtract) . . . 366
9.3.3 Word Sense Disambiguation (LtSensExtract) . . . 379
9.3.4 Inflection Generation (LtFlexer) . . . 379
References . . . 390

Chapter 1
System Design

1.1 Outline

The basic challenge of Information Retrieval has been described as the need to match the conceptual environments of the author of a text and the person who searches for it: They use different concepts, they have different connotations and associative links, they have different intentions when writing/searching texts, they may speak different languages. The challenge for a system is to close this cognitive gap as far as possible and to help users to find the documents that match their intentions best.[1] Since the early days, it was obvious that natural language processing was an important means to support the search, and many researchers worked on natural language as a means to support the retrieval process,[2] in particular in multilingual settings.[3]

The main means to bridge the cognitive gap between text production and text search is the use of *concepts* (called *terms* in the indexing domain), extracted from the texts and collected in an index, which is a structure which will be looked up at search time. Text Content Analysis (TINA) is a family of projects to improve the text analysis and to provide additional information to the index. A closer look reveals properties of terms that need to be considered, if a good index is to be constructed:

---

[1] This has been described in the literature many times, cf. (Salton and McGill 1983), (Ingwersen 1992), (Baeza-Yates and Ribeiro-Neto 1999), (Kowalski 2011), (Sparck Jones and Willett 1997), (Liddy 2005) and many others.

[2] Examples are (Schwarz and Thurmair 1986), (Allan 2000), (Jackson and Moulinier 2002), (Brants et al. 2002).

[3] Cf. (Grefenstette 1998), (Nie 2010) (Peters et al. 2012). Details are given in Chap. 8.

G. Thurmair, *Knowledge-Driven Multilingual Text Analysis and Transparent Information Retrieval*, Cognitive Technologies,
https://doi.org/10.1007/978-3-031-91741-7_1

- Index terms belong to a certain **language**. Multilingualism is a phenomenon which needs to be addressed. Searching a Spanish data base with a Bulgarian query may not return acceptable results.
- Index terms occur in different (inflected) forms, in different spelling, in abbreviations etc.: a first task would be to **normalise** such occurrences, and make the search independent of such phenomena.
- Index terms may be **complex**. A significant part of the search vocabulary consists of multiwords (one term from several words), another part consists of compounds (like in German or Danish) (several terms in one word). There are proper nouns and expressions like dates which need to be generated before they can be searched
- Index terms may also be **ambiguous** (homonyms) and need to be disambiguated.

The task of the TINA project is to tackle such issues and provide better analysis results which should lead to better search results. The search can be supported by additional information for query analysis, query expansion, and query translation.

A first task is to identify the document text. Subtasks are text handling (removing non-textual document parts), language identification, and topic classification. This is presented in Chap. 3.

A second task is to identify the text words. Required tools are sentence splitting, tokenisation and normalisation of the tokens (including spell checking). Chapter 4 will give the details.

A third task is to identify the concepts behind the words. This includes base form production (lemmatisation), compound analysis, treatment of clitics, and lexical analysis. The tools required differ depending on the analysed language, the respective tools are presented first (Chap. 2). The analysers for the different languages are described in Chap. 5.

Special entries are analysed in Chap. 6: Multiword terms and proper names (Named Entities).

Some term candidates may be ambiguous. Their disambiguation requires contextual analysis, shown in Chap. 7: Syntactic categories (part-of-speech annotations) need to be disambiguated by syntactic analysis or tagging; semantic ambiguities must also be disambiguated by context analysis in a word sense disambiguation component.

All analysis results are integrated in a knowledge structure called LtConcepNet and are supposed to support Transparent Information Retrieval (TIR), i.e. searching a foreign language database in users' own language, and returning results again in users' own language. It should be transparent for the users that the documents are not written in their native language. This is presented in Chap. 8.

Finally, Chap. 9 discusses the language resources required for all these components, and tools to produce them.

To address such a task requires knowledge in linguistics and grammars, in computational linguistics, in (multilingual and crosslingual) indexing, in machine translation and in terminology. This may lead to terminological uncertainties: A '*term*' in indexing is quite different from a '*term*' in terminology. The following chapters try to make explicit what is meant in the respective context.

1.2 TINA System Architecture

1.2.1 System Requirements

The architecture of TINA follows design principles which are often cited, like 'modularity' or 'language-independence'. However it is worthwhile to have a closer look on what such principles mean for a multilingual text content and indexing system.

1.2.1.1 Mass Data Analysis

The indexing component must run on large-scale data. They are much larger than the corpus data used in research settings (like Europarl, Reuters, etc.). This has consequences for the TINA architecture.

Of course there are basic **software requirements**, like robustness, no high-priority bugs and the like. As far as speed is concerned, speed is still important, however it does not play the prominent role it used to have. Looking at the last 30 years of development, speed optimization attempts were always overtaken by hardware progress. Instead, TINA can integrate components now which were prohibitive to use with previous hardware configurations.

A much more serious constraint of mass data processing is that **no human intervention** is possible at runtime. The main areas of intervention could be:

- Parameter settings. Many software systems are controlled by parameters which are usually set by humans when the component is launched. This it is not an option in mass data environments: Parameters must be set by system components. The system must decide autonomously in which language an input text is, and what its topic is. If a parameter cannot be set automatically, just defaults must be used.
- User interaction. There are many places in the workflow where user interaction would help, mainly where disambiguation is required. An obvious case is spell checking: Usually the system offers alternatives for a possibly misspelled word, and users decide which one to pick. As this is not possible in mass data environments, the spell checking component can only return corrections which are absolutely safe to execute.
  On the other hand, user interaction is assumed to be necessary in search, e.g. in Transparent Information Retrieval, including query expansion and translation

- Defaulting. In case of out-of-vocabulary words the system must try to go on with some default assumptions on such words, and hope for a high-quality defaulter. Therefore efforts should be spent to develop such defaulters.

As a result, a mass data workflow requires more intelligence than a workflow which can rely on human interaction.

Large scale processing requires a standardised and robust **internal workflow**. In particular in web server contexts where different services reside on different machines[4] the workflow should be stable, with options what to do, if a component fails or is needed/not needed for some language. And special care must be taken to standardise the input to the system, in terms of character code, text handling and format conversion.

It should also be clear what a given component should contribute to the overall analysis result, and how. To support such a requirement, TINA uses a special document model, called LtDoc, as a backbone structure which maintains all information items found so far, delivers the required information to the single components, and integrates their results in the model. This document model is described in detail in Sect. 2.3.1.

1.2.1.2 Modularity

As opposed to other approaches, TINA does not assume that all languages can be analysed with the same software. But if different languages require different tools, it is important that these tools can be switched without harming the overall system. Modularity is therefore a key requirement in a multilingual setting.

Moreover, the complete workflow in TINA consists of a set of components, for example for tokenisation, normalisation, lexical analysis, tagging. Each of these tasks must be encapsulated in a special module, and for each such module there may be several instantiations depending on the language.

Each module is autonomous. There must not be any interrelation between modules except for the communication with the general manager. Modules must be able to be replaced by new versions, extensions to support new languages, or changes in their configuration, without affecting the overall system. Examples will be given where modules need to be exchanged by others depending on the language to be analysed, e.g. switching from 'shallow' to 'deep' processing.

Also, each module is responsible for its own resources. All resources needed must be provided by the module's resource manager. Furthermore, the module is responsible for calling the right worker component; this may depend on the language, on the availability of resources or on the situation in the workflow.

Each module has its own input and output data, a fixed set of parameters, a defined set of data to be used provided by the general manager, and a common error handling structure.

1.2.1.3 Language Independence

It sounds awkward for a multilingual system to require language-independence. But this requirement refers to a consideration how and with which constraints the languages should interact with the software.

[4] PANACEA travelling object, cf. (Toral et al. 2011).

It is a standard requirement that software and language resources should be separated. There are generally two types of software in systems dealing with linguistic content:

- One is completely language-independent and is only concerned with software issues, like web server programming, task scheduling like (McLaughlin and Schmidt 2001), and the like. No knowledge of computational linguistics is needed for this kind of tasks.
- The other one, sometimes called 'lingware', deals with the linguistic content. It is in this section where the requirement of language-independency is relevant.

Language-independent architecture would focus on the following aspects:

- Separation of programs and data. Linguistic resources like dictionaries, tokeniser lists, exception lists and the like should reside in files. In multilingual setups the resource management is responsible to provide the data specifically for those languages and components for which they are needed.
- In cases where several languages need different treatment for some components then at least the basic software formalisms should be available. In TINA there are components for shallow (LtShallow) and for deep processing (LtDeep), described in Sect. 2.2.
- There will always be cases, however, where some languages require special treatment which is not easy to cover by data only. This support should stay as local as possible; for instance there would be language-specific methods within the respective software classes.

Also, if only one language is to be analysed, the system should not be required to load all resources of all languages but should only provide this particular language's resources.

1.2.1.4 Common linguistic description

Language descriptions use different abstractions to describe the phenomena found in their languages: '*onomatopoetica*' would be found in the PKU Chinese tagset (Xia 2001) but not in descriptions of European languages. A comparison of the linguistic descriptions of languages has been collected in the 'World Atlas of Language Structure' (Haspelmath et al. 2005), and on the level of morphosyntactic standardisation efforts like (EAGLES 1996) provide many useful results.

Machine Translation saw an extended discussion which linguistic information was required for translation. Interfaces like Eurotra Interface Structure (Raw et al. 1988), METAL Interface Representation (Thurmair 1990), Transfer Interstructure (Alonso 1990), and above all interlingua-based systems like DLT (Schubert 1986) or UNL (AlAnsary 2014), tried to define structures capturing the results of linguistic analyses, forming an interface to (target) language generation.

A similar effort could not be observed in the field of text analysis/indexing. Many contributions are content with a tagset like the PENN treebank. Other treebanks, like TIGER (Brants et al. 2002), METU-Sabanci (Oflazer et al. 2003), or the Chinese PENN extension (Xia 2001), used their own inventory of linguistic descriptions, designed for monolingual purposes. One of the tasks in TINA was to define a level of linguistic description which covered both the linguistic requirements for an indexing component and a multilingual scenario. Based on this fact, TINA attempts a compromise between language-specific and cross-language descriptions, and tries to provide a compatible framework for linguistic description. It must support the following requirements:

- It must be as language-independent as possible, and allow for the embedding of language-specific phenomena in the general framework.
- The linguistic information must be represented in a common way, in a common categorial space, and in a common description like feature-value annotations.
- The linguistic components need a common design for all supported languages. E.g. there must be common decision on items which stay lexicalised (like lemmata, multiwords) and ones which are represented as features (like number, definiteness, modality). There must also be a common design of the output of linguistic analysis, e.g. the (tree) structures provided for harvesting Named Entities from them.
- It must support different analysis procedures, above all deep and shallow processing, and allow an interrelation of them.
- It also must be able to interact with the Document Model.

The TINA framework is based on a three-tier common tagset, which is shared by all TINA languages. It is designed such that it can support both deep and shallow linguistic processing. The TINA linguistic framework is described in Sect. 2.3.2.

1.2.1.5 Resource Management

Resource management is a central point of attention in TINA. All languages must provide significant sets of resources. Some information items overlap, like (full form) dictionaries for spell checking, for lexical analysis, or for language identification. Such items should be shared as much as possible.

In turn, the single software components should be as autonomous as possible, and take care of their own resources. The resources should be loaded once for a given language but only when a module requires them. As most of them are read-only, they can stay in a common repository as singletons.

An essential aspect is the maintenance of the resources, mainly correcting and adding lexical entries. There must be a single point of intervention for a given language. Instead of providing special purpose coding and maintenance tools, cf. (Bernardi et al. 2001), which would be rather expensive when trying to support all languages and scripts, TINA uses existing software (standard editors working on tab-delimited records) for such tasks, and compiles them into different formats as needed by the respective components.

1.2.1.6 Test and Evaluation

TINA test and evaluation operates on three levels:

- Test of the single components. Each component should be tested and quality optimised in isolation.
- Test of the whole indexing system: which index terms are produced, how are they linked to the multilingual index
- Test of the complete IR system with the TINA multilingual indexing included.

The third type of test cannot be described here as the integration into a proper multilingual IR system still needs to be done; most of the existing IR systems fall short for multilingual requirements.

Of course it would be preferable if tests could be automatised. However, properly annotated test sets are difficult to find, also because of differences in the linguistic definitions used (e.g. different tagsets). TINA uses a bootstrap approach by which the results of the first runs of a component are corrected manually, and used as a comparison for later runs.

Also, to define test criteria for every component is not an easy task. For dictionaries, there are validation criteria defined by ELRA, cf. (Fersøe 2004). For corpus analysis, coverage or out-of-vocabulary percentage is a very important criterion. For single components, evaluation criteria must be revised in the light of the contribution to the overall system. For instance, in language recognition it is not the best evaluation criterion to define the length of a string needed for language identification[5] as the length of the input string is given by the text analysed. Often, the number of correct / incorrect annotations related to the number of analysed items would be sufficient.

If the TINA system is to be tested as a whole, then of course speed and resource consumption is an issue. But more important is that it should be kept in mind that component errors accumulate in a pipeline architecture: Wrong language identification results in unusable analysis results; wrong lexical analysis hampers grammatical analysis or tagging; wrong word sense disambiguation deteriorates the system ontology and the search results. Therefore the interaction of the single components must be tested in a special setup.

[5] This holds for text analysis where documents usually consist of several sentences. Texts from social media, blog posts etc. require different strategies, cf. (Jauhiainen et al. 2018).

1.2.2 TINA System Architecture

1.2.2.1 Pipeline Architecture

The architecture of TINA follows the structure of a pipeline whereby each component builds on the results of the previous ones. The components use a common document model for their interaction: It provides access to the annotation items previously produced, and adds their own findings.

Each component is supposed to be autonomous. There may be different versions for the same task depending on different languages or different technologies; however this should not pose difficulties to the overall system workflow.

The complete overview is given in Fig. 1.1.

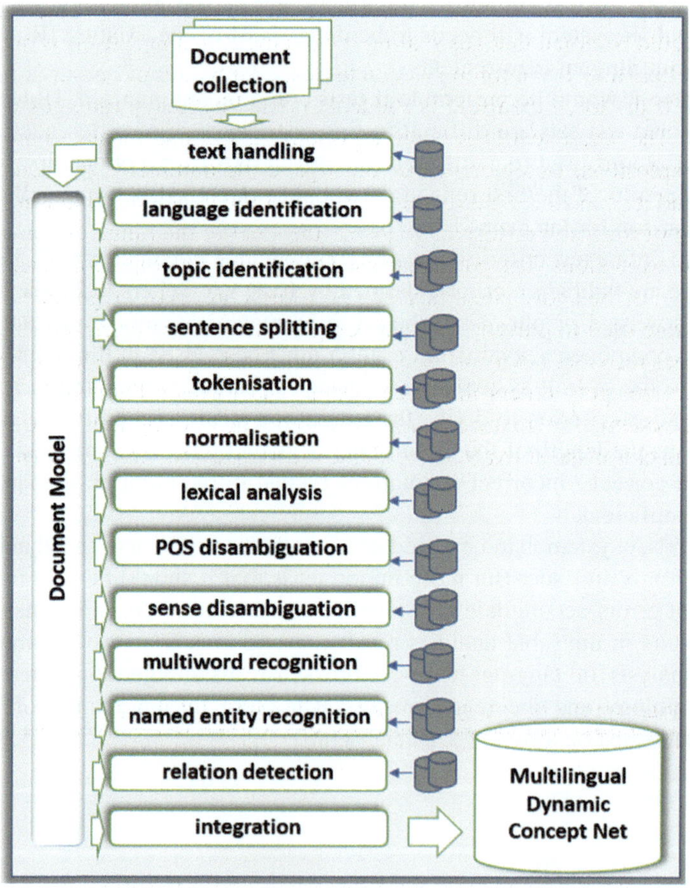

Fig. 1.1 TINA components

The workflow starts with a **Text Handling** component. It extracts the text parts from the external documents, by parsing their markup schemes (HTML, Wikipedia, pdf etc.), identifies relevant layout features and initiates the document model.

According to the result of **the Language Identification** all resources for the following components are loaded; they are all language-dependent.

Topic Identification identifies the topic of the document. It helps in document classification and in word sense disambiguation. It is based on a list of topics which is used throughout the system.

For each paragraph, **Sentence Splitting** returns a list of sentences together with sentence-related annotations. It uses resources like lists of endings or start words, punctuations and the like.

Sentences are split into tokens by a **Tokeniser** component. Different versions are required for the different scripting systems (Chinese does not use spaces for word separation, Arabic script languages have a rather free way of using spaces).

The single tokens can undergo **Normalisation**. This relates to casing, but also to spelling variants or misspellings. The idea is to reduce the complexity of the lexical analysis by eliminating some variations beforehand.

Lexical analysis, also called lemmatisation for some languages, is a key component in the workflow. It is supposed to provide a lexical description for each token in the sentence, in interaction with the dictionary. This component is very language-dependent: While Chinese uses the dictionary already for tokenisation, many European languages can be analysed by a **lemmatiser** using a (full-word) dictionary. Some of them (Italian, Spanish) need to split clitic elements; German asks for a decomposer. Turkish as agglutinative language shows many multi-morpheme tokens. Arabic-script languages require special interaction when tokenising. Access to a lexicon is a key element of this analysis, and the different word forms are analysed into lemmata, part-of-speech, and morphological annotations. For out-of-vocabulary elements some defaulting solution must be provided.

As lexical units (defined on a semantic level) in about 30% of all cases consist of multi-token units a **Multiword Analysis** is an essential part of multilingual indexing. In the end terms must be linked to a multilingual conceptual ontology, be they single or multiwords. Multiword analysis uses part-of-speech patterns as one defining criterion.

Indexing must also include **Named Entity Recognition**. Names are important information items in search. Entities would be person, place and organisation names but also units, dates, and currencies. Again they can be single or multiword units. The analysis of Named Entities can use both shallow or deep techniques. In deep analysis, a harvester would collect the final annotations from the output nodes of the syntactic trees.

Many lexicon units are ambiguous and provide different readings, both on syntactic and on semantic level. **Disambiguation** uses contextual information to do this. For **syntactic** disambiguation, in shallow processing usually a POS tagger is used, in deep processing a full syntactic analyser. The purpose is in the first place to disambiguate the index terms. For **semantic** disambiguation a word sense disambiguation subcomponent must be added here. Again, the purpose is to disambiguate an index term, in order to assign it to the right multilingual conceptual context in the index structure.

A final step in the pipeline consists in the identification of links between the index terms. A special TINA component produces **Term Association** links between terms on the basis of a shallow syntactic document analysis and identifies head–modifier relations of a given term, which are the most general semantic relations.

The whole pipeline of analysis modules is linked to a **document model** by which they communicate. The model collects all analysis results of all modules and makes them available for the following ones.

Of course all components need **linguistic resources**. They are language-specific in nearly all cases. It is the responsibility of the components to load and use them. Most of them are read-only and can be accessed by several components at the same time. The maintenance of resources is an aspect which needs special attention.

Once the index terms are properly built and syntactically and semantically disambiguated they must be integrated into the existing index. The **indexing** procedure in a multilingual setup includes the translation of the index terms, which extends the classical index consisting of terms, frequency, and document pointers by additional links like translations, and transforms the index into a network consisting of index terms and links between them. Unlike other such networks like WordNet (Fellbaum 1998), BabelNet (Navigli & Ponzetto 2012), ConceptNet, cf. (Havasi et al. 2007), (Speer et al. 2017), etc. it is **dynamic**, i.e. providing an index for a given document collection and updated for every new document: The most prominent links in **multilingual** setups are the translations of a term, but other links can be added as well, like semantic links. The result is the extension of the index into a dynamic multilingual conceptual network, called LtConceptNet. It will be described in Chap. 8. Unknown terms (out-of-vocabulary) can remain in the monolingual index, without link connections, and stay searchable this way.

1.2.2.2 Software Embedding

The whole analysis pipeline can be embedded into a global text processing software system in two ways (cf. Figure 1.2):

Inbox/Outbox: There is an inbox into which analysis requests can be placed. The manager/scheduler fetches the requests, processes it and returns the result into an outbox. In high-performance applications, this scheduler will become a bottleneck.

Repository: There is a repository into which a document is placed. Many instances of the scheduler can operate in such a structure in parallel, each one can access the document pool, fetch a task from the agenda, execute it, and place the result back. Such an architecture is much more flexible, scalable, and performant than an inbox/outbox system. An example is the DTS system, cf. (McLaughlin & Schmidt 2001), used in machine translation.

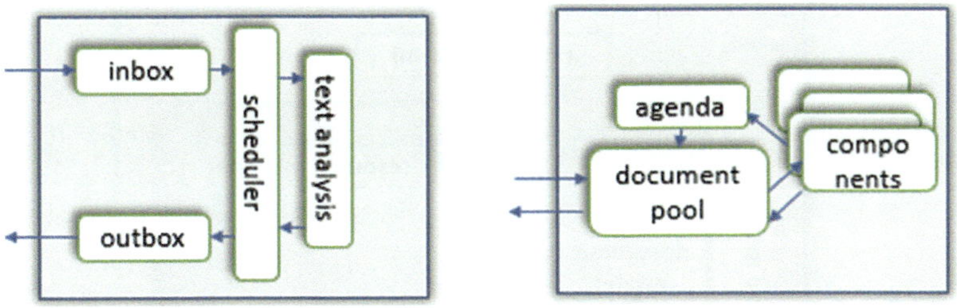

Fig. 1.2 Embedding variants: inbox/outbox, repository

These considerations hold for TINA installations in intranet or enterprise contexts; large internet engines of course require quite different architectures.

1.2.3 TINA Component Architecture

The single components in TINA follow a uniform architecture as far as possible, outlined in Fig. 1.3.

If a component is called by the Global Manager, control is handed to the component manager. This manager is responsible

- to provide the **resources** required for this component; for this purpose a resource manager is called which loads them in case they are not loaded yet
- to provide all **analysis data** required, by accessing the document model for the whole document

With this information, it selects the appropriate worker (there may be several e.g. for several languages). The worker executes its task mostly on a sentence-by-sentence basis.

The results are handed back to the component manager which inserts them into the document model. The component manager is also responsible for error management and for proper shutdown of the component.

There is also an interface for a stand-alone use of the component, be it for integration into other systems, be it for test purposes.

The advantage of such an architecture is an easier maintenance of the software, easier exchangeability of parallel components (like language specific versions), and a better control especially of the resources.

The distribution of single components in a web-based environment is also possible and has been implemented in a project like PANACEA, cf. (Bel et al. 2013): A document (called 'travelling object') is sent from web server to web server, each one being a specialist, e.g. for lexicon lookup, or for tagging, executing its specific

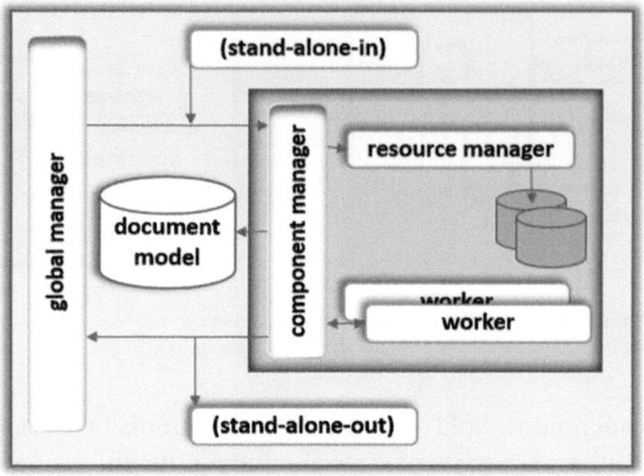

Fig. 1.3 Software architecture of the single components

task, and sending the object to the next server. The analysis status of the object is collected in its header. Such a multi-server architecture is more interesting for research than for industrial environments.

References

AlAnsary, S. (2014). Interlingua-based Machine Translation Systems: UNL versus Other Interlinguas. *The Egyptian Journal of Language Engineering. 1.* https://doi.org/10.21608/ejle.2014.59863.

Allan, J. (2000). Natural Language Processing for Information Retrieval. *Tutorial notes presented at the NAACL/ANLP Language Technology Joint Conf., Washington.*

Alonso, J.A. (1990). Transfer InterStructure: Designing an 'Interlingua' for Transfer-based MT Systems. *Proc. Third Int. Conf. on Theoretical and Methodological Issues in Machine Translation of Natural Language (TMI), 11–13 June 1990, Linguistics Research Center, University of Texas; pp.189–201.*

Baeza-Yates, R., & Ribeiro-Neto, B. (1999). Modern information retrieval. *Addison-Wesley Longman. ACM Press, 1999*

Bel, N., Poch, M., & Toral, A. (2013). PANACEA: Platform for Automatic, Normalised Annotation and Cost-Effective Acquisition of Language Resources for Human Language Technologies. *Proc. Machine Translation Summit XIV: European projects, ed. A. Way, Nice. France.*

Bernardi, U., Gieselmann, P., McLaughlin, St. (2001). A taste of MALT. *Proc. Machine Translation Summit VIII, Santiago di Compostela, Spain.*

Brants, S., Dipper, St., Hansen, S., Lezius , W., & Smith, G. (2002). The TIGER Treebank. *In: Hinrichs, E., & Simov. K., eds. Proc. on First Workshop on Treebanks and Linguistic Theories (TlT) 2002, Sozopol, Bulgaria.*

Brants, Th. (2003). Natural Language Processing in Information Retrieval. *The Clinician 2003.*

Clark, A., Fox, Chr., & Lappin, Sh., eds. (2010). The Handbook of Computational Linguistics and Natural Language Processing. *Wiley-Blackwell.*

EAGLES (1996). EAGLES 1996: Synopsis and Comparison of Morphosyntactic Pnenomena Encoded in Lexicons and Corpora. A Common Proposal and Applications to European Languages. *EAGLES Document Version May 1996* https://www.ilc.cnr.it/EAGLES96/morphsyn/.
Fellbaum, Chr., ed., (1998). WordNet, An Electronic Lexical Database. *MIT Press*
Fersøe, H., (2004). Validation Manual for Lexica. *ELRA.*
Grefenstette, Gr., ed. (1998). Cross-Language Information Retrieval. *Kluwer.*
Haspelmath, M., Dryer, M., Gill, D., & Comrie, B., eds. (2005). The world atlas of language structures. *Oxford University Press.*
Havasi, C., Speer, R., & Alonso, J. (2007). ConceptNet 3: A flexible, multilingual semantic network for common sense knowledge. *MIT Media Lab.* https://www.media.mit.edu/publications/conceptnet-3-a-flexible-multilingual-semantic-network-for-common-sense-knowledge/.
Ingwersen, P. (1992). Information Retrieval Interaction. *Taylor Graham Publishing.*
Jackson, P., & Moulinier, I. (2002). Natural Language Processing for Online Applications. Text Retrieval, Extraction and Categorization. *Amsterdam (John Benjamins).*
Jauhiainen, T., Lui, M., Zampieri, M., Baldwin, T., & Lindén, K. (2018). Automatic Language Identification in Texts: A Survey. arXiv:1804.08186.
Kowalski, G. (2011). Information Retrieval – Architecture and Algorithms. *Springer*
Liddy, E.D. (2005). Automatic Document Retrieval. *In: Encyclopedia of Language & Linguistics. 2. Edition, Elsevier Ltd., 2005, CNLP.*
Maybury, M.T., ed. (1997). Intelligent Multimedia Information Retrieval. *AAAI Press / MIT Press.*
McLaughlin, St., & Schmidt, A. (2001). DTS - A Delivery System for Translation and Translation-Related Services. *Proc. Translating and the Computer 23 (ASLIB), London, UK.*
Navigli, R., & Ponzetto, S.P. (2012). BabelNet: the Automatic Construction, Evaluation and Application of a Wide-Coverage Multilingual Semantic Network. *Artificial Intelligence Vol. 193, 2012, pp. 217–250.*
Nie, J.N. (2010). Cross-Language Information Retrieval. *Synthesis Lectures on Human Language Technologies. Springer*
Oflazer, K., Say, B., Dilek, Z., Hakkani-Tür, D.Z., & Tür, G. (2003). Building a Turkish Treebank. *in: Abeille, A. ed. (2003). Building and Exploiting Syntactically-annotated Corpora, Kluwer Academic Publishers.*
Peters, C., Braschler, M., & Clough, P. (2012). Multilingual Information Retrieval, From Research to Practice. *Springer*
Raw, A., Vandecapelle, B., van Eynde, F. (1988). EUROTRA: An Overview. *INTERFACE, Journal of Applied Linguistics 3.1 (1988), pp. 5–32.*
Salton, G., & McGill, M.J. (1983). Introduction to Modern Information Retrieval. *New York, NY (McGraw-Hill)*
Schubert, Kl. (1986). Syntactic tree structures in DLT. *Utrecht: BSO/Research.*
Sparck Jones, K., & Willett, P., eds. (1997). Readings in Information Retrieval. *Morgan Kauffmann.*
Speer, R., Chin, J., Havasi, C. (2017). ConceptNet 5.5: An Open Multilingual Graph of General Knowledge. *AAAI 31 (2017) 4444–4451.*
Schwarz, Chr., Thurmair, Gr., eds. (1986). Informationslinguistische Texterschliessung. *Olms, 281 S.*
Thurmair, Gr. (1990). METAL: Computer Integrated Translation. *Proc. of a Workshop on Machine Translation, July 1990, UMIST.*
Toral, A., Pecina, P., Way, A., & Poch, M. (2011). Towards a user-friendly webservice architecture for statistical machine translation in the PANACEA project. *Proc 11th European Association for Machine Translation (EAMT), Leuven, Belgium.*
Xia, F. & Palmer, M. (2001). Converting Dependency Structures to Phrase Structures. *In Proc. of the 1st Int. Conf. on Human Language Technology Research.*

Chapter 2
TINA Analysis Strategy

2.1 Rationale

TINA is a multilingual project. It has turned out that a single analysis approach for all languages, including Chinese, Farsi, Turkish, or Spanish, is only a valid option as long as no details of the respective languages come into play. If they do, then quite specific phenomena need to be considered. Then a way must be found to use the best analysis strategies in each case.

There are two main lines of technology, characterised as 'deep' and 'shallow' processing, cf. (Schäfer 2006), differing in the amount of using knowledge bases and in the role of using corpus data. For each language and each component it needs to be decided which technology line is the most adequate solution.

Following the task to choose the best technology for a given component, it is necessary to find a way how the components can interact. This question relates to the technology level but even more to the level of the description of the linguistic content.

The challenge is to find a level of linguistic description which supports both lines of analysis technology. The TINA analysis uses feature value structures, however in different representation forms: While deep processing uses grammar rules to manipulate feature structures, the shallow processing transposes them into tagsets whereby a POS tag represents a special feature configuration. the LtTagset describes POS tags on three levels, from basic tags as used in conventional lemma-based dictionaries to extended tagsets describing fully inflected forms. The extended tagset can be transformed into the respective feature value configurations as used in deep processing.

Finally, the results of the analysis must be made comparable and usable by the other components. This requires a common data structure which every component can use and enrich. The Document Model, called LtDoc, is meant to support such a requirement.

G. Thurmair, *Knowledge-Driven Multilingual Text Analysis and Transparent Information Retrieval*, Cognitive Technologies,
https://doi.org/10.1007/978-3-031-91741-7_2

2.2 Deep Versus Shallow Analysis

Two lines of technology can be followed for linguistic analysis: Linguistic systems in the 80ies and 90ies aimed at full parsing, using grammars and dictionaries. Topics of discussions were certain linguistic phenomena like control and raising, treatment of tense and aspect, anaphora resolution, gapping of constituents, and an important point in discussion were formalisms. Such approaches could be summarised as 'Deep Processing'.

When the technical progress allowed to handle larger corpora a change in analysis turned to a technique called ‚Shallow Processing'. Smaller components based mainly on finite state technology, combined in pipeline architectures rather than in fully integrated systems were developed and applied to many problems in linguistics.[1]

TINA uses common basic technology for all components that use deep processing, called 'LtDeep', and for many components in the line of shallow processing, called 'LtShallow'. The LtDeep description is a derivative of the analysis grammar used in the METAL machine translation system, cf. (Thurmair 1990), which was in use in productive environments from the mid-eighties until recently, covering languages like English, German, French, Russian, Dutch, Danish, and Iberian languages like Spanish, Catalan, Basque and Galician. The LtShallow in TINA branch uses finite-state technology, with an adapted FSA analyser.

The transition or combination of the two lines has been explored e.g. in systems like SPROUT and WHITEBOARD, cf. (Schäfer 2006). It was found that shallow preprocessing increases the efficiency of deep parsing. In TINA most lexical analysis components of the European languages use shallow techniques; components like Named Entity recognition use LtDeep, and components like multiword recognition use LtDeep for analysis but LtShallow for query processing in retrieval.

2.2.1 *Deep Analysis (LtDeep)*

By *Deep Linguistic Analysis* a framework is understood which tries to fully analyse the input on a syntactic and semantic level, by applying linguistic knowledge sources like complex lexicons and grammars which are based on linguistic theories like Transformational Grammar (Chomsky 1965), LFG (Bresnan 2001), HPSG (Pollard and Sag 1994), GPSG (Gazdar et al. 1985), or similar approaches, e.g. (Collins 2003). All of them intend to produce syntactic—semantic interpretations of an input sentence by manipulating complex feature structures. As opposed to shallow processing the intention is to access all available knowledge sources at the same time.

[1] An example of a shallow pipeline is a system like GATE Cunningham et al. (1996), Cunningham et al. (2002), to analyse Named Entities.

TINA uses a special framework for deep analysis, called **LtDeep**. It treats the tokens of a sentence as linguistic objects, not just as strings. Linguistic objects are structures of features and values. As opposed to other formalisms, LtDeep treats feature values as atomic units, i.e. it does not use complex feature-value structures as values of features. Features may be assigned types; in LtDeep the main distinction is between features used only for analysis or also for transfer and/or generation. Also, different languages use different feature sets, with a common kernel as described in Sect. 2.3.2.

LtDeep uses an extended Phrase Structure backbone. In principle, phrase structure grammars are able to model analysis aspects which relate to whole *phrases*, like *definiteness* in Farsi or Turkish: a feature which cannot rely on lexical formatives but requires the analysis of complete node configurations, cf. (Xia and Palmer 2001). But also with a phrase structure analysis the resulting tree forms a dependency structure based on the heads of the phrases; and there are many proposals how phrase structure trees can be converted into dependency trees and vice versa.[2] In addition, a transformational component can handle aspects like discontinuous elements, phenomena like control and raising, gapping but also pro-drop, all cases where deep structure representations differ from surface token lists.

The extension of the Phrase Structure Grammar (PSG) approach as used in LtDeep integrates probabilistic elements, cf. (Charniak 1997), a special feature-handling subcomponent, and transformational operations so that output structures need not match the input sequences.

2.2.1.1 Components of LtDeep

The basic components of LtDeep are a **lexicon** and a parser. The lexicon ideally contains a linguistic description of all tokens of the system, i.e. feature value pairs of the form (Fig. 2.1).

Features may be obligatory or optional. Obligatory features are lemma, POS (part-of-speech), entry-type and others, depending on the component; others may be optional. The POS system follows the Basic Tagset as described below (Sect. 2.3.2) and contains only basic categories.

The **parser** executes grammar rules and creates a structural description for an input string, based on the rules of a grammar. In LtDeep the parser is based on an active chart as described in Winograd (1983) where a chart contains partial parsing results ('well-formed substring table') and an agenda contains new hypotheses; if these hypotheses can be combined with an edge of the chart (and all feature tests

```
feature = value (, value)* ;
```

Fig. 2.1 Feature value description

[2] Cf. Rocchi (2016), Goyal and Kulkarni (2014), Kong et al. (2015).

are positive), a new combined edge is inserted into it, new agenda items will be created, and so on until a terminal symbol is reached which covers the whole input string. This mechanism can be used top-down or bottom-up or middle-out or best-first, depending on the control strategy.

There are extensions to the standard approach in LtDeep:

1. Like in **probabilistic** grammars as described in Charniak (1997) the parser is sensitive to probabilities of analysis structures. Rules provide probabilities, and the parser prefers more likely readings to less likely ones by sorting the more probable hypotheses higher in the agenda, and the edge with the highest probability is processed first. This way a best-first strategy can be adopted.
Rule probabilities should be computed from training data, however such data are not always available, due to the lack of domain-independent data, and due to the lack of matching linguistic descriptions.[3]
2. Beyond the Phrase Structure categories, the parser applies a component for **feature handling**. If the PSG categories match (like '*NP ->Det No*'), then tests on features are carried out to further restrict the combinability of the constituents. This method has the same effect as if a much more fine-grained phrase structure rule set than the standard tagset would be used (like '*noun-dative-plural-feminine-abstract*'). LtDeep provides a set of operators for feature handling because the standard operation which is unification is not sufficient in many cases.
3. A third extension to the standard formalism is a **transformational** component. It is used to manipulate the parse tree, mainly in cases of adjunctions.

A special strategy is used for **pruning**. Even under a probabilistic regime, the parser may fail due to a multiplication of its possibilities before a best-first solution may be found.[4] Therefore a cut is implemented, based on the number of words per sentence and the number of parsing hypotheses.

2.2.1.2 Chart Initialisation: Lexical Insertion

The chart is initialised by looking up all terminal nodes of the sentence in the dictionary, and inserting their feature-value structures into it.

The LtDeep dictionary is a full-form single-word list of entries.[5] Every entry is a list of feature-value pairs, cf. Fig. 2.2.

Lexical features and values are defined in a specification table, used to validate a dictionary.

The dictionary for an LtDeep component is compiled from a basic lexicon in two ways:

---

[3] Most treebanks, like PENN, TIGER, METU etc. use different tagsets and linguistic annotations.

[4] Many parsers restrict the number of rule firings to avoid combinatorial explosions.

[5] Except in Farsi, Pashto, and Turkish where word segmentation is required. This is described in Chap. 5.

```
Lexicon      ::=  entry+
entry        ::=  feavalpair+
feavalpair   ::=  feature '=' value (,value)* ';'
feature      ::=  <a legal feature>
value        ::=  <a legal value for this feature>
```

Fig. 2.2 Format of the LtDeep dictionary

textform=santiago;	lemma=Santiago di Compostela;	etyp=mw3;	POS=NoPL;
textform=di;	lemma=Santiago di Compostela;	etyp=mw3;	POS=MWP;
textform=compostela;	lemma=Santiago di Compostela;	etyp=mw3;	POS=MWP;

Fig. 2.3 Examples of LtDeep dictionaries (Farsi and Turkish)

- single word entries undergo inflections: To create inflected forms from base forms, TINA provides flexers, i.e. inflection generators, for most of its supported languages. They are described in Sect. 9.3.4.
- multiword entries are split into their parts, and every part has its own entry. The multiword head will receive full feature-value description, the POS being the POS and the lemma being the lemma of the multiword; the other multiword parts have also the whole multiword as lemma, a and an artificial POS (like '*MWP*', multiword part[6]). How many parts the multiword has is defined in the '*entrytype*' feature of the head entry. Figure 2.3 gives an example.

The grammar will contain rules which combine multiword parts into the proper multiwords (and remove the useless branches from the analysis tree); this is described in Sect. 6.1.

As lexicon insertion of multiword parts may produce many chart entries for some entries ($_{it}$ '*di*' also in '*Santiago di Chile*', '*Aeroporti di Roma*' etc.; there maybe hundreds of entries with '*di*' in the lexicon), a **lexicon filter** is applied which removes all multiword candidates which cannot provide all their textform parts in the input.

Figure 2.4 gives examples of dictionaries used by LtDeep (for Turkish and Farsi).

2.2.1.3 Grammar Rules

Grammar rules combine terminal feature structures to larger constituents until (one or all) terminal symbols of the grammar ('*S*') are reached.

LtDeep uses a Phrase Structure Grammar as syntactic backbone, and a set of operators for feature manipulation.

---

[6] This prevents the textform from being taken by other multiword hypotheses in the parse.

```
allo=آبرسانی;mi=m;mf=i;bound=f;lemmaid=217;lemma=آبرسانی;cat=NoC;etyp=sw;flex=hA;mwheadid=217;kno=common;
allo=ابرو;mi=a;mf=u;bound=f;lemmaid=218;lemma=ابرو;cat=NoC;etyp=sw;flex=hA;mwheadid=218;kno=common;
allo=آبرو;mi=m;mf=u;bound=f;lemmaid=219;lemma=آبرو;cat=NoC;etyp=sw;flex=hA;mwheadid=219;kno=common;
allo=آبرونی;mi=m;mf=i;bound=f;lemmaid=220;lemma=آبرونی;cat=MwP;etyp=sw;scope=all;mwheadid=220;
allo=آبرودار;mi=m;mf=c;bound=f;lemmaid=221;lemma=آبرودار;cat=Ad;etyp=sw;mwheadid=221;
allo=آبروریزی;mi=m;mf=i;bound=f;lemmaid=222;lemma=آبروریزی;cat=NoC;etyp=sw;flex=hA;mwheadid=222;kno=common;
allo=آبرومند;mi=m;mf=c;bound=f;lemmaid=223;lemma=آبرومند;cat=Ad;etyp=sw;mwheadid=223;
allo=آبرومندانه;mi=m;mf=h;bound=f;lemmaid=224;lemma=آبرومندانه;cat=Ad;etyp=sw;mwheadid=224;
allo=آبرومندی;mi=m;mf=i;bound=f;lemmaid=225;lemma=آبرومندی;cat=NoC;etyp=sw;flex=hA;mwheadid=225;kno=common;
allo=آبرویی;mi=m;mf=i;bound=f;lemmaid=227;lemma=آبرویی;cat=MwP;etyp=sw;scope=all;mwheadid=227;
allo=ابری;mi=a;mf=i;bound=f;lemmaid=228;lemma=ابری;cat=Ad;etyp=sw;mwheadid=228;
allo=آبریز;mi=m;mf=c;bound=f;lemmaid=229;lemma=آبریز;cat=NoC;etyp=sw;flex=hA;mwheadid=229;kno=common;
allo=ابریشم;mi=a;mf=c;bound=f;lemmaid=230;lemma=ابریشم;cat=NoC;etyp=sw;flex=hA,An;mwheadid=230;kno=common;
allo=ابریشمی;mi=a;mf=i;bound=f;lemmaid=233;lemma=ابریشمی;cat=Ad;etyp=sw
```

Fig. 2.4 Examples of LtDeep dictionaries (Farsi and Turkish)

1. Grammar Design

The design of the **Phrase Structure** part basically follows the X-bar theory, cf. Jackendoff (1977), Kornai and Pullum (1997). Modifications of the scheme are required as a consequence of the analysis of the TINA languages. They result in a more ‘conservative’ view of the syntactic structures: No binary branching constraint,[7] no X-bar for functional categories, i.e. VPs instead of TPs, NPs instead of DPs: Persian, Turkish and Russian do not have determiners and express determination by a set of configurational criteria. Moreover, the treatment of multiword entries also affects the X-bar scheme; this is described in the respective chapters below.

In addition to the phrase structure part, LtDeep extends the PSG nodes by **feature structures**, represented as lists of feature-value pairs, and provides **operators** to manipulate them. As opposed to other formalisms, feature unification is not the only operation in feature handling: There are other operations which are needed in practical analysis, like string handling operations, or date manipulation. In addition, all feature traffic is explicitly controlled by the grammar, also to avoid unwanted side-effects of unification A further difference is that values are atomic: strings or lists of strings, but no complex structures like e.g. in HPSG.

Some operators are transformational, changing the parse tree layout. They are used to transform the input tree into a normalised form so that a defined/normalised output structure of the parsing process can be provided. This is important in a multilingual environment as the different languages require different parses, however the following components should be able to apply a uniform treatment:

The **output structures** of the parses should be normalised in order to provide reliable structures for all languages to the tree harvesting components following the parse, like Named Entity Recognition and index term collection.

---

[7] Assumptions like the ‘Binary Branching Constraint’ (cf. Carnie 2013) maybe a nice theoretical assumptions, and enable efficient parsing, however they often do not match the requirements of real texts, like cf. ditransitive verbs with two complements to the verb. Even if all PSG grammars can be converted into binary form (e.g. into Chomsky Normal form Rozenberg and Salomaa (1997)), the resulting parse structures are quite uncomfortable to handle.

```
grammar    ::=   rule+ | comment                    // comment lines start with "=="
rule       ::=   "ID:"        <string>              // a lhs category plus a number: PossLoc-420
                 "PS:"        <lhs> "->" <rhs>+     // a phrasestructure rule: NP -> Det No
                 "Prob:"      <double>              // rule probability
                 "Features:"  featop+               // feature operations
                 "Example:"   <string>              // an input example the rule should cover
                 "Comment:"   <string>              // a comment for this rule
                 "----------"                       // 10 dashes, for historic reasons
featop     ::=   operator arguments                 // e.g.: req 1 nu pl; from 2 nu; set 0 def true;
operator   ::=   < a legal operator >
arguments  ::=   < depend on the operator >
```

Fig. 2.5 LtDeep rule format

2. Rules

The schema of a grammar rule is given in Fig. 2.5. Every rule is identified by an ID, consisting of a category[8] and a number (like 'NP-120'). Then the phrase structure is given, like 'AP ->Av Ad'. The probability of the rule follows, and then the section with the feature operations is given. Finally an example is given, and a comment for this rule for grammar writers is also given.

Apart from the phrase structure part, the core of the rule are the feature operations. They consist of a set of **operators**, all of which return a Boolean value: If it is negative, the rule fails. The operators fall into different categories[9]:

a. **Testing operators**

Featuretests

req	tests if a feature exists and has a specific value	req 1 cat No; req 3 src;
nrq	tests if a feature does not exist or has a different value	nrq 1 gd masc;
intersect	tests if two nodes have a common value; the result is stored in the first constituent	intersect 2.1 nu 2 nu;

Stringtests

equal	String test if a feature has the specified value	equal 1 lemma Prozent;
endswith	String test if a feature ends in the specified value	endswith 1 textform th;
contains	String test if a feature contains the specified string	contains 1 textform @;
grthan	test on string length	grthan 1 textform 4

[8] Categories do not just contain POS elements but also nonterminals, like NP, AP, PossLoc etc.

[9] The numbers refer to the constituent involved: 0 is the new head, 1, 2, 3... refer to the daughters.

Other Tests

leftcorner	tests if a node is left-corner (sentence-initial)	leftcorner 1;
posstime	tests if a digit can be a minute / day / month	posstime 1 textform day;

The *leftcorner* test allows for recursive rules; the *posstime* operator helps in Named Entity time recognition.

b. **Operators for Feature Traffic**

Feature-Operations

set	sets a feature-value pair on the specified node	set 1 nu pl; set 0 netype person;
add	adds a value to the value list of the specified feature	add 0 ca acc;
from	copies a feature-value pair from the specified to the mother node (0)	from 2 offset;

String-Operators

concat	concatenates two strings as value to the specified node	concat 0 lemma 1 textform 2 textform;
concatb	same, just inserting a space between the two	concatb 0 lemma 1 textform 2 textform

The string operations are used in multiword configurations where e.g. a multiword lemma must be constructed from several parts.

c. **Structural/Transformational Operators**

These operators change the tree which the edge maintains. Only three types of operation are implemented:

flatl	deletes the left non-terminal node and raises its daughters	flatl
flatr	deletes the right non-terminal node and raises its daughters	flatr
prune	deletes the specified node and the tree it dominates	delete 1

These operations can be used to normalise a tree. A sentence can be parsed by using a left anchor and recursively right attachment (cf. Fig. 2.6):

The intermediate S nodes are removed by *flatl*, which results in a flat tree[10] (cf. Fig. 2.7). It is a kind of daughter-adjunction plus a deletion of the new daughter's

[10]This is an option if only NE recognition is attempted. For full parsing this strategy is suboptimal.

```
S ->  NamedEntity            // leftcorner
S ->  S    NamedEntity       // recursive parse attaching either NamedEntities
S ->  S    Txt               // or other nodes. Both using flatl
```

Fig. 2.6 Flat tree structures

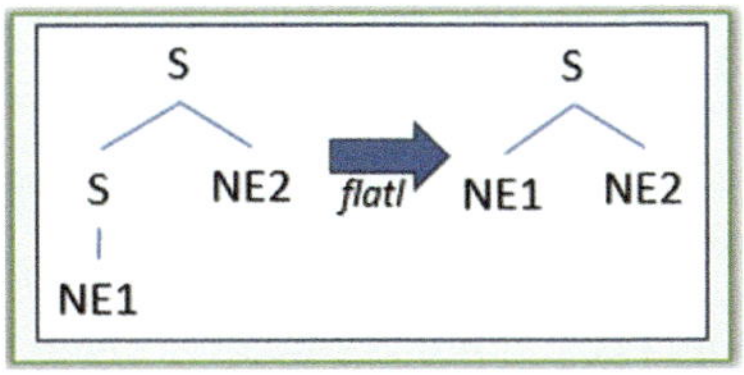

Fig. 2.7 Transformations

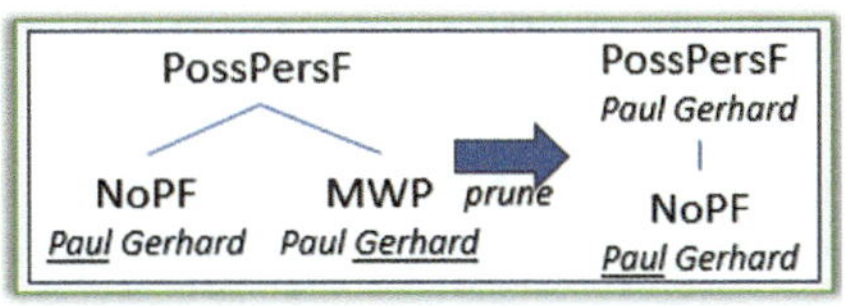

Fig. 2.8 Pruning

sister (after its feature decoration has been percolated to the top node). This way recursive rules on S level are enabled ('*S ->NP S*' and '*S ->S NP*'), and the intermediate S nodes can be removed. With this technique, also the number of recursions can be controlled, by setting a count feature on the S node.

The *prune* operator helps in normalisations. One example is in multiword analysis where the multiword part daughter nodes are removed, the resulting tree is identical to a single word derivation (cf. Fig. 2.8). Another example are cases like definiteness where some languages use lexical material while others (like Russian, Turkish) define it configurationally. In both cases, analysis should result in a feature '*defness=definite;*' on NP level, and then the determiner branch in the case of lexicalised determiners can be pruned away.[11]

Figure 2.9 shows a rule which analyses a two-part multiword, a firstname and a multiword part. Casing, lemma and entrytypes must match. Then onset and offset of the new constituent are defined, and it is marked as a person description with source in the lexicon. The new textform (*Paul Gerhard*) is concatenated from its parts. Finally the MWP path is not needed any more and pruned away.

The whole grammar is organised in packages i.e. rule sets that serve a certain purpose: verb analysis, NE-Person recognition, NP analysis, or multiword analysis.

---

[11] Unfortunately modifiers of the determiner are also pruned away, except if they can be attached to the NP node.

```
ID: PossPersF-101
PS: PossPersF  ->  NoPF   MWP
Prob: 0.86
Features: req 1 casing cp,uc,mx; req 2 casing cp,uc,mx;
                 equal 1 lemma 2 lemma;
                 req 1 etyp mw2; req 2 etyp mw2;
                 from 1 onset; from 2 offset;
                 set 0 src lex;
                 set 0 netype person; from 1 lemma;
                 concatb 0 textform 1 textform 2 textform;
                 prune 2;
Example: We know Paul Gerhard
Comment: lexicalised two-part firstnames
----------
```

Fig. 2.9 Example of LtDeep rule

LtDeep is used in several TINA components: Lexical Analysis for Persian, Multiword identification for Persian, German, and English, syntactic analysis in several languages, Named Entity recognition in English, German, Spanish, Persian and others, Multilingual Linking for these languages, and index term creation.

2.2.2 *LtShallow*

2.2.2.1 Characteristics of LtShallow

The characteristic of shallow analysis is that the analysis of texts is divided into a sequence of different components, each of them with local range, and with no or limited resources. The sequence of components like sentence splitting—tokenisation—lemmatisation—tagging—parsing forms the basis of linguistic analysis.

The very first approaches to text analysis for retrieval applications in the seventies were all based on shallow analysis techniques, simply due to the limited capacities of computers.[12] Analysis was driven by a sequence of different modules: tokenisation—lemmatisation—homograph analysis—phrase analysis—sentence analysis.

The disadvantage of this approach was that lower-level decisions (e.g. a wrong tagger decision), taken with insufficient knowledge, corrupted the decisions of all following components in the analysis sequence. The consequence was to develop deep processing approaches. However, shallow techniques enjoyed a revival since the 1990ies, due to the availability of corpus based techniques and more computational power: Each component could be improved locally, more language

[12] Systems like CONDOR: Fischer (1982), or Systran: Gachot (1986) were designed that way.

resources became available, which not only relativised efforts in data compression[13] but also improved the analysis results. Finite State technologies form the main technology pattern since that time and represent the mainstream technology today, cf. Jurafsky and Martin (2009), Roche and Schabes (1997). Libraries for FS analysis have been provided e.g. by Daciuk and van Noord (2002a, b), and many languages have been investigated in the tradition of Beesley and Karttunen (2003).

LtShallow uses also a sequence of components, with different analysis techniques for each of them. Only two techniques are more widely used in LtShallow: Lemmatisers and Finite State Analysers.

2.2.2.2 Shallow Lemmatisers

As for **lemmatisation**, many of the TINA languages use the LtShallow lemmatiser. The basis of these lemmatisers is a single word full form dictionary with annotations. The annotations consist of a linguistic description, usually a POS tag, and a lemmatisation instruction. Unlike two-level approaches the lemmatisation instruction implements a *cut & append* strategy: A certain number of characters is cut from the word ending, and a different string is appended. The lemmatisation instruction can be created automatically by string comparison.

Details are given below in the section on lexical analysis (Sect. 5.2).

2.2.2.3 FSA Analysers

A second important tool is a **Finite State analyser**. A non-deterministic finite state analyser is used in several components of LtShallow, mainly in lexical analysis of several languages but also for German decomposers.

It allows to define transition conditions on the arcs between two nodes. It allows to specify several conditions per transition, for instance to support cases where a given transition can be used by several POS tags. It also allows for zero transitions; but zero transitions can carry linguistic information. There is an option to add procedures to the transition conditions, e.g. in order to assign linguistic descriptions to zero transitions. The analysis results are kept in a chart.

The arcs collected in a chart undergo a scoring step which aims at finding the 'best' path in the automaton. Different control strategies can be applied: best-first, shortest-path, path with highest probability related to a special computation, path containing a certain POS node etc. This is described in the respective chapters below.

Figure 2.10 shows a simple analyser for Turkish nominal forms like '*ev-ler-imiz-de-ki*' or '*Ankara*'*da-ki*'. The system accepts common nouns (i.e. dictionary marked as nouns) as well as proper nouns (names) which however in Turkish must be followed by a quote. Next there can be plurals, followed by possessive

[13] Like Daciuk and van Noord (2002a, b), Hulden (2009), and others.

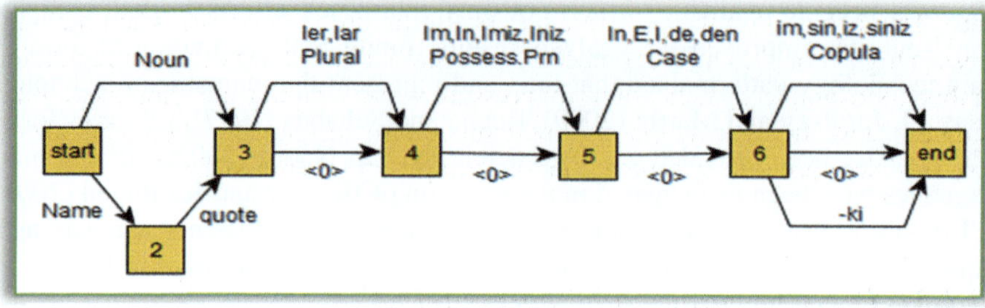

Fig. 2.10 Graphic definition of a simple Turkish noun analyser

pronouns, and a case particle. Finally, there can be a copula (like in '*doktor-um*' '*I am a doctor*') but also a suffix like *-ki* which turns the noun into an adjective ('*Ankara*' *-da-ki profesör*' = '*the professor from Ankara*'), but not both.

Zero transitions allow to analyse simpler forms (like '*profesör*' or '*kedim*' '*my cat*'). But the zero morphemes add linguistic information to the analysis: the <0> between states 3 and 4 annotates *number=singular*, between 5 and 6 it gives *case=nominative*, etc.; so there are different kinds of 'emptiness'.

The definition of the transition tables for FS analysers is a major linguistic task. it is supported by the option of graphically defining an FSA, using a drawing tool like yED.[14] This is a public domain editor which allows for graphically defining FSAs by drawing nodes and arcs and annotations. A compiler converts the yED output into a proper FSA.

However, in many cases the FSAs are too big or complex to allow for easy drawing. In such cases a table-oriented representation is used whereby rows and columns form the nodes and the table cells contain the transition conditions. Again a compiler transforms these tables into proper FS automata.

Beyond lemmatisers and FS analysers, LtShallow uses different approaches for other components like tagging; they are described in the respective chapters below.

Shallow technologies are used in the most TINA components. Lexical analysers in some languages, syntactic analysers and Named Entity recognisers use deep technology.

2.3 Interoperability

Given a situation where different components use different software for analysis it is necessary to integrate all of them into a common system, and to enable their access to all already available analysis results. Interoperability relates to two aspects:

[14] Cf. https://www.yworks.com/. Similar tools are described e.g. in Özenç and Solak (2020).

- A technical way of interaction is given in the form of a **common data structure** to which all components have access. Such a common ground is provided in the **document model**. All components take its content as input parameters (e.g. the document language) and insert their analysis results into their provided sections.
- But it is also necessary to share the **linguistic descriptions** between components: While LtDeep components use basic categorial (POS) information and feature-value structures the LtShallow components nearly exclusively use more elaborate POS information. To find interoperability a **three-tier multilingual tagset** is used such that tags correspond to feature-value configurations, and mapping between the two systems is possible.

Each component of TINA needs to specify which information is required from the document model, which information it can contribute, and which linguistic description level is used.

2.3.1 The LtDoc Document Model

The common backbone of all TINA system components is the document model.[15] The document model, called LtDoc, links all components in TINA to it whereby it serves as a kind of blackboard: All TINA components contribute to the text analysis by using the information collected in the document model so far, and contributing their own results. The document model serves as their common backbone and is described here. The single components are described in the Chaps. 3–7.

2.3.1.1 Document Collections

The Document Model can process a complete corpus. In this case it starts with an optional header collecting the common annotations of the whole data collection, collected from the headers of the single documents, followed by a list of the documents.

The **header** annotations follow the proposals made by XCES, cf. (Ide et al. 2000). They contain annotations like.

- srcloc (location of the original text (URL, file etc.)),
- srccharcode (the original character code of the text),
- srcformat (the original format of the input (html, text, Wiki etc.),
- origlang (the original language, to annotate the fact that e.g. in Europarl English text is produced by non-native speakers).

The header information is relevant only for document collections; for single documents in the indexing task it is less relevant. If required, the header information can be propagated down to the annotations of the<document>tag.

[15]A document model is "the abstract representation of the physical or semantic structure of a given set of documents", according to the digital libraries' definition: https://www.igi-global.com/chapter/opendlib-digital-library-service-system/.

Document	::=	Body, Footer?
Body	::=	para+
Footer	::=	TERMS?, NAMES?

Fig. 2.11 LtDoc top level

Many document collections contain additional useful information, like topic classification (as in Reuters news texts), speakers' native language (as in Europarl speeches), production date etc.; however this information cannot be generalised for all kinds of document collections, and therefore it is not used in the LtDoc core description. It could be collected in a header section.

2.3.1.2 LtDoc: Documents

The basic unit in indexing is the **document**. The whole application originally was called *Document Retrieval* as documents are the main point of reference in search.[16] The TINA Document Model LtDoc is intended to represent all information which can be collected from a single document. It is defined on the basis of an XML schema. In turn, there are information elements which are parts of documents but do not contain linguistically interpretable text (like smileys, formatting instructions, pictures and graphics etc.). Such elements are relevant in systems which must reformat documents (like in Machine Translation), however they need not be considered in indexing (as they do not contain indexable terms), therefore they are just skipped in LtDoc (Fig. 2.11).

1. Document <doc>

Following the traditional way, the single documents have three sections: A **header** containing metadata, a **body** containing the document text and the results of its processing, and a **footer** containing information related to the document as a whole: TERMS for common words, and NAMES for names.

LtDoc stores document text in the <doc> tag. While documents may also contain data which are not part of the text (like formatting instructions, picture of logos etc.) such data will be ignored, and text analysis will remove them in the Text Handling component. Only the text parts will be further processed. LtDoc does not foresee an explicit header tag: Header information will be given as annotations of the <doc> tag.

Document tags in <doc> tag have annotations:

- The **document ID** is required as a reference in retrieval. The ID must be unique in the system. In LtDoc it is an integer, optionally prefixed by the corpus it comes from.

[16] Meanwhile only parts of documents are searched, called *Passage Retrieval* (cf. Karpukhin et al. 2020), however particularly used in question answering research.

- The **language** of the document is an essential piece of information: Depending on the document language the system must load the resources required to analyse it. Instead of initially loading all resources of all supported languages, TINA loads only the resources required by the document language to avoid overhead. The document language is detected by the LtLanguageIdentifier (cf. Sect. 3.2).
- Once the language attribute is set, the **topic** of the document is analysed. The document topic is helpful for semantic disambiguation: Many dictionaries use topic descriptions to disambiguate readings. In TINA the topic information is determined by the LtTopicIdentifier component (cf. Sect. 3.3). This component of course is language-specific and therefore it must run after the language identifier.

As in multi-part documents both language and topic may change, the language and document tags can be set anew on paragraph (and even sentence) level; otherwise they are propagated down from the document tag to the lower level tags.

2. Paragraph <p>

Most document texts are divided into **paragraphs**: <p> .... </p>. Paragraphs are identified by interpreting the document layout. In general, paragraphs are thematic units, consisting of one or more sentences. In multilingual contexts, when texts have to be aligned for parallel corpus collection, aligning texts on paragraph level is more robust than on sentence level. There are multi-topic documents which contain paragraphs of quite different topics, e.g. in news applications (economy, sports etc.).

The annotations of paragraphs (<p> tags) in LtDoc are the following:

- paragraph ID: It is an extension of the document ID, adding the number of the paragraph
- onset and offset: first and last character of the paragraph.
- language: This information is inherited from the document tag; however it can be set anew by running the language identifier for this paragraph (e.g. if too many out-of-vocabulary words are found).
- topic: This information is also inherited from the document tag. Again it can be re-set, if required. The information must be passed down to the sentence analysis.
- flow: Paragraphs can have subflows, like footnotes, references etc. In case discourse structures should be analysed the correct sequence of paragraphs is of importance. The flows will be marked here.
- paragraph type: It is advantageous in linguistic analysis to know if a paragraph is a heading, a normal text, a table cell, or an enumeration: Different paragraph types have different linguistic properties, like capitalisation rules in English headings, partial sentences in enumerations, etc. The analysis could react on such paragraph types, if the text handling component is able to identify them.

3. Sentence <s>

Paragraphs are divided into sentences. Sentences are the basic units in linguistic processing, most components operate on a sentence-by-sentence base. Therefore sentence splitters must be provided to divide paragraphs into sentences (cf. Sect. 4.1). This is of course a language-specific task. The result of sentence splitting is an additional markup in LtDoc, <s> …. </s>

Sentence, like paragraphs, have attributes. The following attributes are used:

- **sentence ID**, an extension of the paragraph-ID, to uniquely identify a sentence, e.g. in case of multilingual alignment. If the global indexing system wants to use references to sentences, then LtDoc can provide their IDs.
- **onset** and **offset** of the characters in the sentence
- **language**, **topic**, and **paragraph type**, percolated down from the paragraph attributes, or set anew by language or topic identifier
- **global casing**, its values being *allupper, alllower, regular.* In languages with case-sensitive scripts it is important for the analysis to know whether a given token (e.g. $_{en}$ *DOG*) is part of an all-uppercase string ($_{en}$ *WALKING YOUR DOG*); if not, it could be an acronym, or even a named entity ($_{en}$ DOG = *Deployable Operations Group* (of *US Coast Guard*)).

The attributes of a sentence markup can be communicated to the following linguistic analysis (LtDeep) by converting them into feature value structures and assigning them to the leftmost node of the linguistic structure during initialisation; this way they are accessible by the analysis grammar.

4. Token <t>

Tokens[17] are the next smaller unit of text analysis. They are built by the tokeniser (cf. Sect. 4.2). The tokeniser creates a <t/> tag for each token of the sentence.

The <t> tag is an empty tag and contains all its information in its attributes.

<t casing = "..", textform = "..", … />

The annotations of the <t> tag belong to two groups: Token-related and lemma-related annotations.

Token-related annotations are produced by the LtTokeniser. They describe the token, like.

- ID: every token has an ID. It is an extension of the sentence-ID, so it can be uniquely referenced in a document.
- onset and offset: first and last character, counted from the document beginning
- textform: contains the textform of the token as found in the input string

[17] The notion of token will be discussed in Sect. 4.2.

- normalform: contains the version of this textform after it has been normalised (e.g. de-capitalised or spell-corrected); this is the task of the normaliser (cf. Sect. 4.3)
- casing: the casing of the textform
- features: features which are collected during tokenisation
- whitespace: In cases where the text needs to be reconstructed this annotation would contain the whitespace(s) following the token. This is however not necessary for indexing.

Lemma-related annotations are contributed by the linguistic analysis components like lemmatisers, taggers etc. All languages must build the basic features of a concept; they are part of the LtDoc kernel. Annotations of this kind are.

- lemma: the lemma of the textform,
- its POS/part of speech, following the POS definition in Sect. 2.3.2,
- its semantic description,
- its entry type (single or multiword).

All these features are described in the chapters on Linguistic Analysis below. Additional information can also be given in the <t/> tag:

- features (additional feature-value pairs).
- head (defining head-modifier relations).

The *features* annotations vary from language to language, or can optionally be added, similar the dependency relations (via phrase heads); some also vary with the application of TINA components.

There is an extension to the document model in several components. For instance, in case of lexical analysis a token can be ambiguous. Then the respective line, instead of only POS and lemma, contains the number of readings, and a *list* of<lemma, POS>pairs. This will later be disambiguated e.g. by a tagging component. Also, for multiword entries and Named Entities there are extensions of the model; this is described in the respective sections below.

5. Tokens and Lemmata

The document model below sentence level is complicated by the fact that tokens and concepts, which are the units of indexing, are not 1:1 related.

There are tokens which contain several lemmata, among those combinations of prepositions and determiners ($_{\text{fr}}$ '*au*' ->'*a le*', $_{\text{de}}$ '*am*' ->'*an dem*'), clitic possessives ($_{\text{it}}$ '*arma-ti*', $_{\text{es}}$ '*vér-lo*', $_{\text{tr}}$ '*ödev-imiz*'), clitic prepositions and determiners ($_{\text{fa}}$ دعراق (d'iraq = in Irak), $_{\text{ar}}$ الفرس (al-fars = Persia)), unit expressions (*50%*, *37°*, *50 kg*), and others (e.g. $_{\text{tr}}$ *gözler-iniz-le* (with your eyes)). Also (closed) compounds like $_{\text{de}}$ '*Schienen-Ersatz-Verkehr*' must be considered here.

token	token-id	token-parts	lemma-parts	lemma
Anna	1			Anna
kommt	2		2,5	ankommen
am	3	1		an
		2		dem
Bahnhof	4			Bahnhof
an	5			--

Fig. 2.12 Token versus lemma

On the other hand, there are lemmata which consist of several tokens. Such multitoken words can be digits (*20 000*), can consist of words which only occur in such expressions ($_{\text{fr}}$ '*parce que*', $_{\text{de}}$ '*gang und gäbe*'), can be words containing spaces for aesthetic reasons ($_{\text{fa}}$ پرداخت ها (par.dāxt.'hā = payments)), or open compounds ($_{\text{tr}}$ '*tatil yeri*' = en '*holiday destination*'), words with gaps ($_{\text{fr}}$ '*ne ... pas*'), verbs with separable prefixes ($_{\text{de}}$ '*kommt .... an*', $_{\text{en}}$ '*write ... off*'). Named entities like '*New York*' could also be mentioned here.

This fact complicates the document model. To replace the tokenlist by a list of lemmata is not the best idea as some of the annotations required in later components are token-specific, not lemma-specific (e.g. onset, casing), and as tokens like separable verb prefixes provide valuable clues for linguistic analysis.

For the multi-lemma tokens, all the lemmata involved need their own description (in terms of lemma, POS etc.). An option could be to give a token not just one description but a *list* of descriptions, the IDs being extended by one level (ID 4 ->4.1, 4.2 etc.), the token-related annotations would remain with the token textform, the lemma-related annotations would go into the subtokens.

For the multi-token lemmata a special feature is required ('lemparts') containing a list of token-IDs which take part in this lemma. The head lemma node would contain such a feature, while the other tokens would have a lemma feature with an empty value (as their lemma is part of the parent lemma). A range representation of multi-token lemmata like in CoNLL-U[18] would not suffice for cases where the lemma parts are discontinuous.

Figure 2.12 gives an example of these options.

The complete body section of the document model is described in Fig. 2.13.

It should be mentioned that the problem of mismatch between tokens and lemmata is not relevant in cases where sentences are lexicalised right from the beginning. This holds for Chinese where segmentation produces words right away, and for some variants of LtDeep: If a rule like {*No ->NoStem+NoFlexive*} is applied, it does not really matter whether there is a space in between or not. This is important for language with Arabic-Persian script where spaces are set much more freely than in Latin script.

[18] https://universaldependencies.org/format.html.

Body ::= doc+				
doc ::= p+				
	docID	::=	{an ID}	
	lang	::=	ar|de|en|es|fa|fr|it|pt|ru|tr	
	topic	::=	{cf. List in Chapter 3.3}	
p ::= s*				
	pID	::=	<an ID>	
	onset	::=	<integer>	
	offset	::=	<integer>	
	lang	::=	{as above}	
	topic	::=	{as above}	
	flow	::=	main | sub	
	paratype	::=	text | heading | enum, | tcell	
s ::= t*				
	sID	::=	<an ID>	
	onset	::=	<integer>	
	offset	::=	<integer>	
	lang	::=	{as above}	
	topic	::=	{as above}	
	globalcasing	::=	regular | allupper | alllower	
	paratype	::=	text | heading | enum, | tcell	
t ::= </>				
	tID	::=	<an ID>	// token-related
	onset	::=	<integer>	
	offset	::=	<integer>	
	textform	::=	<string>	
	normalform	::=	<string>	
	casing	::=	uc|lc|cp|dg|pc|mx	
	whitespace	::=	<string>	
	tokpart	::=	{a tID}	
	lemparts	::=	{a comma-separated list oft tIDs}	
	lemma	::=	<string>	// lemma-related
	POS	::=	{ a POS tag, cf. Chapter 2.3.2}	
	sem	::=	<string>	
	head	::=	{ a tID}	
	feavals	::=	{ feaval*}	

Fig. 2.13 LtDoc: body section

2.3.1.3 Footer Information

The footer section in LtDoc has the task to collect all information found by the analysis of the single sentences, integrate them on the level of the whole document, and prepare them for the backend components. It is here where different variants of terms, different usages of names etc. are joined and document-related computations (like document frequency) are performed (Fig. 2.14).

Footer	::=	TERMS?, NAMES?
TERMS	::=	term*
term	::=	termdef, frequency, atts*, matches+
termdef	::=	lemma, POS, sem
NAMES	::=	(nametype, name*)*
nametype	::=	PERSONS| LOCATIONS | ORGANISATIONS | TIMES |UNITS | COMMEANS | ENTITIES
name		namedef, frequency, atts*, matches+
namedef	::=	lemma, POS, sem
lemma	::=	<string>
POS	::=	< a Btag >
sem	::=	<string>
frequency	::=	<integer>
atts	::=	(feature, value+)+
matches	::=	textform, onset, offset

Fig. 2.14 LtDoc footer section

The footer contains two sections, one containing TERMS, i.e. common words for indexing, the other called NAMES, i.e. named entities. Terms and names share definition, (document-related) frequency, a set of attributes, and a list of matches describing the position of the term/name in the document: the terms in the document may be textforms, i.e. inflected and multipart, while the term in the footer is its canonical form.

Terms and entities are identified by the respective recognisers described in Chaps. 5 and 6, and collected for integration into the LtConceptNet, described in Sect. 8.2. How the footers are produced is explained in Sect. 7.4.

2.3.1.4 Instantiation and LtText

The document model is instantiated when a text enters the system. All external documents are first brought into a standardised format, called LtText.

LtText is intended to allow for standardised text processing:

- The first requirement of LtText is to standardise the **character code** of the input text. LtText uses UTF8 character representation throughout. All text material must undergo character conversion into this representation. The original encoding will be noted in the document header.
- The document model also enables the system to standardise the document processing, independent of the external **text format** in which the documents are written, like HTML, XML, pdf etc. Text Handling is the first component in the TINA processing chain and has the task to instantiate the document model by performing this format conversion (cf. Sect. 3.1).

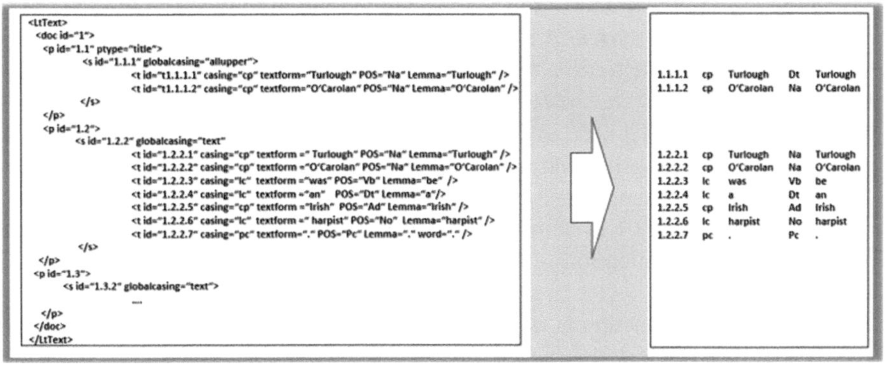

Fig. 2.15 Inline versus standoff formats

The LtText format mirrors the Document model as far as the representation of external information is concerned. It creates an **XML structure** in the tradition of XCES and contributes the following information items:

- the header information,
- the <doc> markups with the *docID* attribute,
- the <p> markups with the attributes *pID*, *onset*, *offset*, *flow*, and *paratype*.

The resulting XML structure is used to instantiate the Document Model for a given document. Once the analysis components (tokeniser, lemmatiser etc.) start working the Document Model switches from an XML-based representation to a standoff annotation scheme as used in many systems, the closest to LtText being the CoNLL-U[19] format. Different annotations are converted into table columns and in turn table rows can be converted into XML markups by making the column rows explicit using the 'column head' as annotation.

XML representations will become complicated, if suprasegmental or overlapping annotations must be represented, like phrase or constituent boundaries (Fig. 2.15).

2.3.1.5 Discussion

As can be seen, the LtDoc document model differs from other such models in several respects.

In general, Information Retrieval systems themselves are not very specific about their document model. A statement like "By *documents* we mean whatever units we have decided to build a retrieval system over"[20] may illustrate this fact, although the document notion influences the retrieval system behaviour (e.g. if document similarity is to be computed).

[19] (CoNLL-2006), http://universaldependencies.org/docs/format.html.

[20] (Manning et al. 2008:4).

Taking LUCENE as an example, their statement is simply: *"Documents are the unit of indexing and search. A Document is a set of fields. Each field has a name and a textual value. A field may be stored with the document, in which case it is returned with search hits on the document.*[21] While the concept of fields is also used for multilingual indexing in Elasticsearch,[22] it does not qualify the textual content of these fields. In contrast, TINA aims at linguistic text analysis and therefore needs a more elaborated document model.

Other than in the IR systems proper, document models and encoding formats are used in many applications, however the focus there is different from the one of LtText.

1. In **software**, the document object model (DOM[23]) is a way to internally represent XML files, organised as a tree of nodes, which allows accessing and manipulating the node content. The structure of the DOM is generic and not related to text processing only. In contrast LtText defines a specific XML structure; as such it is of course processable with DOM technology. However its content is defined and related only to the analysis of text documents.
2. LtText does not aim at the description of texts as in the **Text Encoding Initiative**[24] (TEI), further developed to XCES (Ide et al. 2000). XCES is a XML-based model to describe corpus data. Although there is some overlap, LtText is interested in text encoding only as far as it can be used for linguistic analysis purposes: paragraph structures, text flow, paragraph types, casing. Other text features, like special fonts, or embedded material (like tables, pictures), are not represented in LtText.
3. LtText does not focus on meta-data like in the context of **Digital Libraries**.[25] However, as such elements can provide useful information for search, LtDoc could handle it by extending the document header or the attributes of the<doc>tag. A special text handling application would have to written for such documents.
4. An important difference exists to translation formats like **TMX**[26] or **OTEXT**.[27] Such formats must support reformatting of the text after a translation step: For instance, if the source phrase is highlighted, the target phrase must be highlighted as well; otherwise human postediting will be more cumbersome and time consuming. Therefore the model must provide elements needed for reformatting, like

---

[21] https://en.wikipedia.org/wiki/apache_lucene.

[22] https://www.elastic.co.

[23] "*The W3C Document Object Model (DOM) is a platform and language-neutral interface that allows programs and scripts to dynamically access and update the content, structure, and style of a document.*" https://www.w3schools.com/xml/xml_dom.asp.

[24] Text Encoding Initiative: https://tei-c.org/.

[25] Digital Libraries: https://www.sciencedirect.com/topics/computer-science/digital-libraries.

[26] Savourel (2005); cf. also TMX: https://www.gala-global.org/knowledge-center/industry-development/standards/lisa-oscar-standards.

[27] OTEXT has been used in the IBM translation environmant, cf. Barret (1999).

inline tags (in TMX) or embedded objects.[28] In LtText such inline tags can just be discarded as no reformatting is required. Also, while translation unit tags in TMX are bilingual, LtText does not aim at parallel texts or multilingual sentence representation and therefore does not need the <tu> … </tu> tags of TMX.
5. Compared with standoff representation like **CoNLL-U**, (cf. de Marneffe et al. 2021), the LtDoc approach enables the analysis to use contextual information as collected during the first steps of analysis (topic, paragraph types etc.). In particular topic information may be useful for word sense disambiguation. CoNLL-U focuses on (single) sentence descriptions for treebanks; sentence markups are used as comments rather than analysis contribution; many use corpora which are already sentence-segmented, like the Penn Treebank, cf. Marcus et al. (1993), or WACKy Baroni et al. (2009). Differences can also be seen in the treatment of trailing whitespaces, or range IDs for multi-token words.

As a result, LtText uses text encoding information (like paragraph types) as far as it is useful for linguistic analysis, and it keeps the results of linguistic analysis as far as they contribute to text content analysis and indexing. LtText is designed as a backbone for a series of tools whereby all linguistic analysis components contribute their particular annotations.

2.3.2 *The LtTagset*

The LtTagset is intended to link different linguistic descriptions used for deep and shallow processing into a common room of linguistic analysis: The fact that $_{\text{it}}$ '*boschi*' is a masculine plural noun is independent of deep or shallow processing: Both tools should understand a common linguistic description.

A part-of-speech (POS) tagset is a basic element of linguistic description. All further annotations are based on such a categorial definition for a given word. Categorial definitions like 'noun' or 'verb' are used in conventional dictionaries; they are also the basis of many machine-readable dictionaries for terminology or machine translation. In contrast, conventional tagsets like the PENN treebank tagset and other shallow analysis systems use quite different descriptions, some of which are clearly related to textforms but not to lemmata (like '*inflected participle*'). So some considerations on the status of tags and tagsets may be appropriate.

2.3.2.1 Design

A system like TINA which wants to use both deep and shallow components needs to find a way how interoperability between these different levels of linguistic description can be achieved. In addition, the linguistic descriptions should satisfy multilingual requirements.

[28] How this is done in Machine Translation is described in Thurmair (1990).

1. Monolingual Versus Multilingual Tagset

A tagset is a collection of basic syntactic elements of a language, defined by their distribution: Elements are syntactically identical, if they show a common distribution. So tagsets are syntactically motivated, and as they are related to the syntactic distribution of other tags they are monolingual by definition. An obvious example is the PENN Treebank tagset[29]: It only describes English wordforms.

A multilingual tagset seems to be a contradiction in itself because syntactic structures differ from language to language, and so should do their tagsets. However there are some observations which may justify an attempt towards a multilingual tagset definition.

1. Most conventional dictionaries use a similar basic description inventory for their entries, like ‘*noun*’, ‘*verb*’, and others. A multilingual tagset can be based on such annotations.

2. A second observation is that in bilingual translation dictionaries the POS annotations are surprisingly stable between source and target languages. Only a fraction of entries switches their POS, and this is sometimes again quite regular (e.g. $_{\text{fr}}$ ‘*…-ment*’ -> $_{\text{en}}$ ‘*in a … manner*’). So there seems to be a kind of multilingual understanding of the basic POS tags.

3. Collections like the ‘Word Atlas of Language Structures’ (WALS), cf. (Haspelmath et al. 2005), lists many linguistic phenomena which are shared by all languages, and describes which features are used to express them. Nouns have different ways to express case or definiteness, verbs use tense systems, clauses can be negated by affixes in some languages ($_{\text{tr}}$ ‘*gel<u>me</u>mek*’, $_{\text{fa}}$ ‘نِگذاشتن’), by particles in others ($_{\text{en}}$ ‘*not*’, $_{\text{de}}$ ‘*nicht*’, two-member particles in French: $_{\text{fr}}$ ‘*ne…pas*’). But all of them express the same underlying phenomena (like negation). A multilingual tagset can serve as a basis and capture a part of such common structures.

As a result, a multilingual tagset is a compromise between monolingual descriptions of syntactic distribution and multilingual generalisations. It is a tagset which tries to capture basic syntactic relationships: Nouns can have determiners. Prepositions can precede or follow NPs. Adverbial elements can modify clauses, verbs, nouns, or numerals. All languages use punctuations to organise sentences. Participle clauses can modify nouns, etc. The treatment of such facts can be supported by a multilingual tagset.

Of course such a tagset can only be a general framework, and as a compromise it will contain frictions of different kinds: For instance, not all languages will show all tags. Also, in some languages the selected tags need explanations and denote phenomena which are considered differently in a monolingual point of view, e.g. to describe their system of tense and aspect. As a result, the tag description must specify for each language which tags it provides, and give explanations what is meant by it.

---

[29] Marcus et al. (1993); cf. also (Jurafsky and Martin 2009: 264).

2. Deep Versus Shallow Desription

If both deep and shallow processing are to be integrated in one system, it is of importance that the linguistic descriptions used by both analysis lines can be mapped onto each other.

Deep processing uses information stored in (typed) feature structures. They have a categorial backbone of categories like '*noun*', '*verb*' etc., and annotate them with feature structures. Tags like '*present participle*' are built during the parsing process, as the result of rule application, and are not defined in the dictionary. But the categorial backbone as used in dictionaries should be expressed in a tagset.

In turn, shallow approaches use tagsets describing not just basic categories but *bundles* of features, like '*plural noun*', '*inflected present participle*' and the like. In particular extended tagsets contain morphological information like inflections but also positional properties (pre-/post-modifier), sometimes also semantic information (month names, titles), if they coincide with particular distributional properties.

Interoperability of the two levels of description requires mapping operations between them. The LtTagset approach is to start from a list of basic categories, and extend it to more complex categories, describe them as (extended) tags on one side and on more complex feature structures on the other side. So the mapping is based on a factorisation of the categorial descriptions. It starts from basic categorial POS information and specifies feature-value information for every additional linguistic description, until every POS tag can be described as a configuration of feature-value pairs.

For instance, the PENN Treebank tags '*RBR*' (adverb, comparative) and '*RBS*' (adverb superlative) would share the basic tag 'Adverb' and would differ in the value of their 'degree' feature ('*dg=com*' versus '*dg=sup*'). A tag like '*VBP*' (verb, non-3rd person singular present) assembles a whole cluster of features (person, number, tense) which would be described by feature-value pairs in LtDeep. Btw it is an example of a clearly *mono*lingual tagset, other languages behave quite differently.

There are two consequences of this consideration:

1. The first consequence is that there is not just one tagset but there is a **hierarchy of tagsets**, each of them being more elaborated than the previous one. The transition from one tagset to the next is described by feature-value pairs which can be provided by definition tables. The hierarchy starts with a **Basic Tagset (BTAG)** containing elementary POS information like in dictionaries used in LtDeep. Adding features, mainly syntactic subcategorisations, results in a **Standard Tagset (STAG)**, which is the usual level of description in shallow processing. Adding further feature information, in particular morphological descriptions, results in an **Extended Tagset (XTAG)**. This can be used for tagging, chunking, and other operations of shallow parsing.

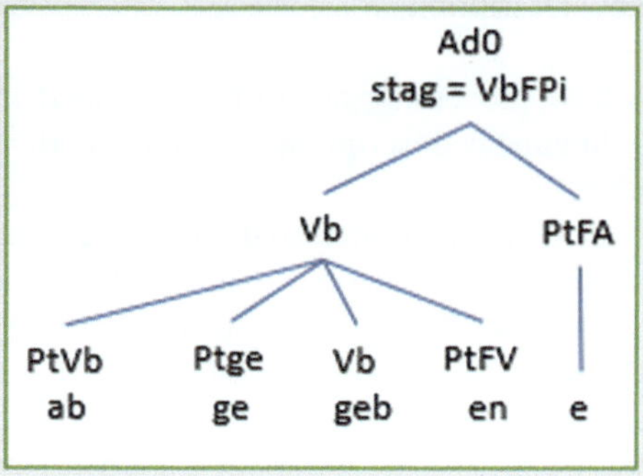

Fig. 2.16 STAG produced by LtDeep

2. The relations between the tagsets are described in **definition tables** which give the used features, their types, their values, and constraints on cooccurrence with other features. They are used to validate dictionary entries and grammar rules for compatibility and well-formedness. It should be noted that not all features needed are elementary (lexical); many of them are produced in the course of derivations (like inflected participles); Fig. 2.16 shows the process of deriving inflected participles by rule applications.
The important issue here is that a formal transition between (extended) tags and feature structures must be available. Only then interoperability between deep and shallow techniques can be envisaged.

Monolingual computational dictionaries usually contain information items beyond the level of description provided by shallow POS based analysis, in particular verb framing, raising, passivation and semantic annotations. Such information will become relevant in later phases of analysis when full parsing and Named Entity Resolution is required.

3. Further Aspects

Some more issues need to be considered in a setup like TINA.
1. The tag system presented here is a framework. The intention is to cover as many linguistic phenomena of the involved languages as possible. Not all languages support every tag; there are languages without articles in the '*Determiner*' POS (like fa, ru), there are languages without gender (en, fa, tr), and there are many languages without '*dual*' as number value, however needed for Arabic. Which tags and which features/values hold for a particular language needs to be specified; this is laid down in the definition table for this language.

2. The definition of tags is not an isolated task. If the whole process flow of text analysis is considered, only tags which can be processed by *all* components can be used, i.e. tags which are not processable by the following components (lexicon lookup, tagging etc.) do not really contribute to the overall analysis. In particular, 'waste-basket' tags like '*residual*', '*other*', '*symbol*' '*nabb*' etc. would collect linguistically quite different phenomena, and grammar rules involving them are hard to imagine. While '*$*', '*€*', '*%*', '°', '*km/h*' are writing variants of nouns and should be treated as such in the dictionary, others like '&' should come out as conjunctions, math operators like '+' '−' '*' would be treated as punctuations of a certain type, and the like. Textforms like '*59%*' are complete noun phrases, the head being the '*%*'=*percent*. Tags like '*nabb*' denote nouns which have an abbreviated form as allomorphs/allographs but behave like nouns, and like other allomorphs the abbreviation has its own feature decoration (e.g. an idiosyncratic plural formation).
3. Multiwords receive the POS tag of their head. '*Nuclear power plant*' would be categorised as noun, '*kick the bucket*' as verb and so on.[30] If the modifiers of the multiword head are invariant, a tag called 'multiword part' could be helpful as it avoids those parts to be used by competing analyses. Multiword treatment is treated in a special chapter (Sect. 6.1).
4. The tagsets discussed here are syntactically based and describe parts of speech (POS). This decision implies that there will be other tagsets to describe other phenomena. For instance, morphological tags as used in the German decomposer analysis must be taken into account, as presented in the respective section below (Sect. 5.2.4). Morphological extensions describing the bound morphemes are also required for the analysis of Turkish and Farsi (Sect. 5.3, 5.4). In any case the tagsets should work both for conceptual analysis (common words) as well as for Named Entity Recognition, which is a sub-case of syntactic analysis.
5. As for the nomenclature for the tagsets, a uniform terminology is intended in LtTagset. Such a decision causes frictions compared to monolingual descriptions, like in cases of tense systems, pronominal elements and the like, but it has advantages in better overall understanding and comparability in multilingual contexts, and it may help to reduce conversion efforts. Idiosyncrasies and deviations will be mentioned in the description of the respective elements in the tagset.

2.3.2.2 Tagsets

LtTagset is a three-tier tagset: Basic tagset (BTAG), standard tagset (STAG), and extended tagset (XTAG).

[30] With the exception of some adverbials: '*in addition*'->Adverb, not Adposition.

1. Basic Tagset (BTAG)

The Basic Tagset follows the conventional linguistic categorisations as used in grammars and dictionaries. It covers basic syntactic distinctions, it is identical for deep and shallow processing and forms the basis of all conversion activities. The naming convention is that BTAG names are capitalised two-letter strings.

Adjectives (Ad)

Adjectives modify nouns. In many languages they stand in agreement wrt morphological features, and in principle they can undergo comparation.

Adpositions (Ap)

In general adpositions are used for case-marking and for marking of semantic relations (temporal, local etc). In some languages they precede the nouns, in others they follow them, many languages have circumpositions in addition. They govern cases ($_{de}$ '*auf+Dat*' is locative, '*auf+Akk*' is directional). Sometimes they have semantic content (e.g. '*he opts for/against a decision*'), sometimes they are semantically empty and are just verb role indicators ($_{en}$ '*he relies on the railway*'). Together with nouns they form prepositional phrases. The heads can also be other categories, like adverbs ($_{en}$ '*they studied until now*', $_{de}$ '*er tendiert nach rechts*').[31]

Adverbs (Av)

Adverbs are elements which modify other constituents. The name is misleading as 'adverbs' do not just modify verbs but also other syntactic constituents: They can modify clauses ('*probably*'), verb phrases ('*yesterday*') but also nouns and adjectives ('*only Peter attended the class*') but also conjunctions ('*except when he was ill*', '*almost without touching the ground*') and numerals ('*he bought about 15 books*'). Which adverbs modify which constituents is a matter of subcategorisation, but for consistency reasons the term 'adverb' is kept in this tagset. Adverbs usually do not inflect.

Conjunctions (Co)

This tag comprises clause-related material. Coordinating, subordinating, infinitive-initiating and other such lexemes are collected here. In some languages there is significant homography to adpositions[32]

[31] Unlike the X-bar theory in TINA the head of the PP is considered to be the noun, not the preposition. In cases of empty prepositions the Ap can be pruned, in other cases it would be converted in a semantic feature like case markers.

[32] To assume just one tag as in the PENN tagset English (IN for prepositions and subordinates) would not be usable in a multilingual setup, cf. below.

Determiner (Dt)

In this tag lexemes are collected which specify nouns, in particular ones related to definiteness, like articles. Some languages use other means to define definiteness, but many have proper lexemes for this purpose; they are collected here.

Numerals (Nm)

All languages use numerals. They can be subcategorised for cardinals, ordinals etc., as described in the standard tagset.

Nouns (No)

This tag is also used by all languages. It covers common nouns, and among many possible subclasses some are relevant for analysis, like time descriptions (month and weekday names), titles, units and the like.

Names (Na)

This tag collects proper names: Persons, locations, companies and institutions and other proper names relevant for Named Entity recognition

Punctuations (Pc)

Most languages use punctuation signs, like commata, brackets, quotes. They need a special POS tag in processing. Many of them are language-specific.

Phrases (Ph)

In all larger dictionaries constructions may be found which are (partial) sentences. They can be idioms, can contain inflected verbs, or can be pragmatically complete utterances (‘*fire!*’). They are difficult to handle in linguistic analysis. For the time being they are collected here under ‘*phrases*’. In this tag also the interjections are collected (like ‘*Holy smoke!*’)

Pronouns (Pn)

Pronouns exist in all languages treated by TINA. They will be divided into personal, possessive, relative etc. pronouns. Some languages use clitic constructions for some of them. Indefinite pronouns sometimes interfere with determiners and/or adjectives, however this tagset takes the more conservative view to collect them under pronouns.

Particles (Pt)

In addition to the yes/no particles, this tag covers negations, comparation-indicating particles ('*more beautiful*'), and a significant amount of bound morphemes. Of those, the affixes are needed to analyse textforms in languages like German (compounds), Turkish (segmentations), or Farsi. Details will be explained in the section on Lexical Analysis (Chap. 5). They are quite language-specific.

Verbs (Vb)

This tag can be found in all languages. Verbs are subdivided in TINA into full verbs, auxiliaries, and modals. Some languages express modality by affixes and other means, some by using lexemes. Such lexemes are collected here as modal verbs ('*can*', '*may*').

Other basic tags are not used in the TINA analysis. Unknowns (UNK) are created for out-of-vocabulary words during analysis, and tried to be resolved by the LtDefaulter component (Sect. 5.2.5). A tag MWP (multiword part) is used as an abbreviation for the description of invariant multiword parts. Tags like '*abbr*', '*foreign*', '*residual*' and the like are not used.[33] Interjections, as used by several tagsets, are treated as subcategorisation of phrases (Ph).

The BTAG tagset is used in machine-readable dictionaries, both mono- and multilingual, and is the basis for POS assignment in LtDeep. For use in shallow analysis it does not contain sufficient details; therefore tagsets with more expressive power are used there.

2. Standard Tagset (STAG)

The Standard Tagset (STAG) is intended to cover the main syntactic subcategories of the basic tagset categories, i.e. it extends the Basic Tagset by defining syntactically/distributionally motivated subcategories. Depending on the language it comprises 50 to 100 tags, and it is the basis for many taggers.

Following the requirement of interoperability, a mapping procedure between BTAGs as used in LtDeep and the STAGs as used in LtShallow must be defined. This mapping adds feature annotations to the BTAGs. These features are mainly syntactic subcategories, named '*s..*' (subcategory_of), completed by the BTAG; so there are *sad=...;*', '*sap=...;*', '*sdt=...;*' etc. features.

The naming convention is that STAGS extend the BTAG name, like '*ApPr*' (prefix), '*ApPo*' (postfix), '*NoCT*' (common noun—title), '*PcQO*' (quote—opening) etc.

[33] There are examples in the Universal-Grammar tag 'X' which are difficult to treat without their tag. However if used, they need a clear instruction how they should be treated on syntactic level.

While BTAGs are used by all TINA languages, STAGs show more language-specific behaviour. Therefore each STAG must specify for which language it holds. This is done in the respective definition files of the single languages.

STAGs for Adjectives (Ad)

Adjectives distinguish the following STAGs, using the feature *sad*=subcategorisation of adjective. If nothing more is known about an adjective, they are assigned just the BTAG (Ad).

AdAt	*sad=attr;*	*attributive*	Regular attributive adjectives, modifying nouns. Often in morphological agreement to the NP head nouns. Numerals (ordinals like ‘*fifth*’) and inflected participles ($_{de}$ ‘*herbeigerufene*’, $_{it}$ ‘*copiato*’) do not belong to this STAG as they show different distributional properties
AdPr	*sad=pred;*	*predicative*	Adjectives in predicative position used in copula clauses: $_{en}$ ‘*she is nice*’, $_{tr}$ ‘*o hasta*’. In languages which drop the copula, like Arabic or Turkish, the adjective serves the predicate function in providing person, number, and tense. Also in resultative clauses ($_{en}$ ‘*he painted the wall blue*’)
AdNa	*sad=name;*	*nationality*	Adjectives denoting names, like nationalities, groups (‘*Catholics*’): In many languages they are capitalised ($_{en}$ ‘*French*’), often they show special inflectional behaviour ($_{de}$ ‘*Frankfurter*’). In Russian, many person names are categorised this way ($_{ru}$ базедовой)
AdAv	*sad=adv;*	*Adverbial*	Adjectives in adverbial use ($_{de}$ ‘*er schläft lange*’). In some languages they are morphologically marked ($_{en}$ ‘*safe-ly*’, $_{fr}$ ‘*formidablement*’, $_{it}$ ‘*pluviamente*’)
AdNS	*sad=nst;*	*non-standard*	Adjectives in non-standard position. Adjectives in non-standard position to the NP head are marked here: pre-ordered in Roman languages ($_{it}$ ‘*caro, certo*’, $_{fr}$ ‘*joli, mauvais*’), post-ordered (and uninflected) in German ($_{de}$ ‘*das Röslein rot*’). They need special consideration in tagging

Abbreviated adjectives are treated as allomorphs of their non-abbreviated forms. No special tag is provided for them.

STAGs for Adpositions (Ap)

Adpositions take NPs or AvPs ($_{en}$ ‘*he came from behind*’) as complements, and are case-governing. The subcategorisation of adpositions (using the feature *kap*=kind of Ap) is based on positional properties. The following STAGs are distinguished:

ApPr	*sap=prep*	*Preposition*	Prepositions are $_{de}$ '*vor*', $_{en}$ '*on*', $_{fr}$ '*pour*' etc. Some languages (Arabic, Farsi) attach clitic prepositions to nouns. As not all prepositions share this property, the ones which are able to take a proclitic position must be marked (by a feature clitic=yes) in the XTAG
ApPD	*sap=pdet*	*preposition-determiner*	In some languages there are morphemes which combine prepositions and determiners into one lexeme, like $_{de}$ '*am*', '*im*'.[34] $_{ar}$ 'بهذه' (with this)
ApPo	*sap=post*	*postposition*	Postpositions follow the NP. Some languages use only postpositions ($_{tr}$ '*için*', '*sonra*'), some have both pre- and postpositions ($_{de}$ '*gemäß*': '*gemäß der Regel*' – '*der Regel gemäß*')
ApPoL	*sap=loc*	*locative postposition / localiser*	In Turkish there are postpositions with special properties. They are noun derivations and can bear case information. Examples are $_{tr}$ '*alt*', '*ön*' cf. '*altına*' '*önünde*'. Similar phenomena can be found with Chinese localisers which also are closely related to nouns, ch. (Cheung 2016: 279)
ApC1	*sap=cir1*	*circumposition first part*	
ApC2	*sap=cir2*	*circumposition non-first part*	Circumpositions enclose the NPs. Their first part is coded ApC1, their second part ApC2. Both parts form the lemma ($_{de}$ '*um ... willen*', $_{en}$ '*for ... sake*', $_{fr}$ '*à ... près*'). Both parts can themselves be formed by multiwords ($_{ps}$ 'د په باب', about)

STAGs for Adverbs (Av)

Adverbs are modifiers of other constituents. They use the feature *sav* (= 'subcategories of adverb') for subcategorisation. If nothing more is known, adverbs stay with the BTAG 'Av'. Otherwise they can be subcategorised according to several aspects:

From a distributional point of view, they can modify different constituents, like nominal, verbal, and sentential constituents. Nominal modifiers are distinguished in relation to their position (initial—final). In the XTAG features it is stated which constituents they modify (nouns, adjectives, numerals). Verbal modifiers are the classical 'adverbs', clausal modifiers include proposition-related ($_{en}$ '*he was possibly ill*') or discourse-related ($_{en}$ '*therefore he did not say anything*') adverbs.

From a semantic point of view, adverbs would be subcategorised into negations, queries, pro-PPs, verbal adverbs etc., shown below.

Not all of these subcategories will be found in all languages. However the analysis of their distribution behaviour is essential for proper tagging and parsing.

[34] They cannot be simply replaced by '*an dem*', ,*in dem*' usw, as there are many idiomatic expressions like '*im Augenblick*', '*am besten*' etc.

AvMI	*sav=nomodi*	*nominal modifier in initial position*	These adverbs modify nominal constituents in initial position ($_{en}$ '*quite*', '*only*'). Modifications of adjectives ($_{en}$ '*quite green*') and numerals ('*around 10*') are also covered here, distinguished by a feature in XTAG
AvMF	*sav=nomodf*	*nominal modifier in final position*	Here are post-nominal adverbs ($_{en}$ '*he was brave enough*')
AvVb	*sav=vbmod*	*verbal modifier, 'adverb')*	These are the classical adverbs which modify VPs, like '*yesterday*', '*usually*' etc. Most of the members of BTAG Av belong to this category
AvCl	*sav=clmod*	*sentence adverbs - clausal modifiers)*	Adverbs modifying clauses. They can be conversational, i.e. used to create coherence in a discourse, like '*therefore*', '*consequently*', '*so*' etc., relevant if discourse structures are analysed. They can also be modal, expressing the authors point of view in relation to the propositional content, like '*perhaps*' or '*possibly*', cf. (Huddleston and Pullum 2002:576)
AvPn	*sav=pnmod*	*(pronominal adverbs, pro-PPs)*	These adverbs consist of adpositions and pronominal elements, and form pro-PPs in analysis. Examples are: $_{en}$ '*there-of/for*', $_{de}$ '*da-mit/zu/bei*'
AvPr	*sav=avmod*	*predicative adverbs, verbal adverbs)*	These are adverbs which govern clauses, like in Russian (должны, можно, надо: *it must be / is possible, is necessary*) in Turkish verbs on '*-erek*', '*-ince*' and others[35]
AvRl	*sav=rel*	*relational adverbs*	These adverbs express links to other (mainly clausal) constituents but differ from conjunctions in their distribution. Examples are $_{en}$ '*he was so upset about all this that he ...*', $_{it}$ '*como*'. In Farsi the elements before *ke* (like ' طوری که ', 'چنان که': '*so ... that*') are categorised as AvRl
AvNg, AvNg1, AvNg2	*sav=neg*	*negational adverbs*	These are negation adverbs: $_{en}$ '*not*' de '*nicht*'. Two part negations ($_{fr}$ '*ne ... pas*') are marked *AvNg1* and *AvNg2*. Some languages (Turkish, Farsi) express negation by particles ($_{tr}$ '*gelmiyor*')
AvQu	*sav=qu*	*(query adverbs)*	Query adverbs (wh-queries) like $_{en}$ '*where*', '*when*', $_{fr}$ '*où*', $_{tr}$ '*neden*', occur in all TINA languages except Chinese[36]

[35] In Turkish the classification is done using particles like *-ince*, *-ip*, *-ken*, *-erek* etc.

[36] In Chinese the sentence-final ne 呢 is used to express wh-queries, cf. (Cheung 2016).

STAGs for Conjunctions (Co)

These lexical units govern clauses. In English there is significant overlap between prepositions and conjunctions, however this does not hold in a multilingual context: Many languages use quite different words. Conjunctions can be multiwords: $_{\text{fr}}$ '*parce que*'.

If nothing is known about a given conjunction, then just the BTAG *Co* is assigned.

CoCo	*sco=coord*	*main clause conjunction*	Such conjunctions ('*and*', '*or*') exist in all TINA languages
CoSu	*sco=subord*	*subclause conjunction*	Conjunctions initiating subordinate clauses also exist in all TINA languages. They could further be subdivided into temporal, local modal conjunctions, however this would not be necessary for the standard tagset
CoIn	*sco=inf*	*infinitive conjunction*	Infinitive clauses are also introduced by conjunctions by many languages ($_{\text{en}}$ '*to*', $_{\text{de}}$ '*zu*', $_{\text{es}}$ '*a*')
CoKo	*sco=comp*	*comparative clause conjunction)*	Special conjunctions are used in comparative constructions ($_{\text{en}}$ '*than*', $_{\text{de}}$ '*als*', $_{\text{es}}$ '*que*': '*he is smarter than I had expected*'). The head of such clauses must contain a comparative construction
CoC1	*sco=coc1*	*multipart conjunction first part*	
CoC2	*sco=coc2*	*multipart conjunction non-first part*	Multipart conjunctions can be found in many languages: $_{\text{en}}$ '*neither...nor*', $_{\text{fr}}$ '*ni... ni*', $_{\text{tr}}$ '*ya ... ya da*'). The first element is coded as CoC1, all following parts as CoC2
CoFi	*sco=fin*	*final conjunctions*	Such conjunctions indicate the end of a clause, often in connection to an enumeration. Such conjunctions are $_{\text{en}}$ '*etc.*', $_{\text{de}}$ '*usw.*'. They help in parsing and tagging
CoSuC	*sco=subcop*	*subclause conjunction with personal pronouns*	There is a special case in Arabic where subordinating conjunctions can take suffixed pronouns (e.g. أنّك), cf (Abu-Chakra 2018:26

STAGs for Determiners (Dt)

The subcategorisation of determiners follows the standard classification. While some languages like Russian, Turkish etc. do not have articles, they still use demonstratives ($_{\text{tr}}$ '*bu ev*' this house), indefinitives ($_{\text{tr}}$ '*bir ev*' a house) or possessives ($_{\text{ru}}$ 'моя машина' my car) for determination, which imply definiteness information.

Some subcategories of determiners are parallel to subcategories of pronouns, some tagsets[37] revert them and define for instance '*indefinites in determiner/pronoun position*'. However in later parsing a different treatment must be re-opened again. Therefore they are (also) tagged as subcategories of determiners.

DtDd	*sdt=def*	*definite determiner*	
DtDi	*sdt=indef*	*non-definite determiners*	Many languages use special lexemes to characterise an NP as definite ($_{\text{en}}$ '*the*', $_{\text{de}}$ '*das*', $_{\text{it}}$ '*il*') or not ($_{\text{en}}$ '*a*', $_{\text{de}}$ '*ein*', $_{\text{it}}$ '*una*'). They are marked as *DtDd* and *DtDi* respectively
DtDm	*sdt=demon*	*demonstrative determiners*	Demonstratives are $_{\text{en}}$ '*this*', $_{\text{tr}}$ '*bu*', $_{\text{zh}}$ '这 zhè', $_{\text{fa}}$ اینin, cf. این خانه *in xâne=this house*
DtPo	*sdt=poss*	*possessive determiners*	Possessives are $_{\text{en}}$ '*my*', $_{\text{it}}$ '*mio*', zh '我的 *wǒ de*'. Some middle-Eastern languages use possessive clitics instead of determiners ($_{\text{tr}}$ *ev-im*, $_{\text{fa}}$ 'خانهی من' *xâneje man: my house*)
DtQu	*sdt=qu*	*query determiners*	Such determiners are en '*which*', $_{\text{de}}$ '*welche*', $_{\text{tr}}$ '*hangi*', $_{\text{zh}}$ '哪' nǎ'
DtRl	*sdt=rel*	*relational determiners*	This type of determiner occurs in relative clauses with changing subject, and is used to express the link to the matrix NP. Examples are $_{\text{en}}$ '*whose*', $_{\text{de}}$ '*dessen*', $_{\text{it}}$ '*cuo*'. While $_{\text{fr}}$ '*dont*' precedes the determiner, $_{\text{it}}$ '*cui*' follows it, and $_{\text{de}}$ '*dessen*' and $_{\text{es}}$ '*cuyo*' replace it: $_{\text{it}}$: L'uomo la cui casa è bella; $_{\text{fr}}$ *L'homme dont la maison est belle*, $_{\text{de}}$ *Der Mann, dessen Haus schön ist*, $_{\text{es}}$ *El hombre cuya casa es hermosa*

STAGs for Numerals (Nm)

Beside the known cardinal and ordinal subcategories there are other types of numerals with require quite different descriptions. Some are NP modifiers (like distributives or multiplicatives), others are sentential adverbs (adverbial numbers like $_{\text{en}}$ '*firstly*', $_{\text{de}}$ '*erstens*') or verbal modifiers like multiplicatives ($_{\text{en}}$ '*twice*', $_{\text{it}}$ '*primo*'). Therefore more subcategories are required for numerals. In some languages they are characterised by morphological means, in others they are built compositionally. The LTtagset defines the following subcategories:

[37] Like in the Universal Dependencies definition: https://universaldependencies.org/u/pos/

NmCa	*snm=card*	*cardinal number*	Cardinals like $_{\text{en}}$ '*one*', $_{\text{ru}}$ 'два', $_{\text{it}}$ '*tre*', $_{\text{tr}}$ '*dört*' exist in all TINA languages
NmOr	*snm=ord*	*ordinal number*	The same holds for ordinals, like $_{\text{en}}$ '*first*', $_{\text{de}}$ '*zweite*', $_{\text{ru}}$ 'третий'
NmDg	*snm=dg*	*digits*	Digits like *1, 2, 3* are mostly cardinals but they are prima facie underspecified, and require interpretation
NmAv	*snm=av*	*adverbial number*	They are used in initial position in phrasal enumerations. Examples are $_{\text{de}}$ '*erstens*', $_{\text{it}}$ '*primo*', $_{\text{fr}}$ '*secondement*'. They are sentential adverbs
NmCl	*snm=coll*	*collective numbers*	These numerals address collectives of individuals: $_{\text{en}}$ '*threesome*', $_{\text{de}}$ '*zu dritt*', $_{\text{it}}$ '*in tre*'. They have adverbial function
NmDs	*snm=dist*	*distributive numbers*	They count groups of individuals: $_{\text{en}}$ '*three each*', $_{\text{de}}$ '*je drei*' $_{\text{tr}}$ '*beşer*'. They also function as nominal modifiers[38]
NmFr	*snm=fract*	*fractions*	The denominators of fractions are specially marked: $_{\text{de}}$ '*Drittel*', $_{\text{fr}}$ '*tiers*', $_{\text{it}}$ '*terzo*'
NmMu	*snm=mult*	*multiplicatives*	Multiplicative numerals are $_{\text{en}}$ '*twice*', $_{\text{de}}$ '*dreimal*'

For the recognition of NmDg no lexical analysis is required, members of this tag are delivered by the tokeniser. Most of the other subcategories could be covered under the BTAG Av as most of the other tagsets do.

STAGs for Nouns (common nouns) (No, NoC)

Except for the classifier and deverbal nouns, the subcategorisation of common nouns follows semantic criteria: They are all required in Named Entity analysis. As the Content Analysis comprises both common nouns, i.e. terms, and proper nouns, i.e. named entities, the tagset is supposed to reflect this fact.

No, NoC		*common noun*	nouns with no further subcategorisation
NoCC	*sno=class*	*classifier nouns*	The are most important in Chinese; each noun would specify which classifier to use. But they also occur in other languages albeit in rudimentary form. Examples are zh 朵, tr 'tane', fa 'خوشه' (xuše, bunch)
NoCD	*sno=day*	*day names*	Day and month names are given in all TINA languages. They are needed in the analysis of dates in Named Entity recognition. Day names also comprise '*Christmas*' '*Thanksgiving*' etc

[38] While they precede the head in German NPs (*Teams von je 4 Personen*) they are in split position in English, with a post-head element: *teams of 4 people each.*

NoCM	*sno=month*	*month names*	Day and month names are given in all TINA languages. They are needed in the analysis of dates in Named Entity recognition
NoCN	*sno=nat*	*nationalities*	Inhabitants of states (*'Albanians'*), regions (*'Bretons'*) or members of ethnicities (*'Inuit'*, *'Beloch'*) are collected here. They are not considered proper names. In some languages these words are capitalised
NoCT	*sno=title*	*titles (of persons)*	Titles are often abbreviations. They cover specific positions in NPs, and indicate person names in Named Entity recognition
NoCU	*sno=unit*	*unit names*	Unit names are also mostly abbreviations, like *'cm'*, *'kg'*, *'sec'* but also *'$'*, *'€'* etc. They are needed in Named Entity recognition. The type of a unit (e.g. *'currency'*) is given in the XTAG

Acronyms (NoCA) would not be particularly subcategorised as they have the same distribution as regular nouns. It is not straightforward to decide whether an acronym is a common (No) or a proper noun (Na) (cf. *'EBF=European Banking Federation'* is the Name of a legal entity, versus *'ERP=Electronic Road Pricing'* is a common noun)[39] however the respective treatment is like the other nouns. Acronyms are treated as allomorphs of their non-abbreviated forms and share their lemma; the fact that they have special inflection paradigms are expressed as features of these allomorphs in the XTAG.

STAGs for Names (proper nouns)

Proper nouns are key elements in Named Entity Recognition. Their subcategories are also found in the LT resources and tagsets.

No, *NoP*		*proper name*	if no additional information is given, such a word is just a name
NoPC	*sna=comp*	*company names*	This subcategorises companies, like *'AT&T'*, *'Microsoft'*, *'Stada'*. In borderline cases the legal definition is consulted. Company indicators, like *'Ltd'*, *'GmbH'*, *'SA'* are categorised as (common) nouns
NoPI	*sna=inst*	*institutions*	Institutions are all other kinds of organisations: Government institutions, private organisations like football clubs, also organisations like *'Mafia'*, *'Catholic Church'* etc. would be classified as institutions. Further distinctions can be given in the XTAG

[39] There are also homographs: *BMW* as a car is a common noun but *BMW* as a company would be categorised as a proper noun.

NoPL	*sna=loc*	*location names*	Location names ('*Asia*', '*Paris*', '*Lake Van*') are language specific and need to be translated in multilingual contexts. Subclasses (countries, cities etc.) are given in the XTAG
NoPF	*sna=fname*	*first names*	Lists of first names help in Named Entity recognition as the last names of persons are often unknowns. There are limitations of course: In Turkish, nearly every noun can be used as first name (*Hasret, Sevgi, Özlem, Yılmaz, Gül, Güneş* etc.)
NoPN	*sna=lname*	*last names*	The same holds for last names (*Hendersson, Taylor, Dubois*). The most frequent ones may be lexicalised and need a tag description then
NoPP	*sna=pers*	*person names*	Most of the complete names are (two-part) multiwords. But there are also single word person names like '*Napoleon*' '*Jesus*'

Like in the common noun subcategorisation, proper nouns are not subcategorised for acronyms. Those acronyms that are proper names (*UNICEF*) are treated like the other proper names of the respective subcategory.

STAGs for Punctuations (Pc)

Although this BTAG has only a few members, it needs to be subcategorised extensively because each punctuation has different functionality. Attempts to build larger classes (all brackets, all sentence-final punctuations) were abandoned because each punctuation requires different treatment in parsing.

PcBO	*spc=ob*	*opening bracket*	
PcBC	*spc=cb*	*closing bracket*	All forms of brackets are meant: [({ < and> })], as well as two-byte versions in Chinese: 〈, 《, 『, 【, 》, 〉, 』, 】
PcEO	*spc=eo*	*exclamation mark opening*	*Spanish ¡*
PcEC	*spc=ec*	*exclamation mark closing*	*regular !*
PcIO	*spc=io*	*interrogative/ question mark opening*	*Spanish ¿*
PcIC	*spc=ic*	*interrogative/ question mark closing*	*regular ?* The opening tokens ¡ and ¿, i.e. inverted exclamation and question marks, are needed in Spanish
PcPd	*spc=pd*	*period*	*single as well as multiperiods:., …*
PcCm	*spc=cm*	*comma*	*regular, and inverted form ، in Farsi*
PcCl	*spc=cl*	*colon*	

PcSc	*spc = sc*	*semicolon*	Commata and semicolons are also found in Arabic scripts (، and ؛). The interpretation is language-specific, the tokeniser has to take care of it
PcHy	*spc = hyph*	*hyphen*	Several forms of hyphens are covered by this subcategory. They need language-specific special treatment in analysis. It needs to be decided if they are word boundaries or not
PcQS	*spc = sq*	*single quote*	All forms of single quotes (smart quotes, opening / closing quotes) are collected here
PcQD	*spc = dq*	*double quote*	All forms of single quotes (smart quotes, opening / closing quotes) are collected here
PcSy	*spc = sy*	*symbol*	This subcategory collects mathematical operators (+ - / * 2) but also other non-textual elements, like smileys. Symbols need subcategorisation in the XTAGs
PcDS	*spc = ds*	*date separator*	In Arabic there is a special type of colon used for separating the elements of dates

STAGs for Phrases (Ph)

Phrases are just treated as a whole, without further analysis. However, a subclass of the phrases are the interjections.

Ph		*Phrase*	
PhIj	*sph = intj*	*Interjections*	Most tagsets use a special tag for interjections. Here it is a subcategory of the phrases

More detailed analysis of such phrases could be helpful.

STAGs for Pronouns (Pn)

The pronouns in the LT tagset follow the conventional subcategories.

PnDm	*spn = dem*	*demonstrative*	
PnIn	*spn = indef*	*indefinite*	
PnPo, PnPoC	*spn = poss*	*possessive*	Possessive pronouns with special forms occur also in clitic positions in some languages ($_{tr}$ '*ev-im*' my house). The special (clitic) tag PnPoC is required as there are pronoun forms which can be used only in clitic position
PnPs, PnPsC	*spn = pers*	*personal*	Such pronouns have also forms to be used in clitic positions where they represent copula clauses with dropped copula (tr '*hasta-yım*' *I am sick*)
PnQu	*spn = int*	*interrogative*	

PnRl	*spn=rel*	*relative*	
PnRx	*spn=rfx*	*reflexive*	
PnRz	*spn=rez*	*reciprocal*	These units are pronouns in some languages ($_{tr}$ '*birbiri birbirimizi birbirimizden*' etc.), with number and person information. In other languages reciprocal relations are expressed by particles (PtRz): $_{en}$ '*one another*', $_{de}$ '*einander*'

STAGs for Particles (Pt)

The definition of the tagset for particles is a challenge in a multilingual tagset. All languages have idiosyncratic small particles, and only some of them can be grouped into a multilingual representation. Moreover all bound morphemes in the LT analysis are classified as particles. They are also quite language-specific. As a consequence, only some particles are defined in the STAG tagset, and others are left to language-specific extensions.

PtIn	*spt=inf*	*infinitive particles*	Free particles are $_{en}$ '*to*', $_{de}$ '*zu*', bound particles are e.g. $_{de}$ '*zu*' (*an-zu-zeigen*), $_{tr}$ '*mek/mak*' (*gel-mek-ten*)
PtKo	*spt=komp*	*komparative particles*	They are used in comparative constructions: $_{en}$ '*more*', '*most*', $_{es}$ '*más*', '*menos*', $_{it}$ '*più*', $_{tr}$ '*daha*'
PtNg	*spt=neg*	*negations*	Negations in free position are $_{en}$ '*not*', $_{de}$ '*nicht*', in bound position $_{tr}$ '*me/ma*' (*gel-me-mek*), $_{fa}$ 'ن *na*' (نیامد nay-amad)
PtPr	*spt=prec*	*precursors*	These are 'empty' pronouns using as precursors for other constituents: $_{en}$ '*there*', $_{de}$ '*es*', $_{it}$ '*ci*' ('*C'era molta gente*')
PtQu	*spt=qu*	*query particles*	While yes/no queries in many languages are expressed by positional means, some languages use lexicalised means, like $_{tr}$ '*mi/mı/mü/mu*', $_{fa}$ 'مگر magar', 'آیا aya'
PtRz	*spt=rez*	*reciprocal particles*	These particles express reciprocity. Examples are $_{en}$ '*each other*', $_{de}$ '*gegenseitig*', $_{fa}$ 'یکدیگر yekdigar'
PtVb	*spt=vb*	*(separable) verb particles*	Several languages know separable verb particles, like $_{en}$ '*off*', '*over*' (in '*stand off*', '*roll over*'), $_{de}$ '*an*', '*ab*' ($_{de}$ '*sagt … an*'), $_{fa}$ 'بر *bar*', 'فرو *foru*' (in برمیگردد *bar-mi-gardad*). They need special treatment in lemmatisation as they are part of the verb lemma

Of course there are other particles, like $_{fr}$ '*en*', $_{tr}$ '*ki*' $_{zh}$ '者' or '偏巧', $_{fa}$ 'را *rA*', specific for those languages. They are collected in language-specific extensions of the Pt tag.

Bound morphemes to be used in Lexical Analysis e.g. for Turkish and Farsi need a categorial description as well. They are categorised as particles (Pt) with specific extensions: the naming convention is that particles are identified along their position (prefix—infix—suffix—inflection) and the BTAG they occur in (noun—verb—adjective), followed by language-specific information. For instance, *PtSN* describes nominal suffixes, *PtFV* verbal inflections. Details are collected in language-specific extensions of the tagset.

STAGs for Verbs (Vb)

Verbs are subcategorised in a conventional way (feature *svb*), distinguishing full verbs, auxiliaries and modals. In addition they are divided according to their forms (feature *vform*), distinguishing finite verbs, infinite verbs, gerunds and participles. Most languages have two participles, the present active forms are called *gerunds* here, the past mainly passive forms are called *participles*. Of course all these forms have different positions in the systems of their respective languages, and also different linguistic descriptions in their XTAG features; therefore the naming here is only an approximation, indicating common aspects.

In some languages gerunds and participles can be inflected ($_{\text{it}}$ *'fabricanti'*, *biforcati'*, $_{\text{de}}$ *'laufend-en'*, *'informiert-er'*) and resemble adjectives. However distributional differences to real adjectives exist, in particular in framing: While adjectives may have just one complement (*'proud of'*), participles can provide the complete verb frame, albeit in transformed form: $_{\text{de}}$ *'das dem Schüler vom Lehrer (gestern) geschenkte Buch'*[40]; so they must be dominated by a clausal constituent.

VbFF	*svb=full* *vform=fin*	*finite full verbs*	
VbFI	*svb=full* *vform=inf*	*infinite full verbs*	Many languages have two infinitives: a 'regular' one and a 'nonregular' one. Nonregular infinitives comprise: Short infinitives in Turkish (*'oturma'*) and Farsi ('رست' *rast*, 'مرد' *mord*), infinitives with insertions in German (*'heraus-zu-kommen'*), infinitive pessoal in Portuguese (*'estud-ares'*)
VbFG/ *VbFGi*	*svb=full;* *vform=grd*	*gerund full verbs (inflected)*	Examples for inflections are $_{\text{de}}$ *'kommende Woche'*
VbFP/ *VbFPi*	*svb=full;* *vform=par*	*participle full verbs (inflected)*	Several languages allow the participle forms to be inflected: $_{\text{es}}$ *'refugiada'*, $_{\text{de}}$ *'verbesser-t-es'* $_{\text{tr}}$ *'gel-en-ler'*
VbAF	*svb=aux;* *vform=fin*	*finite auxiliaries*	
VbAI	*svb=full;* *vform=inf*	*infinite auxiliaries*	Nonregular infinitives are not known, and not covered in the tagset
VbAG	*svb=aux;* *vform=grd*	*gerund auxiliaries*	
VbAP	*svb=aux;* *vform=par*	*participle auxiliaries*	There are very few forms but still inflected forms can be found, e.g. $_{\text{de}}$ 'der gewesene Minister'

[40] In English the phrase would be positioned *after* the head: *the book donated to the pupil by the teacher yeaterday*.

VbMF	*svb=mod; vform=fin*	*finite modals*	
VbMI	*svb=mod; vform=inf*	*infinite modals*	No nonregular modal infinitives are known
VbMG	*svb=mod; vform=grd*	*gerund modals*	
VbMP	*svb=mod; vform=par*	*participle modals*	Inflected modal participles ($_{de}$ '*die gekonnte Übung*') indicate the function of a full verb rather than a modal

Major differences between the languages will be found in the feature decoration, e.g. tense systems, types on infinite forms etc. All this needs to be treated in the XTAG schemata.

3. Extended Tagset (XTAG)

Design

It has been explained above that the TINA tagsets is a three-tier hierarchy, each tier expanding the previous level: Starting from a basic tagset (BTAG) as used in standard dictionaries, the first extension relates to syntactic subcategorisations of these tags, described in the Standard Tagset (STAG). Further extensions can be introduced on the basis of the syntactic subcategories, referring to the morphological descriptions of the STAG members. This is the level of the Extended Tagset (XTAG). Many STAG members can be morphologically extended.[41]

The morphological descriptions are specified in the form of feature value pairs; the tagset tries to be as generic as possible to support all TINA languages. This implies compromises in some cases, e.g. in the description of tense systems in different languages. Some features (like *gender*) are not available in some languages, others use different value sets (like *case*). There are also differences in the distribution of morphological features to STAGs: For instance, Arabic needs inflection information for some subordinative conjunctions, Portuguese needs '*person*' information in the *Infinitivo Pessoal*.[42]

The XTAG tagset is only relevant for languages which use fully inflected forms in their analysis. It does not provide information for Chinese as Chinese characters are not changed by morphological information; and it does not support Turkish as the analysis of this language does not use fullform dictionaries because of the

[41] In a system of typed features, STAG features would be of type '*synt*' while the XTAG features would be of type '*morph*'.

[42] (Hundertmark-Santos Martins 1998:171ff).

complexity of its agglutination capabilities, cf. Sect. 5.3. A special case is Persian: An XTAG version of Farsi has been produced based on canonical writing, to be used for speech understanding dictionaries; however in analysis this cannot be used due to the spelling and whitespacing variations, and a different way of analysis is used (cf. Sect. 5.4).

It should also be noted that the XTAG system, as well as the STAGs describes only feature configurations which can be assigned to single word textforms. But many morphological descriptions are compositional, like degree information ($_{\text{en}}$ '*more likely*'), auxiliaries for verbs ($_{\text{es}}$ '*ha comido*', $_{\text{de}}$ '*ist gekommen*') or definiteness markers ($_{\text{fa}}$ 'خانه را' *hane rA* = the house (akk)). Such cases need to be handled by syntactic analysis and are not subject to XTAG descriptions. So the XTAG is just a subset of a complete description of the morphology of a language.

Moreover, the XTAG features do not cover phenomena which are relevant in word formation, for instance the use of different stems to build different verb forms ($_{\text{es}}$ '*contar—cuente*', $_{\text{ru}}$: 'читать'—'чита') or plural forms (broken plural in Arabic and Farsi: عصب pl. اعصاب = *nerve/s*). Because the analysis always deals with a given textform the formation of this textform is not in its focus. In the Basic Lemma Dictionary of course such information items need to be represented (cf. Sect. 9.2.1).

XTAGs: Morphological Extensions

The XTAG descriptions use the following morphological features:

- **Gender**: In addition to the standard values feminine (f), masculine (m), neuter (n), a feature undefined (u) is used for cases of pronouns ($_{\text{it}}$ '*quale*', '*colui*', $_{\text{fr}}$ '*leur*', '*quelque*').
- **Number**: uses singular (s) and plural (p), for Arabic dual (d) in addition
- **Case**: nominative (n), genitive (g), dative (d), accusative (a) in many languages, in addition oblique (o), locative (l) and ablative (b) in Turkish, instrumental (i) and prepositional (p) in Russian.

These features are mainly used for nominal constituents: nouns, adjectives, pronouns. Adjectives have degree information in addition:

- **Degree** with values for positive (p), comparative (c), superlative (s), and elative (e)

Verbal constituents use the following features:

- **Person** with 1, 2, 3
- **Mood**: Values are indicative (i), imperative (m), subjunctive (s), conditional (c)
- **Tense**: The values give present (p), imperfective (i), perfective (r), future (f), pluperfect (q). However the different languages provide different nuances

XTAG:	Tense value P	Tense value I	Tense value R	Tense value F
ar	imperfect / ongoing	perfect / past	--	--
de	Präsens	Imperfekt	--	--
en	present	past	--	--
es	presente	préterito imperfecto	pretérito indefinido	futuro imperfecto
It	present	imperfetto	passato remoto	futuro presente
fr	present	imparfait	passé simple	futur1
pt	present	pretérito imperfeito	pretérito perfeito simples	futuro simples
ru	präsens	präteritum	--	futur

Fig. 2.17 Variations in XTAG tense feature

behind these tags. An example for the values of the tense features is given in tab. 2–17; they would have to be set into relation with the other tenses in the respective languages.

- **Aspect** is coded for Russian (perfective (p) and imperfective (i)) and Farsi (durative (d))
- **Voice** with active (a) and passive (p) is coded for Russian and Turkish

Negation is a Boolean feature used everywhere, be it in verb morphology or in adverb descriptions.

As explained above, these values refer only to cases where full wordforms must be described. Of course complete paradigms exist in the single languages, many forms being built compositionally. So the XTAG annotations form only a subset of the complete morphological descriptions (Fig. 2.17).

Figure 2.18 table shows the morphological features and values used in XTAGs.

The number of XTAGs differs strongly between the languages, as shown in Fig. 2.19. No XTAGs are given for Turkish and Chinese, Farsi XTAGs are result of the fullform generation as explained above.

As for the notation, the TINA XTAG extensions begin with a hyphen, followed by a description of features and values, the features being single-letter uppercase letters, the values lowercase single-letters. Examples are given in Fig. 2.20.

This notation has the advantage that a feature can have several values (like '*Gfm*' *gender=feminine or masculine*), and the features are position independent, compared to notations as e.g. in Multext where the tags have a predefined set of features in predefined positions.[43]

As for processing, different POS taggers using Extended Tagsets have been compared e.g. in (Allauzen and Bonneau-Maynard 2008), showing that taggers are sensitive to the tagsets which they process.

[43] Cf. the examples in Erjavec (2016).

Feature	ar	de	en	es	fa	fr	it	pt	ru	tr
Gender	f,m	f,m,n	f,m	f,m,u	--	f,m,u	f,m,u	f,m,u	f,m,n	--
Number	s,p,d	s,p	s,p	s,p	s,p	s,p	s,p	s,p,u	s,p	s,p
Case	n,o	n,g,d,a	n,g,o	--	--	--	--	--	n,g,d,a,i,p	n,g,d,a,l,b
Degree	--	p,c,s	p,c,s	p,c,e	p,c,s	p,c	p,c,e	p,c,e	p,c,s	p,c,s
Person	1,2,3	1,2,3	1,2,3	1,2,3	1,2,3	1,2,3	1,2,3	1,2,3	1,2,3	1,2,3
Mood	i,m	i,m,s	i,m	i,m,s,c	i,m,s	i,m,s,c	i,m,s,c	i,m,s,c	i,m	--
Tense	i,p,f	i,p	i,p	i,p,f,r	i,p,r	i,p,f,r	i,p,f,r	i,p,f,r	i,p,f	--
Aspect	--	--	--	--	d	--	--	--	i,p	--
Voice	--	--	--	--	--	--	--	--	a,p	a,p
Neg	t	t	t	t	t	t	t	t	t	t

Fig. 2.18 XTAG features and values

language	ar	de	en	es	fa	fr	it	pt	ru	tr
no. XTAGs	216	715	264	332	293	306	332	353	818	--

Fig. 2.19 XTAG: size of tagsets

Language	Textform	XTAG	Lemma
de	heiler	NoCo-GmNpCa	heiler
	heiler	NoCo-GmNpCg	heiler
	heilers	NoCo-GmNsCg	heiler
	heilerfolge	NoCo-GmNpCa	heilerfolg
en	oakwoods	NoCo-NpCn	oakwood
	oakwood's	NoCo-NsCg	oakwood
	maladministered	VbFF-MiTiNpP1	maladminister
es	podía	VbMF-MiTiNsP1	poder
	debería	VbMF-McNsP1	deber
	glándulas	NoCo-GfNp	glándula
fa	توانستنها	NoCo-Np	توانستن
	توانگر	NoCo-Ns	توانگر
	توانگری	NoCo-NsDi	توانگر
	توانگران	NoCo-Np	توانگر
	میگرداند	VbFF-MiTiAdP3Ns	گرداندن
	میگشتند	VbFF-MiTiAdP3Np	گشتن
ru	желай	VbMF-MmNsP2Va	желать
	желали	VbMF-MiTiNpVa	желать

Fig. 2.20 XTAG examples

2.3.2.3 Comparisons

The definition of the linguistic environment is a central point in text analysis: All components and tools will have to use it, and inadequate linguistic descriptions will propagate through the whole system. An accurate setup of the basic linguistic data, esp. tagsets and feature/value annotations, is a key requirement; therefore a more detailed review may be justified.

1. Monolingual Tagsets

PENN treebank

The tagset used in the Penn treebank[44] is one of the best-known tagsets. It is monolingual as it is applicable to English only, and it is an extended tagset as it does not just cover syntactic but also morphological distinctions (like tense, number, degree).

The tagset is a shrink-version of the Brown corpus tagset.[45] Compared to this tagset the Penn treebank tagset does not represent tags like qualifiers (in pre and post-head position), and auxiliaries, which were excessively tagged in the Brown corpus (more than 5 tags for each '*be*' and '*have*'). And it collapses prepositions and subordinating conjunctions into one tag ('*IN*'). However a reduced tagset requires more effort to express syntactic phenomena which otherwise would be described more easily, e.g. by using an AUX tag; it also produces more ambiguities in the description of such phenomena in a corpus.

Some decisions could be worth discussing: The creation of a special tag for nouns like '*#*' or '*$*'; a '*list item*' tag which rather belongs to the deformatting component; the omission of an AUX node; a reduced set of pronoun subcategorisation; no differentiation in the tag '*verb baseform*' (*VB*) between infinitive and imperative use, and so on.

For a multilingual setup as tried in the LtTagset, the main difficulty with the Penn treebank tagset is that it is completely monolingual, optimised for English analysis. The distinction of verb forms between '*3-ps-singular*' and '*non-3-ps-singular*' is only valid in a system with rather poor morphological markers. Merging prepositions and subordinating conjunctions is hardly possible in other languages as different lexemes are used there. The analysis intended by using pre- and post-determiner tags mirrors a specific kind of NP analysis, following (Quirk et al. 1972), which is not used in the grammars of other languages. The tag for possessive endings ("s') is also English-specific and requires more elaborate tagging in other languages. Finally the morphological extensions of the tagset would have to be elaborated more in morphologically richer languages.

In the LtTagset, most Penn treebank tags can be mapped to STAG elements (cf. Fig. 2.21[46]). The morphological phenomena would be covered in the XTAG set, some tags like '*FW*' (foreign words) or '*LS*' (list item markers) would be dropped, the possessive endings ('*POS*') would be merged with clitic possessives in other languages, prepositions and subordinating conjunctions would be split (with many homograph lexemes), and the nominal modifiers would be regrouped. Such a

[44] Marcus et al. (1993).

[45] Cited in Jurafsky and Martin (2009).

[46] * = The morphological differentiations are covered in the XTAG set. ** = The punctuations are grouped differently in the STAG, cf. above. In Penn, the sentence-final '.' covers :?!, mid-sentence ',' covers:; …, – – *** = The list items are part of the text handling component.

BTAG	Penn	STAG	meaning
Ad			
	JJ	AdAt	Adjective
	JJR	AdAt*	comparative
	JJS	AdAt*	superlative
Ap			
	IN	ApPr / CoSu	preposition or subordinating conjunction
Av			
	RB	Av	Adverb
	RBR	AvPr*	Adverb, comparative
	RBS	AvPr*	Adverb, superlative
	WRB	AvQu	wh-adverb
Co			
	CC	CoCo	coordinating conjunction
Dt			
	DT	DtDd/i	determiner
	PDT	AvNo	predeterminer
	WDT	DtWh	wh determiner
Nm			
	CD	NmCa	cardinal number
No			
	NN	NoC	noun, singular or mass
	NNS	NoC*	noun plural
Na			
	NP	NoP	proper noun sg
	NPS	NoP*	proper noun plural

BTAG	Penn	STAG	meaning
Pc			
	$ # l“ r“ ( ) ; . :	Pc_	Punctuations **
Ph			
	UH	PnIj	interjection
Pn			
	PP	PnPs	personal pronoun
	PP$	PnPo	possessive pronoun
	WP	PnQu	wh-pronoun
	WP$	PnPsC	Possessive wh-pronoun
Pt			
	EX	PtPr	existential there
	POS	PtPoC	possessiv ending
	RP	Pt_	particle
	TO	PtIn	to
Vb			
	VB	VbFF	verb base form *
	VBD	VbFF	past tense *
	VBG	VbFG	verb, gerund or pres. participle
	VBN	VbFP	verb, past participle
	VbP	VbFF	verb, non-3rd-p sing. present *
	VbZ	VbFF	verb, 3rd-p sing. present *
	MD	VbMF	modal
XY			
	FW	--	foreign word
	LS	--	list item marker ***
	SYM	PcSy	Symbol

Fig. 2.21 PENN tagset and STAG. (* with XTAG extensions, **with STAG extensions, *** not used)

1. Nouns (N)	7. Adverbs (ADV)
2. Verbs (V)	8. Conjunctions (KO)
3. Articles (ART)	9. Adpositions (AP)
4. Adjectives (ADJ)	10. Interjections (ITJ)
5. Pronouns (P)	11. Particles (PTK)
6. Cardinal numbers (CARD)	(12. others)

Fig. 2.22 STTS tagset

conversion can be done automatically to a large extent. In turn, the LtTagset would add all POS tags needed for the description of Named Entities.

STTS

The Stuttgart-Tübingen Tagset' (STTS), cf. (Schiller et al. 1995), is a monolingual tagset for German. It is used in many German treebanks like NEGRA and TIGER. It is also the basis of further developments and adaptations, cf. (Simon 2004).

The STTS defines a list of basic tags (as given in Fig. 2.22).

This list is similar to the BTAGs as defined above. The two differences lie in the definition of determiners versus articles, and in the definition of cardinals instead of numerals. The former is the result of the definition of attributive pronouns (indefinite, possessive etc.): PPOSAT (‚*mein Haus*') as attributive versus

PPOSP (‚*meins*') as nominalised. Attributive pronouns are classified as Dt In STAG. The latter results from the definition of other types of numerals (ordinals, multipliers etc.) as adjectives, so only cardinals are left in the numeral tag (CARD).

The STTS is a detailed and knowledgeable description of the German parts of speech. Most of the STTS tags can be converted into STAGs quite easily, and provide a good basis for a multilingual tagset.

From a multilingual point of view, two issues would need reconsidering.

- First, the adjective tag comprises not just adjectives but also some numerals like ordinals, and participles (both past and present): Non-inflected participles are marked as predicative (ADJD), inflected ones as attributive (ADJA) adjectives. This may be due to the fact that these tags show the same inflectional pattern. However this is not the case in all languages; moreover participle constructions should clearly be described as verbal constituents in syntax: They would have a VP head, and undergo verb framing in more complex constructions ($_{de}$ ‚*das gestern von ihm für Otto gekaufte Buch*' -> $_{en}$ ‚*the book which was bought yesterday by him for Otto*'). For this reason the LtTagset deviates from the STTS and contains participles as special tags, both in non-inflected and inflected form, as well as more extended numerals, esp. ordinals which can take modifiers like '*around*' which normal adjectives cannot take.
- Second, the treatment of adverbs in STAG differs from the STTS treatment. Adverbs in the LtTagset are differentiated on the basis of the constituents they modify: This can be nominals (adjectives, nouns: ‚*Nur die Vögel haben überlebt*'), adverbs (‚*er arbeitet fast umsonst*'), verbs (‚*er ist gestern gekommen*') or clauses (‚*er ist möglicherweise krank*'). This distinction is valid for many languages. As a consequence the LtTagset uses this subcategorisation.

Other differences, like punctuations, are marginal.[47]

Peking University (PKU) Level 1 tagset

The general Chinese dictionary of TINA contains about 63 K entries, tagged according to the PKU (Peking University) tagset, cf. (Yu et al. 1998). This tagset is also used in 'The PKU 863 Chinese-English Parallel Corpus' project, cf. (Sun et al. 2002), as given in Fig. 2.23. It is a monolingual tagset and was developed to POS-tag Chinese corpora. It bears the challenge to map this tagset to the tagsets used in multilingual indexing, like the LtTagset, to streamline the annotation framework.

The LtTagset tries to incorporate the specific Chinese tags into its own framework, for two reasons:

[47] For instance, imperatives must be put under finite forms, as they show person, number, and tense.

PKU tag	meaning	STAG	PKU tag	meaning	STAG
a	adjective	Ad	n	common noun	No
b	non-predicate adjective	AdAt	o	onomatope	PnOn
c	conjunction	Co	p	preposition	Ap
d	adverb	Av	q	classifier	NoCC
e	interjection	PnIj	r	Pronoun	Pn
f	directional locality	ApPL	s	space word (on earth, border area,suburb)	No
g	morpheme	Pt	t	time word (moment, last year, december)	No
h	prefix	PtP	u	Auxiliary	VbA
i	idiom	PnId	v	Verb	Vb
j	abbreviation	- **	w	symbol and non-sentential punctuation	PcSy
k	suffix	PtS	x	unclassified items	PtSy
l	fixed expressions	Pn	y	modal particle	Pt
m	numeral	Nm	z	state (smooth, enlighted, foggy, golden)	Av

Fig. 2.23 PKU tagset

- As POS information is a constitutive element in concept descriptions, similar concepts would require compatible tag information. Therefore it is important to be able to establish systematic relations between the tagsets of different languages, especially in the area of content words. Bilingual dictionaries would map the POS information from source to target language. The more similar the tagsets are the less difficult this operation will be.
- This also holds for tasks like producing bilingual parallel corpora, or running linguistic tools like taggers on the languages at hand. The less different the tagsets are the easier the task will be.

Both reasons are particularly relevant for the open word categories, and the TINA dictionaries use these categories nearly exclusively.

In the attempt to map the PKU tagset to the STAG tagset, every tag of PKU must be linked to its STAG counterpart. According to (Cheung 2016), there are common POS tags between Chinese and Western tagsets, like Noun (‘*n*’), Verb (‘*v*’), Adjective (‘*a*’), Adverb (‘*d*’), Adposition (‘*p*’), Conjunction (‘*c*’), and Numerals (‘*m*’); the main differences are seen in the tags for classifiers, for localizers, and for sentence-final particles.

In the STAG framework, such tags are tried to be covered by the tagset. PKU tags like interjections (‘*e*’), idioms (‘*I*’), fixed expressions (‘*l*’), and onomatopoetica (‘*o*’) would form subcategories of the ‚phrase-related’ Ph tag; morphemes (‘*g*’), prefixes (‘*h*’), and suffixes (‘*k*’) would form subclasses of particles (Pt_). The ‘Chinese-specific’ POS tags ‘classifier’ (‘*q*’) and ‘localizer’ (‘directional locality’: ‘*f*’) would be mapped into the STAGs NoCC and ApPL respectively, and the sentence-final particles would stay under the ‚Pt_’ BTAG, with a special flavour for Chinese. Space (‘*s*’) and time (‘*t*’) words[48] seem to be semantic subcategories of nouns or adjectives.

[48] Including month names, which would undergo separate treatment in the STAGs.

1	ADJ	7	NOUN
2	ADP	8	PRON
3	ADV	9	PRT
4	CONJ	10	. [Pnct]
5	DET	11	VERB
6	NUM	12	X

Fig. 2.24 Google universal tagset

Proper nouns are subcategorised by PKU in a similar way to STAGs (persons '*nr*', places '*ns*', organisations ('*nt*') and other proper nouns ('*nz*')). Abbreviations ('*j*') would be collected under the tags of the non-abbreviated forms, and the unclassified terms ('*x*') would be treated as special kinds of particles (*PtSy*).

The purpose is to represent all relevant distinctions of the PKU tagset in the STAGs, to support interoperability between the tagsets and the tools based thereon. An alternative would be to create a special tagset for Chinese in the way it is proposed for the Penn Chinese treebank, cf. (Xia 2000), which again would be just a monolingual tagset.[49]

2. Multilingual Tagsets

The Google Universal POS tagset

This tagset, given in (Petrov et al. 2012), is multilingual, its scope is a basic tagset. It was constructed by mapping existing treebank tagsets to this basic set (Fig. 2.24).

It resembles the LtTagset BTAG, except for their POS'*X*' (a kind of 'catch-all'), the POS 'noun' also covering names, and a missing '*Phrase*' POS as used in STAG. Compared to other tagsets it is unique as the POS tags are the only information that is provided for corpus annotation. However this tagset has turned out to be too generic to be of use in linguistic analysis; this was the reason why it has been extended by all the other tagset constructions: The more generic the tagset is the more disambiguation effort is needed in parsing. Therefore the effort would go into a more expressive tagset which also covers the multilinguality requirement.

The EAGLES recommendations

The specifications of the EAGLES project[50] aim at a tagset which is multilingual and extended, i.e. including morphological specifications. It defines multilingual morphosyntactic descriptions. The goal is to provide a platform which can serve

[49] Xia (2000) also contains a mapping from their tagset to the PKU tagset in the annex, all of them are monolingual.

[50] EAGLES: http://www.ilc.cnr.it/EAGLES96/home.html.

1	N	[noun]	8	C	[conjunction]
2	V	[verb]	9	NU	[numeral]
3	AJ	[adjective]	10	I	[interjection]
4	PD	[pronoun/determiner]	11	U	[unique/unassigned]
5	AT	[article]	12	R	[residual]
6	AV	[adverb]	13	PU	[punctuation]
7	AP	[adposition]			

Fig. 2.25 EAGLES tagset/obligatory items

as an interchange between different tagsets. Such an 'interset' would be linked to existing tagsets by converters, or tagset drivers, cf. (Zeman 2008).

The EAGLES recommendations divide the data into obligatory and recommended information items, and define two types of extensions, namely optional generic and optional language-specific attributes.

Obligatory items are required by any tagset that wants to be linked to the Intermediate tagset. The only obligatory items are the basic parts of speech, defined as in Fig. 2.25.

All other information items are recommended, or extensions.

The **recommended** information items refer mostly to morphological phenomena, like number, gender, case, tense, etc. For syntactic subcategories similar to the STAG definition, '*type*' features are used in several POS tags, they present recommend values, like '*type=preposition*' for adpositions.

On top of the recommended values, **extensions** are defined. Such extensions may be generic or language-specific.

Generic extensions may be subcategories of adverbs, or of nouns, like '*netype*', if nouns are named entities, as used in Freeling.[51] Examples of **language-specific** extensions are the subcategories of the '*unique*' POS, i.e. the particles in the different languages: Infinitive markers (de, en), second negative particles ($_{\text{fr}}$ '*pas*'), existential particles ($_{\text{en}}$ '*there*', $_{\text{de}}$ '*es*') and others.

Technically the features take fixed positions in the tag definition lattice, their values are of length one on the respective position.

Derivations of the EAGLES recommendations can be seen e.g. in the tagsets of Freeling or in SketchEngine.[52]

While most of the tags of the LtTagset can be mapped from and into the EAGLES Intermediate tagset, some decisions in the EAGLES tagset need a closer look:

As for '*residuals*', they comprise formulae, foreign words but also symbols, acronyms, abbreviations and other material. However, abbreviations should be

---

[51] Freeling: https://freeling-user-manual.readthedocs.io/en/latest/tagsets/.

[52] Sketchengine: https://www.sketchengine.eu/.

Open class words		Closed class words		Other	
ADJ	adjectives	ADP	adpositions	PUNCT	punctuations
ADV	adverbs	AUX	auxiliaries	SYM	symbols
INTJ	interjections	CCONJ	coordin.conjunctions	X	(others)
NOUN	nouns	DET	determiners		
PROPN	proper nouns	NUM	numerals		
VERB	Verbs	PART	particles		
		PRON	pronouns		
		SCONJ	subord. conjunctions		

Fig. 2.26 UDF tagset

tagged like their non-abbreviated parts,[53] and acronyms are very often nouns, common or proper nouns, and should be tagged as such.

Also, to merge different phenomena into the '*unique*' tag may let the tagset look more elegant, however for the following analysis components the tag is not really helpful; the parser would nevertheless have to look into the single items (like '*infinitive marker*'). The same holds for merging the attributive and the nominal pronouns into one tag ('*PD*'): At the end of the day it is still necessary to split it into '*prontype*' and '*dettype*' features as the parser must react quite differently.

In general, it should be remembered that assigning tags to word forms is not a purpose in its own; instead, it is just one element in the analysis chain, and the definition of the tagset should reflect the following analysis steps. For instance, even if an inflected participle has a quite similar distribution to an adjective, the parser would have to build a VP from it, as explained above, so overall it is advantageous to split the POS. The same holds for the distinction between common and proper nouns, here belonging to the same tag: They would have to be treated quite differently later on. Similarly, if Arabic words are tagged 'DTJJ: adjective with the determiner "Al" (ال)' or 'DTNN: noun, singular or mass with the determiner "Al" (ال)' etc.[54] it complicates the parsing process, if such clusters of tags must be handled. As a consequence, the tagset design should keep in mind that defining tags is just the beginning, not the goal of the analysis.

The LtTagset tagset reflects also the requirements of the following analysis steps, esp. parsing and Named Entity recognition. Unlike EAGLES' definition, the distinction between required and optional features is not kept,[55] instead the STAG set contains all tags which are needed for the analysis components. Also the interoperability between LtDeep (using BTAG) and LtShallow (using STAG and XTAG) is an essential requirement.

[53] At least on STAG level; their morphology can be different.

[54] SketchEngine, Arabic tagset: https://www.sketchengine.eu/stanford-arabic-parser-tagset/. The treatment of clitic elements is not envisaged in most of the tagsets discussed here.

[55] Maybe the distinction is required, if more languages than just the TINA languages must be considered.

Universal Dependencies annotations

The **Universal Dependencies**[56] project aims at developing annotated tree banks in many languages, and in this context defines multilingual annotation schemata for words as well as for dependency relations between them. As this is not the goal of the TINA project some phenomena need not be annotated, e.g. typos are corrected instead of annotated. Also the languages covered by TINA would not require many annotations defined in the UD project. However there is significant overlap in the area of POS annotations and the morphosyntactic features.

Like the other projects, the UD project defines a basic tagset, as given in Fig. 2.26.

This tagset can easily be compared to the BTAG system, only three changes are requied: SCONJ and CCONJ are subcategorisations of BTAG Co (conjunction), AUX is part of the verb descriptions there, and the interjections are a subcategory of phrases (Ph).

The UD project does not provide an intermediate level of syntactic description with subcategorisations like the STAG level. In LtTagset the feature annotations are typed: 'Syntactic' features take care of syntactic subcategorisations, 'morphological' features cover morphology, 'semantic' features would cover 'animacy' but also 'adverb type', 'pragmatic' features would cover 'politeness', etc. The Universal Dependency offers a significant set of features and values below the POS level which can be used by annotators if required. To compare the STAG tags the closest level of information would be the 'type' feature provided for many POS tags. Differences like subcategorisation of adjectives, only given in STAG, subsumption of some numerals under adverbs in UD without further annotation, the option to provide both auxiliaries and modals under the AUX node, or a subcategorisation for SCONJ into finite and infinite subclauses, would indicate different analysis strategies in TINA compared to the UG approaches.

While the BTAG is an essential part of a concept the STAG is a means of disambiguation in case of homographs. Full parsing would require full linguistic descriptions as in XTAG, plus additional syntactic-semantic information like verb frames.

References

Abu-Chacra, F. (2018). Arabic, An Essential Grammar. *Routledge*

Allauzen, A., & Bonneau-Maynard, H. (2008). Training and Evaluation of POS Taggers on the French MULTITAG Corpus. *Proc. 6th Int. Conf. on Language Resources and Evaluation (LREC'08), Marrakech, Morocco.*

[56] https://universaldependencies.org/, cf. de Marneffe et al. (2022).

Baroni, M., Bernardini, S., Ferraresi, A., & Zanchetta, E. (2009). The WaCky wide web: a collection of very large linguistically processed web-crawled corpora. *In: Language Resources and Evaluation 43.3, pp. 209–226.*

Barret, A. (1999). Otelo and the Domino translation object. *Proc. Machine Translation Summit VII, pp. 177–179, Singapore, Singapore*

Beesley, K. & Karttunen, L. (2003). Finite-State Morphology. *CSLI Center for the Study of Language and Information*

Bresnan, J. (2001). Lexical-Functional Syntax. *Oxford: Blackwell Publishers.*

Carnie, A. (2013). Syntax: A Generative Introduction. *John Wiley & Sons*

Charniak, E. (1997). Statistical Parsing with a Context-free Grammar and Word Statistics. *Proc. AAAI'97/IAAI'97, Providence, Rhode Island*

Cheung, C.H.C (2016). Chinese: Parts of speech. *In: Chan, S.W., ed., The Routledge Encyclopedia of the Chinese Language. New York: Routledge*

Chomsky, N. (1965). Aspects of the Theory of Syntax. *MIT Press*

Collins, M. (2003). Head-Driven Statistical Models for Natural Language Parsing *Computational Linguistics, 29:4, pp. 589-637*

Cunningham, H., Wilks, Y., & Gaizauskas, R. (1996). GATE – a General Architecture for Text Engineering. *Proc. 16th Int. Conf. on Computational Linguistics (COLING) 1996 Volume 2*

Cunningham, H., Maynard, D., Bontcheva, K., & Tablan, V. (2002). GATE: A Framework and Graphical Development Environment for Robust NLP Tools and Applications. *Proc. 40th Annual Meeting of the Association for Computational Linguistics (ACL), 2002, Philadelphia, PA, USA.*

Daciuk, J., & van Noord, G. (2002). A Finite-State Library for NLP. *Proc. 12th Computational Linguistics in the Netherelands (CLIN)*

Daciuk, J., & van Noord, G. (2002). A Finite-State Library for NLP *In: Bruce Watson, Derick Wood (eds) (2002). Implementation and Application of Automata. Lecture Notes in Computer Science. pp. 65–73*

de Marneffe, M., Manning, Ch., Nivre, J., & Zeman, D. (2021). Universal Dependencies. *In: Computational Linguistics, vol. 47, no. 2, pp. 255–308 (ISSN 1530–9312)*

Erjavec, T., ed. (2016). MULTEXT-East Morphosyntactic Specifications. V. https://nl.ijs.si/ME/Vault/V5/msd/html/

Fischer, H.G., ed. (1982). Information Retrieval und natürliche Sprache. Integrierte Verarbeitung von Daten und Texten im Modell CONDOR. *K.G.Saur*

Gachot, D.A. (1986). The SYSTRAN Renaissance. *Proc. Machine Translation Summit 1986, Munich, pp. 60–65*

Gazdar, G., Klein, E.H., Pullum, G.K., & Sag, I.A. (1985). Generalized Phrase Structure Grammar. *Oxford: Blackwell, and Cambridge, MA: Harvard University Press.*

Goyal, P., & Kulkarni, A. (2014). Converting Phrase Structures to Dependency Structures in Sanskrit. *Proc 25th Int. Conf. on Computational Linguistics (COLING) 2014: Technical Papers, Dublin, Ireland, pp. 1834–1843*

Haspelmath, M., Dryer, M., Gill, D., & Comrie, B., eds. (2005). The world atlas of language structures. *Oxford University Press*

Huddleston, R., & Pullum, G.K. (2002). The Cambridge Grammar of the English Language. *Cambridge Univ. Press*

Hulden, M. (2009). Foma: a Finite-State Compiler and Library. *In Proc. of the Demonstrations Session at EACL 2009, pages 29–32, Athens, Greece.*

Hundertmark-Santos Martins, M.T. (1998). Portugiesische Grammatik. *Tübingen (Niemeyer)*

Ide, N., Bonhomme, P., & Romary, L. (2000). XCES: An XML-based Encoding Standard for Linguistic Corpora. *Proc. 2nd Int. Conf. on Language Resources and Evaluation (LREC 2000), Athens, Greece*

Jackendoff, R. (1977). X-bar Syntax: A Study of Phrase Structure. *MIT Press, Cambridge (MA)*

Jurafsky, D. & Martin, J.H. (2009). Speech and Language Processing. *Pearson Int. Edition (2nd ed.).*

Karpukhin, V., Oguz, B., Min, S., Lewis, P., Wu, L., Edunov, S., Chen, D., & Yih, W.T. (2020). Dense Passage Retrieval for Open-Domain Question Answering. *Proc. 2020 Conf. on Empirical Methods in Natural Language Processing (EMNLP), pp. 6769–6781, Online.*

Kong, L., Rush, A., & Smith, N. (2015). Transforming Dependencies into Phrase Structures. *Human Language Technologies: The 2015 Annual Conf. of the North American Chapter of the ACL, pages 788–798, Denver, Colorado*

Kornai, A., & Pullum, G. (1997). The X-bar Theory of Phrase Structure. *Language 66,1*

Manning, C.D., Raghavan, P., & Schütze, H. (2008). Introduction to Information Retrieval. *Cambridge University Press*

Marcus, M.P., Santorini, B., & Marcinkiewicz, M.A. (1993). Building a Large Annotated Corpus of English: The Penn Treebank, *in Computational Linguistics, Vol. 19, No. 2 (June 1993), pp. 313--330 (Special Issue on Using Large Corpora).*

Özenç, B., & Solak, E. (2020). Visual Modeling of Turkish Morphology. *In Proc. of the 12th Int. Conf. on Language Resources and Evaluation (LREC), pp. 3984–3990, Marseille, France.*

Petrov, Sl., Das, D., McDonald, R. (2012). A Universal Part-of-Speech Tagset. *Proc. of the 8th Int. Conf. on Language Resources and Evaluation (LREC'12), Istanbul*

Pollard, C., & Sag, I. (1994). Head-driven phrase structure grammar. *Chicago: Unversity of Chicago Press.*

Quirk, R., Greenbaum, S., Leech, G., & Svartvik, J. (1972). A Grammar of Contemporary English. *Longman*

Rocchi, L. (2016). Definiteness vs. Indefiniteness in the Turkish language. *Rivista Internazionale di Tecnica della Traduzione/Int. Journal of Translation 2016 (18), pp. 185–205*

Roche, E., & Schabes, Y. (1997). Finite-State Language Processing. *MIT Press*

Rozenberg, G., & Salomaa, A. (1997). Handbook of Formal Languages. Volume 1: Word, Language, Grammar. *Springer*

Savourel, Y., ed. (2005). TMX 1.4b Specification. https://www.gala-global.org/tmx-14b

Schäfer, U. (2006). Integrating deep and shallow natural language processing components: representations and hybrid architectures. *Diss. Saarbrücken*

Schiller, A., Teufel, S., Thielen, C., (1995). Guidelines für das Tagging deutscher Textcorpora mit STTS. *Technical Report, IMS-CL, University Stuttgart, 1995*

Simon, B.P. (2004). Ein Annotationsschema für deutsche Sätze – Anleitungsvorschlag zur Verwendung mit @nnotate. *Diss. TU Berlin*

Sun, L., Xue, S., Qu, W., Wang, X., & Sun, Y. (2002). Constructing of a Large-Scale Chinese-English Parallel Corpus. *Proc. 19th Int. Conf. on Computational linguistics (COLING '02): The 3rd Workshop on Asian Language Resources and International Standardization*

Thurmair, Gr. (1990). METAL: Computer Integrated Translation. *Proc. of a Workshop on Machine Translation, July 1990, UMIST*

Winograd, T. (1983). Language as a Cognitive Process, vol. 1: Syntax. *Addison-Wesley*

Xia, F. (2000). Chinese The Part-Of-Speech Tagging Guidelines for the Penn Chinese Treebank (3.0), *IRCS Technical Reports Series, no RCS-00–07 University of Pennsylvania.*

Xia, F. & Palmer, M. (2001). Converting Dependency Structures to Phrase Structures. *In Proc. of the 1st Int. Conf. on Human Language Technology Research.*

Yu, S., Zhu, X., Wang, H., Zhang, Y. (1998). The Grammatical Knowledge-base of Contemporary Chinese, A Complete Specication (in Chinese). *Tsinghua University Press, 1998.*

Zeman, D. (2008). Reusable Tagset Conversion Using tagset Drivers. *Proc. 6th Int. Conf. on Language Resources and Evaluation (LREC'08), Marrakech, Morocco*

Chapter 3
Text Analysis Preprocessing

Pre-processing consists of

- text handling, i.e. conversion of text into the LtText format,
- language identification, and
- topic identification.

First the textual elements of a document need to be isolated and standardised by converting them into normalised character code and common format annotations. The target structure of text handling is the LtText format.

Next, the most important task is to determine the **language** of the document. Only if its language is known, the system can load the resources required to analyse it.

The identification of the document **topic** is used to decide whether or not a document should become member of a document collection in a given domain.

These components are called before proper linguistic analysis is possible because the required resources are not yet in place. A consequence is that they must deal with (non-normalised non-lemmatised) textforms, including inflections, capitalisations etc.; this affects the resources needed e.g. in topic identification.

The result of the preprocessing step is an instantiated LtDoc Document Model, with the tags for language and topic instantiated, and more tags set as far as the information is provided by the external documents, like paragraphs and their types.

3.1 Text Handling (LtText)

The purpose of the Text Handling component is to instantiate a Document Model for the respective document. The relevant information elements are the markups (<doc> and <p>), their attributes (*ptype* and *flow* for <p>), and the text (*onset* and

G. Thurmair, *Knowledge-Driven Multilingual Text Analysis and Transparent Information Retrieval*, Cognitive Technologies,
https://doi.org/10.1007/978-3-031-91741-7_3

offset of the text per<p>). In case headers are supported in the Document Model instance the relevant items would be the location (URL), character code, original format and other useful information (like '*creation date*') of the respective document.

There are elements in a document which require different analysis techniques (like figures, graphics, tables etc.); they will be removed from the document texts by the texthandling component.

The text data to analyse are enterprise or intranet data in most cases, but also open source (like governmental) data, or results of internet searches. Internet documents can be collected using crawlers, which are initialised by seed URLs; in cases of focused crawlers also sets of seed terms are used to thematically restrict the search results. Such crawlers are described in Sect. 9.3.1.

Depending on their origin, the text data are encoded in different formats. Each of these formats has its specific challenges.

TXT formatted texts require special attention to three issues:

- The **paragraph** boundaries<p>must be detected. Such a detection can use empty lines, indents or other means.
- Some txt files use **hard returns** at the end of each line. If hard returns would be used as<p>markers (as usual), then this would break the text flow and leave defective text chunks. This happens for instance when pdf text is converted into txt format. In such cases the real paragraph boundaries must be detected, and the connected text parts must be rebuilt.
- Some **list items** are not straightforward to detect. In particular if capital letters are used as list items, they could be ambiguous and also belong to the text flow (e.g. abbreviated first names). Only contextual clues can decide whether they are list items or text flow elements.

XML documents need to be parsed, and their markups need to be interpreted. Their definition files (DTDs or schemata) need to be examined to find the<p>elements and their attributes. Several solutions exist, e.g. the Pub2TEI project which aims at transforming a variety of publisher XML format to the same TEI XML format.[1] For other XML formats special converters need to be written.

It should be noted that XML does not support all characters: If the proper XML characters occur in texts, they must be escaped: < & > ' &quot. Most control characters below \u001F are not allowed either.[2]

pdf documents need to be converted into editable text. Several converters exist, for instance the GROBID software with a particular focus on technical and scientific publications.[3] Once converted into txt, the pdf documents must be further processed to find the markups and attributes needed by LtText, as described above.

[1] https://github.com/kermitt2/Pub2TEI.

[2] https://en.wikipedia.org/wiki/valid_characters_in_xml

[3] https://cwiki.apache.org/confluence/display/tika/GrobidJournalParser.

HTML documents need to be parsed and the relevant markups need to be interpreted: The relevant HTML markups are the ones which relate to special text elements[4]:

- <p>tags are taken over into LtText as<p>. If the document contains<header>and<body>, these elements are also retained.
- headings (<h1>...<h6>) receive the<p>attribute *'paratype = "heading"'*. Headings in English have special capitalisation rules, therefore the identification of headings is needed in analysis: Terms in headings can also receive a higher weight in indexing.
- list items (<li>) are attributed with *'paratype = "enum"'*. List items often contain defective sentences
- tables<table><td><th><tr>set the attribute *'paratype = "tcell"'*
- inline tags like<b><i><u><href>
<strong>are not retained but just ignored.

Many other tags can be deleted or ignored.

As HTML documents contain many parts which are irrelevant for text analysis, like links, navigation elements, advertisements etc. These 'boilerplate' elements are eliminated when HTML documents are cleaned. Details are given in Kohlschütter et al. (2010) or Papavassiliou et al. (2013).

Wikipedia documents have their own annotation schemata, called Creole (Markup).[5] It is a simplified annotation possibility for light formatting of Wikipedia texts. Several parsers exist for analysing Wiki texts.[6]

TMX (Translation Memory Exchange Format) is a format in which translation memories are stored.[7] It is a bilingual repository consisting of pairs of sentences (source sentence and target equivalents). The TMX converter would extract texts in just one language. A drawback of TMX is that there is no<p>element, i.e. either every sentence is also a paragraph or all sentences belong to just one paragraph.

TMX can also contain inline tags, mainly for formatting. They are ignored by the LtText converter. This is a difference to a Machine Translation converter where such elements need to be retained for proper reformatting of the target/translated text.

For all these formats, and also some additional ones (like emea (Tiedemann 2009), Europarl etc.) special converters are applied to standardise the input text into the Lt Document model. They are called HTML2LtText, WIKI2LtText, TXT2LtText etc.

There are some common components of the format converters:

---

[4] https://developer.mozilla.org/en-US/docs/Web/HTML/Element.

[5] https://de.wikipedia.org/wiki/Creole_(Markup).

[6] E.g. https://github.com/fhellwig/creole-parser.

[7] https://www.gala-global.org/tmx-14b.

- The different **character codes** need to be standardised. LtText only deals with UTF-8 Unicode, so external documents must convert their character codes hereinto. There are libraries like ICU to support this task, cf. Prager (1999).[8]
- Some **special characters** need to be normalised. This relates for instance to the different types of spaces (zero-width non/breaking etc.), of quotes (smart, left/right quotes etc.), hyphens (of different length). They all are normalised into standard whitespaces, quotes, and hyphens. This measure frees the later linguistic analysis from taking care of each of these variants.
- The computation of **onsets and offsets** is based on the <p> tags. After every <p> tag the onset is recomputed. An alternative would be to compute the onsets on a <doc> basis instead of <p>.

Each converter analyses its source format, and provides a properly instantiated Document model. The original format is stored in the header annotation 'srcformat=...'.

3.2 Language Identification (LtLanguageIdentifier)

The purpose of the Language Identifier is to identify the language of a given document (<doc>), or paragraph (<p>), and to instantiate the <language> attribute of those tags. A special component called LtLanguageIdentifier, was developed for this purpose.

3.2.1 Setup

3.2.1.1 Requirements

a. The LtLanguageIdentifier operates on **documents** to be used for indexing. Its purpose is to identify and prepare loading the language resources for these documents. As a consequence, the available text length which can be used by the identifier differs from a whole document (if the <lg> attribute is to be assigned to the <doc> tag), to a paragraph and to a sentence (attributing <p> or <s> tags respectively).[9]
A second aspect is that the character code for the whole component would already be normalised to UTF8 by the Text Handling, so byte-oriented approaches dealing with different character codes as in (Brown 2012) would not have to be followed either.

---

[8] https://unicode-org.github.io/icu/userguide/conversion.

[9] Efforts to identify the languages of very short texts (cf. Tromp and Pechenizky (2011)) or even tweets (cf. Zubiaga et al. (2014) are intended for different applications.

A third aspect is that the input data will not be clean and contain significant amounts of noise, which is a challenge for the LtLanguageIdentifier, as will be seen below.
b. As for the **languages** to be covered, it is clear that a small set of about 10 languages (as in McNamee (2005) or Mathur et al. (2017)) may not be sufficient, on the other side 900+languages as in (Brown 2012) would not need to be supported either. Of course all languages analysed by the TINA tools must be covered, and there should be an open perspective in case new languages must be added.
Overall the LtLanguageIdentifier recognises about 40 languages. These are (with regard to the scripts used):

- Latin script:
  - Norwegian (no), Swedish (sv), Danish (da), Finnish (fi), Estonian (et), Latvian (lv), Lithuanian (lt)
  - English (en), German (de), Dutch (nl)
  - French (fr), Italian (it), Portuguese (pt), Romanian (ro), Spanish (es), Catalan (ca), Galician (gl)
  - Czech (cs), Slowak (sk), Slovene (sl), Polish (pl), Croatian (hr)
  - Albanian (sq), Basque (eu), Hungarian (hu), Turkish (tr), Maltese (mt)

- Greek script: Greek (el)
- Kyrillic script: Russian (ru), Bulgarian (bg), Ukrainian (uk), Macedonian (mk)
- Hebrew script: Hebrew (he)
- Arabic-Persian script: Arabic (ar), Farsi (fa), Pashto (ps), Urdu (ur)
- CJK script: Chinese (zh), Japanese (ja), Korean (ko)

There are clusters of 'close' languages among these languages, like Spanish-Catalan-Galician, or Czech-Slovak-Slovene, notoriously difficult to distinguish, showing a drop in recognition rate like in the TweetLID data set, cf Zubiaga et al. (2014). Overall, the amount of languages recognised in comparable to tools like Cybozu or with the Google API, at least as far as European languages are concerned.[10]
c. A third aspect of Language Identification is the **computational cost**: It is just the first component in a series of tools, and should not require much computation. In particular in the context of focused crawlers (as described in Sect. 9.3.1) efficiency is a key requirement. This aspect is explicated e.g. in Ahmed et al. (2004).
d. Sometimes recognition results produce close results, and an option would be to present all recognition scores to an evaluator. However in the production line of indexing only one result can be used, and questionable cases must be decided right away.

---

[10] https://github.com/shuyo/language-detection/blob/master/src/com/cybozu/labs/langdetect/

e. The result of language identification is the instantiation of the<lg>attribute in the document model. This attribute is related to the<doc>,<p>, or<s>tag, the value is a (2-letter) ISO definition of the language. Even if the identifier returns alternatives, the Document Model just keeps one value for the<lg>attribute.

3.2.1.2 State of the Art

There are several surveys of Language Identification techniques,[11] documenting the progress from first attempts, cf. Dunning (1994), Beesley (1999), to commercial packages e.g. by Shuyo (2010) or Sites (2013). Two basic techniques are identified (cf. Grefenstette 1995): word-based and n-gram-based. While word-based techniques identify a language by looking at their most typical/frequent words, n-gram recognition collects typical n-grams from a training corpus. The n-grams used range from 2 to 7, with longer n-grams (trigrams to pentagrams) showing better results than shorter ones. N-gram creation requires normalisation of the training data (removal of digits, acronyms, punctuations etc.) and design decisions (n-grams with/without white spaces, normalisation of casing).

Both approaches seem to produce similar results; their efficiency also seems to depend on the languages involved, cf. Mathur et al. (2017), word-based approaches seem to come out better with close languages according to Jauhiainen et al. (2018). For non-Latin scripts, word-based techniques were claimed to be more appropriate e.g. by Panich (2015). And Prager (1999) states that a combination of word-based and n-gram-based techniques shows the best results.

As for the classifiers used, the Naive Bayes approach seems to be favoured by many identifiers. Other approaches, like the Cumulative Frequency Addition Classifier (Ahmed et al. 2004) also perform well, according to Panich (2015).

Overall, Mathur et al. (2017) report accuracies for language identification between 79% (Yandex) and 89% (Google Translate API), with 95% for their LIDE component. Of course this depends on the number and kind of languages involved: For close languages like Spanish/Catalan/Galician the recognition drops to 75%, cf. Zubiaga et al. (2014).

3.2.2 *LtLanguageIdentifier*

3.2.2.1 Approach

The LtLanguageIdentifier tries to use all available resources for recognition:

[11] E.g. Garg et al. (2014) or Panich (2015), Jauhiainen et al. (2018).

- Sometimes the recognition of a **script** is sufficient for language identification, like Greek or Hebrew.
- Next, recognition of languages based on their **most frequent words** is attempted.
- The third knowledge source is an **n-gram** analysis.

The data for word and n-gram based components are taken from Wikipedia dumps.

Each heuristics is run independently, produces their own results, and notes it in a kind of white-board architecture. To make them comparable, and to compute the best overall result, a simple scoring system is set up into which the component results are mapped.

The Language Identifier returns a value for the<lg>attribute in the<doc>(or<s>) tag in the Document Model.

3.2.2.2 Resources

As just stated, the LtLanguageIdentifier uses three criteria for language recognition:
1. **Script analysis:** A significant help in language identification is the analysis of the script, i.e. the characters used. Script analysis is sufficient in cases where just one language must be recognised:

- Greek script clearly identifies Greek as the language, and so does Hebrew.
- Arabic-Persian scripts limit the languages to identify to Arabic, Farsi, Pashto, and Urdu.
- CJK scripts are used by Chinese, Japanese and Korean, whereby the Korean characters can be clearly identified, and Hiragana and Katakana characters can be used to recognise Japanese.
- Cyrillic scripts are used by Russian, Ukrainian, Bulgarian and Macedonian.
- All other languages analysed by LtLanguageIdentifier use Latin scripts.

Within the respective scripts, further distinctions can be made, looking at special characters indicating languages: The occurrence of '*ß*' in a Latin script points to German,; '*ê*' is only used in Romanian and French, '*è*' only in Lithuanian, etc. Similar heuristics can be applied in other scripts: e.g. the occurrence of' ړ' points to Pashto as text language.

For each script, all 'non-standard' characters were collected and tagged with the language(s) in which they occur.[12] This information was taken from the UTF8 code page descriptions and stored in form of triplets of<character, language, weight (in this language)>.[13]

---

[12] https://www.unicode.org/charts/

[13] The weight could be computed on the basis of the frequency of the character in that language; in LtLanguageIdentifier is was just approximated, and a value between 5 (characters pointing to several languages) and 15 (for characters pointing to just 1 language) was assigned to indicate how specific the character was for a given language.

2. **N-gram analysis:** As such script-based heuristics have clear limitations, esp. in Latin scripts where languages like English or Dutch do not have any idiosyncratic characters, an n-gram based resource was created for each language. The data were again collected from Wikipedia.

For each language supported, dumps of Wikipedia were analysed, converted to LtText format, and the first 10 K documents were used for n-gram building. The number of tokens analysed is between 1.2 and 6.9 Mio, depending on the language. The documents were normalised: Punctuations, digits, and acronyms were removed.[14]

N-grams were built (2, 3, 4, 5-gram) from the tokens, and their frequency was counted. N-grams containing white spaces were not used. The result was one ngram list per language, giving<length, ngram, frequency>. Next, for each length the most frequent n-grams were taken, down to a threshold where the sum of the frequencies of the selected n-grams covered 25% and 50% respectively of the total frequencies of the given language. As a result, between 1000 and 6000 n-grams per language were collected. N-grams below the threshold were discarded.

3. **Word analysis:** Even if the score of an ngram reflects its importance for a given language, it is felt that ngrams forming whole words should also be used for the identification process: Humans would identify languages by looking at their most frequent words: If an input candidate contains many words of a certain language, the text may have been written in this language. This heuristics was found a good alternative to n-grams both in Latin and in non-Latin scripts.[15] It is modelled by using a word list.

A first attempt was to re-use spell checker wordlists. But it was abandoned as it does not contain frequency information, in favour of a frequency-based attempt: The same Wikipedia resources as for n-grams were used for the **word-based** resource. The Wikipedia data were tokenised and then frequency-annotated. High-frequency tokens were extracted and merged into a final resource for a given language. Like for the n-grams, the criterion to include a word into the final list is its contribution to the coverage of a language: All words needed to cover 50% of a text were included in the word list, starting from the most frequent ones downwards. This way, between 400 and 7000 words per language were selected, 2500 words on average.[16]

3.2.2.3 Classification

As the LtLanguageIdentifier uses three different knowledge sources with different content a common ground had to be found for classification. A common scoring system was therefore adopted for the three resources, assigning scores between 2 and 5. It is based on language coverage:

[14] Casing was not normalised as it is significant in German as well as in Named Entity analysis.

[15] Panich (2015), useful also in very similar languages.

[16] For the CJK languages no word based resources were built due to the lack of tokenisation in this state of analysis.

Characters, n-grams (of length 2–5), and words occurring in>50% of the texts get scores>4, if occurring in 25% get scores>2, all the rest gets scores = 0. Tetra- and pentagrams have higher scores than bi- and trigrams, cf. Ahmed et al. (2004).

The classifier itself uses a kind of Cumulative Frequency Addition Classifier, just using scores instead of frequencies, Each method reports its scoring result, and from there the final decision is computed, using the same weight for each knowledge source. In cases where the final score clearly favours a language this language is output. In cases where several rather equally likely languages are detected, all these languages are returned. Results falling below a lower bound are returned ‘unknown’.

3.2.2.4 Test

The test was designed as follows: Five sentences of each supported language were collected, taken from Wikipedia, related to the same topic. The test would have to assign the correct language to them. In addition, five sentences each were added (Gaelic, Assamese, Bahasa) for the language identifier to return ‘unknown’. Overall the test suite consisted of about 220 sentences.

The results are given in Fig. 3.1:

For the 40 languages supported by the LtLanguageIdentifier, only one incorrect assignment was returned for this test set, which means an accuracy on 99.5%.[17]

For the three additional languages (Gaelic, Assamese, Bahasa), all Assamese examples came back as ‘unknown’: Assamese has its own script. Together with Gaelic and Bahasa, both Latin script, the accuracy of the ‘unknown’ languages was 60%.

Overall, the accuracy of the LtLanguageIdentifier reached 97.4, which is a competitive result. Each of the three heuristics contributed to the recognition result.

3.2.2.5 Assessment

1. The main determining factors for language identification seem to be the number of languages to be recognised, and the closeness, or similarity, of these languages. It is easy to detect a language which has its own script: If Russian is the only language with Cyrillic script, its detection is easier than in cases where languages like Bulgarian, Macedonian etc., all using Cyrillic script, are also in the test set. In turn, close languages like Spanish—Catalan—Galician or Slovak—Slovene—Czech are difficult to distinguish and have lower recognition rates, cf. Zubiaga et al. (2014), Jauhiainen et al. (2018).

[17] The only incorrect assignment was a recognition of Galician for Spanish input, a notoriously difficult distinction (cf. Panich 2015).

norm	recognised	Text
ar	ar	اللغة العربية ذات أهمية قصوى لدى المسلمين، فهي لغة مقدسة
bg	bg	След войната Байерн става член на Южната Дивизия на немската Първа Дивзия, която е била разделена на пет части по това време, наричана още Оберлига.
ca	ca	El curs de la guerra, tot i ser-li favorable als primers anys, es tornà en una derrota inexorable que el conduí a suïcidar-se
da	da	Arbejdsvilkårene kan give påvirkninger, der for nogle mennesker kan være belastende og for andre stimulerende.
el	el	Εβραίοι ονομάζονται τα μέλη του εβραϊκού λαού,
es	es	Autocracia es un concepto político.
es	gl	Participación en asociaciones de carácter político (partidos políticos, sindicatos)
et	et	Washington asub halduslikult Columbia ringkonnas.
fa	fa	ر ۲۱ دسامبر ۱۹۰۷ مادرش کلارا هیتلر با یک مرگ دردناک بر اثر سرطان سینه در سن ۴۷ سالگی فوت کرد. هیتلر از طریق دادگاهی در لینز
fi	fi	Verdunissa hyökättiin sen takia, että se oli lähellä Saksan rautateitä.
gl	gl	Xunto a estes existe unha pequena minoría de lingua alemá.
it	it	L'importanza della Cina nel ventunesimo secolo[7][8] si riflette in virtù del suo ruolo come seconda potenza economica per prodotto interno lordo;
ja	ja	1904年、ヒトラーはシュタイアー市の中等学校（レアルシューレ）に再入学するように家族から勧められたが、
ko	ko	히틀러 평전에 따르면 히틀러는 자아가 강하고 상급학교 진학 이후의 학업성적은 전혀 우수하지 못하였다.
lt	lt	Šią grupę sudarė trys pagrindinės valstybės:
mk	mk	Меѓународна соработка за промовирање на колективната безбедност потекнува од Концерт на Европа кој се разви по Наполеоновите војни во 19 век, во обид да се задржи статус кво меѓу европските држави и така да се избегне војната.
pl	pl	Uznawany jest przez większość historyków za osobiście odpowiedzialnego za politykę rasową nazistowskich Niemiec,
ps	ps	د لومړی نړیوالي جګړي وروسته المان ډېر لوی قرض پر اوږو درلودو او په میلیونونو خلک بې بې کاره وو.
ro	ro	În conformitate cu o legendă a întemeierii națiunii cehe, regina Libuše s-a căsătorit cu un plugar umil pe nume Přemysl, fondând Dinastia Přemyslid.
ru	ru	Один из внешних символов медицины с конца XX века — шестиконечная «Звезда жизни»
sl	sl	Nemška vojska ga je pri tem podprla in si s tem zapečatila svojo usodo.
tr	tr	Bu sayfayı yaratmak için aşağıdaki metin kutusunu kullanınız.
ur	ur	جرمنی کے سب سے بڑے صوبے باویریا کا صدر مقام ہے۔ میونخ برلن اور ہمبرگ کے بعد جرمنی کا تیسرا بڑا شہر ہے۔
zh	zh	兩德统一，原屬東德的图林根三個行政區，

Fig. 3.1 Language identifier results. Only Spanish-Galician is incorrectly recognised

2. The recognition results of LtLanguageIdentifier unfortunately dropped significantly when applied to 'real' data, in this case a system to validate the language of Translation Memory sentences.[18] One of the reasons for this fact may be that the data are not clean. They contain foreign language segments like company and person names, or address data like foreign language street names. Named Entities in foreign languages also influence the recognition of the sentences, and would require a more robust recognition approach. In particular the character-based script analysis seems to need more robustness.[19] Also it may be better to focus on larger segments like<doc>or<p>tags instead of sentences (<s>tags) where headings like '*information*' would not be safely recognisable.
3. Also the purpose of Language Identification needs to be revisited: It is used to decide which language resources the system should load to process a given text. Now, if a segment in an English Translation Memory contains only the address of a Swedish business partner, the identifier would probably return '*Swedish*' as language, causing the system to unload the English resources and load the Swedish ones, just for one small segment. This is not what is intended. The fallback solution of the LtLanguageIdentifier in these cases is to use the language ID of the previous segments as an additional input parameter, and keep this information for cases where the language of a segment is not safely detectable, where a segment is rather short, and where the number of out-of-language segments is rather small.
4. Quality improvement and domain adaptation is possible in LtLanguageIdentifier by just adding additional words to its words list.

3.3 Topic Identification (LtTopicIdentifier)

Topic Identification is the assignment of a document to one of the topics of a given domain. Domains, or taxonomies, are structured into different topics, or classes. A new document is classified into one or several of such topics. The classification is based on the analysis of features of the incoming document, usually key terms. The classifier identifies the best matching topic based on these features. An example is the assignment of incoming emails to different departments of an enterprise.

Unlike other approaches, the LtTopicIdentifier put its focus on the content side (identification of good features), less on the classification side.

3.3.1 Rationale

There are several reasons to include document classification to a text analysis chain:

---

[18] The task was to detect non-translated segments in such files.

[19] E.g. by factorising the contributions of the three heuristics in the overall scoring, and reducing the weight of the script analysis.

language direction	term	sense in general	topic	sense in topic
de > fr	Ausdauer	persévérance	*[sports]*	endurance
fr > de	auditeur	Zuhörer	*[economy]*	Rechnungsprüfer
en > fr	flasher	exhibitionniste	*[automobile]*	clignotant
fr > en	Stupeur	astonishment	*[medicine]*	stupor
en > it	assembler	montatore	*[technology]*	assemblatore
it > en	piastrina	plate	*[military]*	badge
en > es	hash	picadillo	*[IT]*	aleatorización
es > en	comportamiento	behaviour	*[finance]*	performance
en > pt	injury	ferida	*[law]*	infração
pt > en	recurso	resource	*[medicine]*	remedy
de > en	Abschuss	shooting	*[sports]*	kick
de > en	Mast	pole	*[agronomy]*	Fattening
			[nautics]	mast
en > de	adhesion	Haftung	*[physics]*	Adhäsion
	plane	Fläche	*[nechan.engin]*	Hobel
			[botanics]	Platane
			[physics]	Ebene
			[transport-air]	Flugzeug

Fig. 3.2 Word senses/translations depending on topic selection

The first reason relates to document collection. Many indexing tasks originate from an initial document collection covering a new domain of interest. Such a collection can be created with the help of a topic identifier. It separates documents which are relevant for a given profile of interest from others which are not. Sometimes such topic identifiers are part of a crawler. Such a 'focused crawler' searches for documents which belong to a given topic.[20] Focused crawling is a kind of web search which collects documents belonging to a specific topic, like 'health & safety'. The crawler, when inspecting a document candidate, must have a quick method to decide whether to add it to the domain-relevant document collection or to discard it. Detailed analyses are too much of effort for such a decision, so a fast and robust technique must be applied here.

A second reason to include a topic identifier in text analysis is the requirement to do semantic analysis of the index terms. Semantic disambiguation is an important requirement in multilingual indexing as the index terms must be linked to terms in other languages with the same meaning. Figure 3.2 gives examples where the target language links of terms point to quite different entries depending on the topic of the document.

This fact is not just relevant in multilingual indexing but of course in text translation as well: As topic information obviously is essential in translation. Previous MT systems like METAL, cf. Thurmair (1990) or Personal Translator, cf. Aleksić

[20] An example of a focused crawler is described in Papavassiliou et al. (2013).

and Thurmair (2011), offered the option to set the topic as a global parameter. But as many users did not use it they deteriorated the translation quality, and it was decided to set the topic automatically, by using a topic identifier.

3.3.2 *LtTopicIdentifier*

Topic Identification, or text classification, is one of the most researched fields in Natural Language Processing, starting in the Eighties, cf. Panyr (1986), up to proposals of unsupervised learning e.g. Janai and Vijayarani (2021) or Ko and Seo (2003). An overview is given by Gasparetto et al. (2022). The focus in research is on the classifiers and their properties, sometimes also the type of features, from terms, cf. Azam and Yao (2012) or Sarkar et al. (2014), to N-grams as in Cavnar and Trenkle (1994).

However it is doubtful if the selection of the classifier is the most influential factor in topic detection, as other factors come into play in reality, like the shape of the taxonomy, or the form and linguistic status of the features. In an analysis chain like TINA a significant factor is the position of the topic identifier in the processing sequence.

3.3.2.1 Embedding

Unfortunately the position of the Topic Identifier in the Text Analysis chain influences its shape significantly:

- If used in the initial phase in document collection, a quick and cheap decision must be attempted as not much processing time per document is available in the crawling procedure. The Topic Identifier, although always language-specific, should be designed as a low resource component, such that not much more effort than list lookup in the respective language would be required.
- If used in semantic disambiguation, more elaborate linguistic processing is available, and more resources can be consulted, leading to better recognition results, cf. Goller et al. (2000).

In any case a topic identifier is a language-specific component, and it needs to build resources for each participating language.

The LtTopicIdentifier in TINA is called before more elaborate text analysis is tried. As it is known that the quality of topic identification can be increased by using linguistic analysis tools like lemmatisers and decomposers, cf. Goller et al. (2000), the LtTopicIdentifier needs resources which overcome such deficiencies to reach a comparable level of quality. These resources must cope with the disadvantage of having less analysis options and must improve quality by expanding the resources and including textform and normalisation variants of relevant classifier terms.

3.3.2.2 Taxonomy

Real world applications cannot deliver topic classification systems with an 'empty' taxonomy; such an approach would not be accepted by customers. Therefore a kind of 'generic' taxonomy which can be used as a starting point for adaptations and modifications must be provided with the software.

It should be noted that the taxonomy itself is one of the factors which influence the quality of a Topic Identifier, i.e. the number and closeness of the different topics which it includes. A taxonomy consisting of only two classes (like '*sports*' and '*laser technology*') would be rather easy to support while a rather narrow domain (like '*dental technology*' versus '*dental surgery*') with numerous subtopics clearly poses more challenges. This fact, however, is seldom mentioned in the literature.

There are several options to set up taxonomies for topic identification:

1. Most of them are user-specific, and divide the document collection into subgroups each of which serves specific user interests. Examples are email distribution systems, automotive maintenance systems which classify user comments along the car components they refer to, police systems which identify different types of crimes, and many more.

The LtTopicIdentifier must be able to support such scenarios by offering tools to define such user-specific taxonomies for a document collection. Such tools, from crawlers to term extraction and linguistic analysis, are described in Chap. 9.

2. However there are cases where special-purpose taxonomies are not sufficient, and general taxonomies are required. An example is Machine Translation Systems where it is unknown which texts the users may want to translate. General taxonomies would start from a kind of top-level 'upper' ontology as proposed e.g. by Niles and Pease (2001), and define lower-level topics underneath. However, a general classification of topics is not easy to find: Most document collections, even large ones like Europarl (Koehn 2005), are already domain specific.[21] Existing classification systems like Dewey's Decimal Classification show a clear predominance of social sciences over natural science and engineering.[22] Balanced corpora like the British National Corpus, are rather small and focus on different things, esp. linguistic phenomena.[23] A resource covering rather many domains is Wikipedia. It claims to cover all areas of human knowledge, and indeed Wikipedia has frequently been used for classification purposes, for instance by Hassan et al. (2012). Other proposals use thesauri as starting points, e.g. the Kadokawa thesaurus, cf. Kang et al. (2001).

3. The TINA LtTopicIdentifier follows a different tradition, which originated in terminology research. The definition of a term includes a description of its position in a taxonomy, according to e.g. (Wüster 1991); therefore it was an early interest of

[21] This can easily be proven by comparing just the top frequent 300 terms of such collections.

[22] Dewey: https://en.wikipedia.org/wiki/Dewey_Decimal_Classification.

[23] http://corpora.lancs.ac.uk/bnc2014/

the terminology science to define a general classification, to be able to embed single terms in it. Several such taxonomies were proposed; the LtTopicIdentifier taxonomy is based on the DANTerm system proposed by Engel and Nistrup-Madsen (1985) and adapted by Hohnhold (1984).

This taxonomy is a semi-hierarchical structure, dividing the topics into 'common technical vocabulary' (CTV) for the technical and 'common social vocabulary' (CSV) for the social sciences, both under 'general vocabulary' which contains words which are not considered to be terms of a special domain.[24] Below CTV and CSV, the respective topics are aligned in a rather flat order; some of them have subdomains themselves. The taxonomy is given in Fig. 3.3.

4. There are also attempts not to use classifications but define free topics and to do Topic Modelling in an **unsupervised** way, without relating the documents to an existing classifications. This approach became a reseach field for Machine Learning. Topics are defined differently here, e.g. "*In statistics and natural language processing, a topic model is a type of statistical model for discovering the abstract "topics" that occur in a collection of documents.*"[25] This distorts purpose and means of topic identification, and defines a topic to be the output of some statistical operations. However it is known that such an output creates 'topics' which cannot be easily interpreted by humans and only partially correspond to human intuition. This problem is then discussed as 'labelling', trying to answer what the terms in clusters resulting from statistical techniques could have in common. For real world applications such techniques are of only limited use as topics are defined according to practical necessities and are prior to the documents potentially used for topic detection.

There are however, applications without predefined or with fast changing topics, like in social media. They require special attention anyway due to the constraints of such media (very short messages, colloquial speech etc.), cf. Lo et al. (2017). But even the news domain provides already a predefined topic hierarchy, e.g. the Reuters classification.[26]

3.3.2.3 Features

Documents are assigned to topics of a taxonomy by using features, i.e. terms which describe the respective topic. All topic identification components crucially depend on the presence of terms indicating a specific topic. The more such terms can be found the more precise the topic assignment can be. Such a consideration follows human intuition, if humans should decide on topic assignment. While most papers in the literature on topic identification focus on different classification

[24] For this reason they stayed outside of the DANterm taxonomy.

[25] https://en.wikipedia.org/wiki/Topic_model.

[26] https://liaison.reuters.com/tools/topic-codes.

Tag	Meaning	Tag	Meaning	Tag	Meaning
		GV	general vocabulary		
CTV	common technical vocabulary			CSV	common social vocabulary
bio	Medicine and Bio Sciences			art	Art & Culture
-biol	Biology			-arch	Architecture
-food	Nutrition and Food			-aud	Audio Culture
-med	Medicine			-lit	Literature and Theatre
-pharm	Pharma Industry			-vis	Visual Arts
tech	Natural Sciences and Technology			econ	Economy
-car	Car Industry			-agr	Agriculture
-chem	Chemical Industry			-fin	Finance
-civeng	Civil Engineering			-ind	Productive Industry
-electr	Electrical Engineering			-soc	Labour and Social Welfare
-energ	Energy Industry			-trade	Trade
-env	Environmental Issues			-transp	Transportation and Logistics
-geo	Geo Sciences			hum	Humanities and Social Sciences
-it	Information Technology			-hist	History
-mat	Manufactures			-ling	Linguistics
-glas	Glass Industry			-paed	Education
-met	Metal Industry			-phil	Philosophy
-pap	Paper Industry			-psych	Psychology
-plast	Plastic Industry			leis	Leisure and Sports
-tex	Textile Industry			-cook	Cooking
-wood	Wood Industry			-games	Games
-math	Mathematics			-sport	Sports
-mecheng	Mechanical Engineering			soc	Society and Social Issues
-min	Mining and Oil Industry			-adm	Administration
-photo	Optics and Photography			-cust	Customs
-phys	Physics			-law	Law and Legal System
-qual	Quality Assurance Issues			-crime	Criminality
-tel	Tele Industry Telephony			-media	Media
-transp	Transportation and Traffic			-mil	Military Issues
-rail	Railway Issues			-arms	Military Arms
				-pol	Politics
				-rel	Religion

Fig. 3.3 TINA domain classification

techniques, the focus should also be directed to finding more and better terms for the description of topics.

The form of the features is also a matter of consideration. Terms taken just as they occur in texts (textforms) seem to perform not as well in classification as terms which underwent linguistic processing (lemmatisation, decomposition); such measures, which imply aspects of normalisation, tend to increase the classifier quality, cf. Goller et al. (2000).

1. LtTopicIdentifier uses two resources to collect the features for the classifier:

   - As mentioned above, the LtTopicIdentifier taxonomy is derived from work in terminology. So it is not surprising that many dictionary resources contain domain annotations. In particular bilingual dictionaries both for human lookup and for MT use such annotations for translation disambiguation, c.f. the examples in Fig. 3.2. The dictionaries contain many entries with topic annotations; they can be harvested and added to the feature sets of the respective topics. As they originate from dictionaries they are usually in lemma form.
   - However, dictionary terms are usually not systematic and often do not cover a topic as a whole; so it turned out to be useful to support the topic feature collection in a corpus-based way. In LtTopicIdentifier, training corpora were built for each of the topics, with about 30 documents per topic. As this is a time-consuming task, tools as described in Sect. 9.3 to support the corpus creation were built: a focused crawler to collect relevant documents, and a term extraction tool to extract relevant terms from them.

     The resulting term candidates must be manually validated; noise terms which do not indicate the topic must be removed. It could be noted that there is a discrepancy in how a term is viewed: *Is 'Afghanistan' a drug term?* From the point of intrinsic semantics, it is not but simply a place name; however from the point of drug term topic identification it functions like a drug term.[27]

     If corpus-based collections alone are used for feature collection, it also holds that some terms describing the topic do not occur in the corpus data. For specialised domains this is even rather frequent as such terms often describe proprietary knowledge. Therefore the dictionary and the corpus-based sources, both contributing to the topic description, have to be merged into one term list per topic.

2. Two more operations are required to complete the feature resource for the LtTopicIdentifier. The first one is the assignment of **probabilities** to terms. For terms resulting from corpus analysis, standard weights like term and inverted document frequency (tf·idf) can be used. However terms collected from the dictionary

[27] This is because during the American presence Afghanistan became a leading country in opium production.

en term translation for de *‚Schlagfrequenz'*	Dictionary Source Quality	Corpus Frequency
blow sequence	high	34
beat frequency	mid	7830
impact frequency	mid	0
hammer-blow sequence	mid	3
blow frequency	low	118
lateral runout frequency	high	0
stroke frequency	low	1160

Fig. 3.4 Topic terms: dictionary vs. corpus evidence

don't have frequency information (cf. Fig. 3.4). To be able to make use of them, and allow users to edit the feature lists, a scale of 1 to 5 points was created to annotate the terms: The corpus weights are mapped onto this scale, the dictionary terms are weighted using a manual 'preferred' annotation.[28]

The second operation is a consequence of placing the topic identification before the linguistic analysis in the processing chain: The input documents can only be analysed on textform level, not on lemma or concept level, as such tools are not yet available in the processing chain. Therefore the topic feature list must be expanded and turned into **inflected forms**. This can be done either by using a flexer (cf. Sect. 9.3), or by doing manual extension (not all inflected forms of a term will be relevant).

The result is a list of terms consisting of a textform, a score, and a topic to which the term belongs. For each language such a topic list needs to be built. Some terms can be machine-translated.[29]

Such a topic feature list combines dictionary-based and corpus-based methods. Of course it requires manual work, however this work creates a static resource which can further kept in maintenance mode.[30] The idea is to have more and also more accurate topic features, and therefore achieve better identification quality.

3. The collection of features for the Topic Identification usually starts with the collection of example documents for each topic, and extracts the topic-specific features from there.[31] The LtTopicIdentifier follows such a different approach, by first defining some seed terms and then searching documents based on them: Just one big corpus was used. Some seed terms (like "*sports football hockey racing*" ->*sports*) are collected for each topic. In an interactive bootstrapping procedure,

---

[28] This annotation is described in the Sect. 9.2 below.

[29] If the MT system is sensitive for topics…

[30] It is surprising how many terms are not available from corpus tools (like term extraction) as compared to domain specific dictionaries, and in turn, how many dictionary entries have no evidence in the corpus data.

[31] Details are described e.g. in Goller et al. (2000).

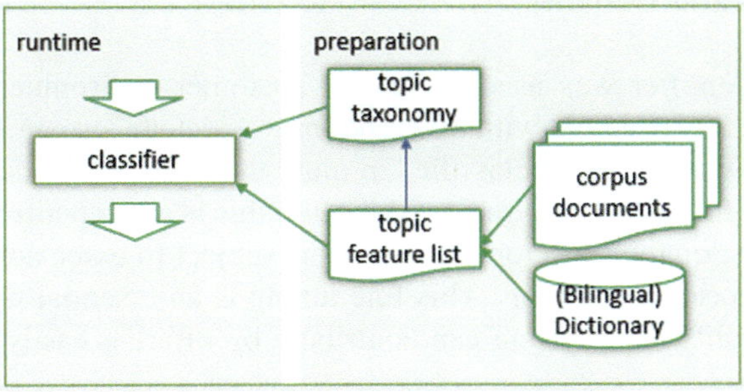

Fig. 3.5 Topic Identifier components

after correlation analysis the best correlated terms (both single and multiword terms) are returned and inspected by human experts, and added to the term list if supporting the respective topic. This procedure is repeated several times. In the result, several thousand candidates per topic were inspected, and between 200 and 700 terms per language were collected for each of the about 60 topics.

3.3.2.4 Classification/Topic Identification

The classifier applied here at runtime is a modified Cumulative Frequency Addition Classifier: The document under examination is deformatted and analysed by a simple tokeniser.[32] The resulting tokens are compared with the topic feature list and a simple scoring mechanism is applied: For each term found in the input a topic instance is created. For each instantiated topic a score is computed consisting of the number of terms times their score. The topic with the best score wins.

The output of the topic identifier is handed to the Document Model, where the<topic>feature is instantiated on the<doc>tag. Of course the topic annotation can be propagated down to the<p>level, alternatively the identifier can be called again on<p>level in case of multi-topic documents, like news (Fig. 3.5).

As explained above, the focus in LtTopicIdentifier is not on improving the statistical machinery but on improving the topic features, and providing a fast identification option for a focused crawler. Significant improvements can be achieved by adding linguistic preprocessing: using concepts instead of text forms; however not yet available in this early state of the processing chain.

Also, this approach has the advantage that it is transparent: Users can easily understand why certain decisions are made, and can easily improve it as the topic feature list is user edible. This is not the case for other approaches.

[32] The simple tokeniser just strips punctuation marks from the text forms.

3.3.2.5 Tests and Usability

The LtTopicIdentifier was used in several customer environments. For the MT setup, a test was carried out with 500 randomly selected German and English documents. They were correctly classified in more than 80% of the cases, and no false positives were found.[33] For a 'real world' setup this is a competitive result.

The baseline quality of a topic identifier is subject to extensions and modifications in real world installations. This fine tuning is an essential part of the overall component quality. The system can contribute by offering easily usable tools for such adaptations.

References

Ahmed, B, Cha, S.-H., & Tappert (2004). Language identification from text using n-gram based cumulative frequency addition. *Proc. Student/Faculty Research Day, CSIS, Pace University (2004), S. 12–1*

Al Jabani, S., Salman, M.A., & Mohammed, M. (2019). Pragmatic Text Mining Method to Find the Topics of Citation Network. Proc. Int. Conf. on Big Data and Networks Technologies 2019: 190–205

Aleksić, V., & Thurmair, Gr. (2011). Personal Translator at WMT 2011. *Proc. Workshop on Statistical Machine Translation (WMT 2011), Edinburgh*

Azam, N., Yao, J. (2012). Comparison of term frequency and document frequency based feature selection metrics in text categorization. Expert Systems with Applications 39(5): pp. 4760–4768

Beesley, K. (1999). Language Identifier: A Computer Program for Automatic Natural-Language Identification of On-line Text. Proc. 29th Annual Conf. of the American Translators Association, Berkeley, Calif.

Brown, R. (2012). Finding and identifying text in 900+ languages. *Digital Investigation 9, 2012, pp. 34–43*

Cavnar, W., & Trenkle, J. (1994). Mi. N-gram-based text categorization. Proc. 3rd Annual Symposium on Document Analysis and Information Retrieval (SDAIR), Las Vegas, Nevada, 1994.

Chen, H.H., Ku, L.W. (2002). An NLP & IR approach to topic detection In: Allan, J. (eds) Topic Detection and Tracking. The Information Retrieval Series, vol 12. Springer, Boston, MA. https://doi.org/10.1007/978-1-4615-0933-2_12

Dunning, T. (1994). Statistical identification of language. Technical Report CRL MCCS-94--273, Computing Research Lab New Mexico State University, 1994.

Engel, G., & Nistrup-Madsen, B. (1985). Danterm. *TermNet News (1985) 12, p. 8*

Garg, A., Gupta V., & Jindal, M. (2014). A Survey of Language Identification Techniques and Applications. *Journal of Emerging Technologies in Web Intelligence 6 (2014), Nr. 4, pp. 388–400*

Gasparetto, A. , Marcuzzo, M., Zangari, A., & Albarelli, A. (2022). A Survey on Text Classification Algorithms: From Text to Predictions. Information 2022, 13, 83. (https://doi.org/10.3390/info13020083)

Gasparetto , A., Marcuzzo, M., Zangari, A., Mathur, P., Misra, A., Budur, E. (2017). LIDE: Language Identification from Text Documents. https://doi.org/10.48550/arXiv.1701.03682

[33] Cf. Thurmair (2005).

Goller, Chr., Löning, J., Will, Th., & Wolff, W. (2000). Automatic Document Classification: A thorough Evaluation of various Methods. *In: Knorz, G., Kuhlen, R., eds. : Informationskompetenz – Basiskompetenz in der Informationsgesellschaft. Proc. des 7. Int.en Symposiums für Informationswissenschaft (ISI 2000), Darmstadt*

Grefenstette, Gr. (1995). Comparing Two Language Identification Schemes. Proc 3rd Int. Conf. on Statistical Analysis of Textual Data (JADT 1995), 1995, pp. 263–268

Hassan, M., Karray, F., & Kamel, M:S. (2012). Automatic Document Topic Identification using Wikipedia Hierarchical Ontology. *Proc. 11th Int. Conf. on Information Science, Signal Processing and their Applications, (ISSPA 2012), Montreal, Quebec, Canada, 237–242*

Hazen, T., & Zue, V. (1993). Automatic Language Identification using a segment-based approach. Proc. 3rd European Conf. on Speech Communication and Technology (Eurospeech 1993), Berlin, Germany

Hohnhold, I., (1984). Die Terminologie-Datenbank TEAM im Sprachendienst Siemens München *BDÜ-Mitteilungsblatt für Dolmetscher und Übersetzer 30 (1984) 3, p. 7.*

Janani, R., & Vijayarani, S. (2021). Automatic text classification using machine learning and optimization algorithms. Soft Computing, volume 25, pages 1129–1145 (2021)

Jauhiainen, T., Lui, M., Zampieri, M., Baldwin, T., & Lindén, K. (2018). Automatic Language Identification in Texts: A Survey. arXiv:1804.08186

Kang, S.J., Li, H.F., & Lee, J.H. (2001). Semi-Automatic Construction of Language Independent and Practical Ontology for Word Senses Disambiguation (R). https://doi.org/10.3115/1118220.1118226

Ko, Y., & Seo, J. (2003). Automatic Text Categorization by Unsupervised Learning. https://doi.org/10.3115/990820.990886

Koehn, Ph. (2005). Europarl: A parallel corpus for statistical machine translation. *Proc. Machine Translation Summit X, Phuket, Thailand.*

Kohlschütter, Chr, Fankhauser, P., & Nejdl, W. (2010). Boilerplate detection using shallow text features. *Proc. 3rd ACM Int. Conf. on Web search and data mining, (WSDM '10), New York City, NY, USA, pp. 441–450*

Lo, S.L., Chiong, R., & Cornforth D. (2017). An unsupervised multilingual approach for online social media topic identification. *Expert Systems with Applications, Volume 81, 2017, pp. 282-298*

Mathur, P., Misra, A., & Budur, E. (2017). LIDE Language Identification from Text Documents. https://doi.org/10.48550/arXiv.1701.03682

May, Ch., Ferraro, Fr., McCree, A., Wintrode, J., Garcia-Romero, D., & Van Durme, B. (2015). Topic Identification and Discovery on Text and Speech. In Proc. of the 2015 Conf. on Empirical Methods in Natural Language Processing, pages 2377–2387, Lisbon, Portugal. Association for Computational Linguistics.

McNamee, P. (2005). Language identification: a solved problem suitable for undergraduate instruction. *In: Journal of Computing Sciences in Colleges 20 (2005), Nr. 3, S. 94–101*

Meddeb, A., & Romdhane, L.B. (2022). Using Topic Modeling and Word Embedding for Topic Extraction in Twitter. 26th Int. Conf. on Knowledge-Based and Intelligent Information & Engineering Systems (KES 2022) Procedia Computer Science 207 (2022) 790–799

Niles, I., & Pease, A. (2001). Towards a Standard Upper Ontology. *Proc. of the 2nd Int. Conf. on Formal Ontology in Information Systems (FOIS-2001pp.2–9, Ogunquit, Maine, USA,*

Panich, L. (2015). Comparison of Language Identification Techniques. *BA thesis, Univ. Düsseldorf*

Panyr, J. (1986). Auromatische Klassifikation und Information Retrieval. Tübingen (Niemeyer)

Papavassiliou, V., Prokopidis, Pr., Thurmair, Gr. (2013). A modular open-source focused crawler for mining monolingual and bilingual corpora from the web. *Proc. 6th Workshop on Building and Using Comparable Corpora (BUCC), pp. 43–51, Sofia, Bulgaria, 2013.*

Prager, J. (1999). Linguini: Language Identification for Multilingual Documents. *Journal of Management Information Systems, 1999, S. 1–11*

Sarkar S.D., Goswami, S., Agarwal, A., & Aktar, J. (2014). A novel feature selection technique for text classification using Naive Bayes. In: International scholarly research notices, 2014

Shuyo, N. (2010). Language Detection Library. Cybozou Labs

Sites, D. (2013). Google compact language detector 2, 2013. https://github.com/CLD2Owners/cld2

Thurmair, Gr. (1990). METAL: Computer Integrated Translation. *Proc. of a Workshop on Machine Translation, July 1990, UMIST*

Thurmair, Gr. (2005). Improving MT Quality: Towards a Hybrid MT Architecture in the linguatec 'Personal Translator'. *In: Proc. of Machine Translation Summit X: Invited paper, Phuket, Thailand.*

Tiedemann, J. (2009). News from OPUS – A collection of Multilingual Parallel Corpora with Tools and Interfaces. *in: Nicolov, N., Bontcheva, K., Angelova, G., & Mitkov, R., eds. (2009). Recent Advances in Natural Language Processing (vol V) (Benjamins)*

Tromp, E., & Pechenizkiy, M. (2011). Graph-based n-gram language identification on short texts. *In: Proc. 20th Machine Learning conference of Belgium and The Netherlands, S. 27–34*

Wüster, E. (1991). Einführung in die allgemeine Terminologielehre und terminologische Lexikographie. *Bonn 1991 (3rd ed.)*

Zubiaga, A., San Vicente, I., Gamallo, P., Pichel, J., Alegria, I., Aranberri, N., Ezeiza, A., & Fresno, V. (2014). Overview of tweetlid: Tweet language identification at sepln 2014. *Proc. SEPLN 2014: Tweet-LID@ SEPLN (2014)*

Chapter 4
Text Segmentation

All components following the Preprocessing (Text Handling, Language and Topic Identification) are language specific, and need resources in the supported languages. The first group of analysis tools is intended to create word candidates ('tokens') from texts (paragraphs) as produced by the text handling: Sentence splitting, tokenisation, and normalisation.

4.1 Sentence Splitting (LtSSplit)

The first task in text analysis is to split a text into linguistically processable units, i.e. sentences. Input of the component is a text as given by the Document Model within the<p>and</p>tags, output is the text with inserted<s>and</s>tags and their annotations.

4.1.1 Rationale

4.1.1.1 Impact

The way a text is split into sentences has significant impact for many applications:

- Parallel/Bilingual sentence production requires proper alignment, otherwise applications like bilingual term extraction or machine translation are less precise than they could be.

G. Thurmair, *Knowledge-Driven Multilingual Text Analysis and Transparent Information Retrieval*, Cognitive Technologies,
https://doi.org/10.1007/978-3-031-91741-7_4

- Computer aided translation (CAT): In translation-memory-based (human) translation sentence segmentation has an immediate impact on the translation quality, and even on the payment of translators as it depends on the number of sentences translated.
- In pipeline-based systems like TINA improperly segmented texts cause significant consequences for the components following in the pipeline. Especially parsing in deep processing is affected.

4.1.1.2 Standard Approach

Most sentence splitters follow the approach to define a list of punctuations which indicate a sentence break (like period, exclamation mark, question mark) and a list of exceptions (mostly abbreviations[1]) which do not cause a period to split a sentence.

This approach has a significant weakness:

- Abbreviations are not a fixed class. They are not just language-dependent but also domain-dependent: Fig. 4.1 shows examples of abbreviations in different domains. So every domain would need their own abbreviation list. The Topic Identifier could be used to tell the sentence segmentation component which abbreviation list to use; however the creation and maintenance of such lists is a significant task.
- Abbreviations have different distribution with respect to sentence splitting. Some indicate there should never follow a sentence boundary (like '*Mr.*', '*Prof.*', '*Dipl. Ing.*', '*St.*'); some strongly indicate a sentence end (like '*etc.*', '*usw.*'and similar), some are only part of a multiword abbreviation (de '*F.D.P.*').
- Abbreviations can even be created ad-hoc. For instance, an SAP writing instructions manual defines: „*Space restrictions may necessitate the use of abbreviations in dialog boxes, for field names in the applications, for lines in the menu bar or pull-down menus and for function key names. Use Webster's Ninth New Collegiate Dictionary as the reference for abbreviating words.*".[2] Such dialog boxes and field names need to be translated…
- Punctuations can also be parts of words, a famous example being '*Yahoo!*'. The system should not break a sentence every time this token occurs in the input.

As a result, abbreviation lists will always contain flaws. If used as the only criterion for sentence splitting, they will mislead the analysis sufficiently often.

---

[1] By abbreviation, strings ending in a period are meant. Acronyms like UN, OPEC etc. are not included.

[2] https://experience.sap.com/files/guidelines/References/abbrev_en.htm. German abbreviation rules are given by Duden.

	Medicine		Electrical Engineering
a.c.	Before meals.	cont.	Continuous quantity
b.i.d.	Twice daily.	e.m.f.	Electromotive force
h.s.	At bedtime.	inst.	Instantaneous quantity
p.o.	By mouth.	p.u.	Per unit
p.r.n.	As needed.	Ref.	Reference
q.d.	Each day.		Theology
q.i.d.	Four times daily..	al.	alii, alias ("others")
	Civil Engineering	Appatis.	Approbatis ("approved")
R.B.W	Reinforced brick work.	Cod.	Codex (Manuscript)
U.S.C	Ultimate stress of concrete	Card.	Cardinalis ("Cardinal")
W.S.D	Working stress design	I gr.	Igitur ("Therefore")
	Chemistry	Conf.	Confessor
alc.	#alcohol(ic)	El.	Electio, Electus ("Elect")
alk.	Alkaline	Exe.	Excommunicatio
anhyd.	Anhydrous	hebd.	Hebdomada ("Week")
app.	Apparatus	mph.,mp.	Metaphysics
atm.	Atmospheric	s.d.	sine data ("no printing date")
av.	Average	Lte.	Licite ("Lawfully")
c.d.	current density		
dilg.	Diluting		

Fig. 4.1 Abbreviations in different domains

For such reasons the progress in computational linguistics went into the development of better technologies: Regular Expressions and Finite State approaches were developed, e.g. by (Karttunen et al. 1997), an example can be found in (Jurayfsky and Martin 2009: 105).

There was even a standard defined on this basis, called **Segmentation Rules eXchange format** (SRX): „*The purpose of the SRX format is to provide a standard method to describe segmentation rules that are being exchanged among tools and/or translation vendors*".[3] The progress of this standard is that it does not just consider the string preceding the punctuation but also the context following it. However the main content of an SRX file would still be an abbreviation list described by regular expressions, which is a hard task once hundreds of abbreviation need to be defined.[4]

[3] SRX 2.0, 2008, defined in https://www.ttt.org/oscarStandards/srx/

[4] (Mitkowski and Lipski 2011) report on successful tests with SRX and the GENIA corpus. However this corpus uses periods nearly exclusively for sentence splitting http://www.geniaproject.org/genia-corpus/.

A closer look at sentence boundaries reveals that in languages using the Latin script, case information is at least as helpful as abbreviations. In strings like '*The return of the Yedi This movie was one of the most successful …'*, the capitalisation indicates a sentence boundary, even if no punctuation is available. Such phenomena of missing sentence boundary indications frequently occur in cases where HTML titles are converted into text.[5] In turn, in cases like '*He mentioned dragons, giants, fairies* etc. *in his talk'* the lowercase continuation indicates that the abbreviation does not end the sentence.

Such facts led to a change in the paradigm of sentence splitting in TINA; it can also be found in other publications like (Mikheev 2002).

4.1.2 LtSsplit

4.1.2.1 Approach

1. In LtSSplit, sentence boundary detection is based on **configurations of token annotations**. Relevant annotations are:

- Words which are usually in **lowercase** (mainly function words). If they show up in capitalised form, they indicate a sentence boundary before. Such words, called 'startwords', can be found by analysing the sentence-initial words in corpus data. In principle they comprise all words except proper names, but in fact mainly function words plus a certain amount of content words (identifiable by frequency lists) are relevant here.
- Words which usually occur in **sentence-final** position, like separable prefixes. If words like *'off'*'or *'by'* 'occur before period, this indicates a sentence boundary. Again, the members of this class can be identified by analysing corpus data.
- Words often used bevor **colons**. Sentences like '*The number of channels (default: 4) can be determined by a special switch*' should not be broken after '*default*'.
- **proper names**: Their capitalisation would not indicate a sentence boundary. The collection of proper names in TINA is part of the Named Entity recognition component. Once available there they can also be of use here.
- **abbreviations**. In unclear cases, periods after abbreviations would indicate a sentence boundary. As stated above, these lists are subject of updates and domain-adaptation measures.
- Abbreviations which never indicate a sentence boundary. Words like ‚*Prof.*', ‚*Mr.*', ‚*St.*' or abbreviated first names, as well as multipart abbreviations would be collected here.
- and of course, digits and punctuations deserve closer attention.

[5]As the HTML handler removes the soft return when converting it into text.

LtSSplit has a special resource where all these forms are collected, and annotated with the respective class indicator. Such a resource must be built for all supported languages.

2. Sentence splitting is based on the recognition of **patterns** formed from these classes. Examples are:

- Capitalised words after period indicate a sentence boundary if they are members of the startword list, but they do not if they are names.
- Lowercase words after period indicate a non-sentence-break, even after abbreviations.
- Colon words after parenthesis and before colon cause the colon not to break a sentence.

Such heuristics become more complicated when parentheses and quotes come into play. As already mentioned these heuristics are language-dependent: For instance in German, periods after digits (like $_{\text{de}}$ '5.') are used for forming ordinals and must not be treated as sentence boundaries (except in front of startwords…). Also, as nouns are capitalised, they cannot function as startwords in German.

3. The implementation of the sentence splitter does not use regular expressions. Instead the parser looks at the single tokens and just sets flags for the elements found. These flag patterns are then interpreted to decide if a boundary needs to be set.

This analysis method needs only one pass to analyse a sentence, no parsing cycles are needed as in cascaded rules like SRX, so it is sufficiently fast (Fig. 4.2).

LtSsplit is called with language and a text buffer (from<p>to</p>) as parameters. It first calls a simple tokeniser which only worries about word-initial and word-final punctuations. Next it parses the token list and sets the pattern flags accordingly. At some point (usually the first non-punctuation token after a sentence-break candidate), the flag pattern is evaluated by comparing it with a list of sentence-boundary-indicating patterns, and a<s>tag is inserted into the text, if the pattern is found there. LtSSplit returns a textbuffer with<s>and</s>tags inserted.

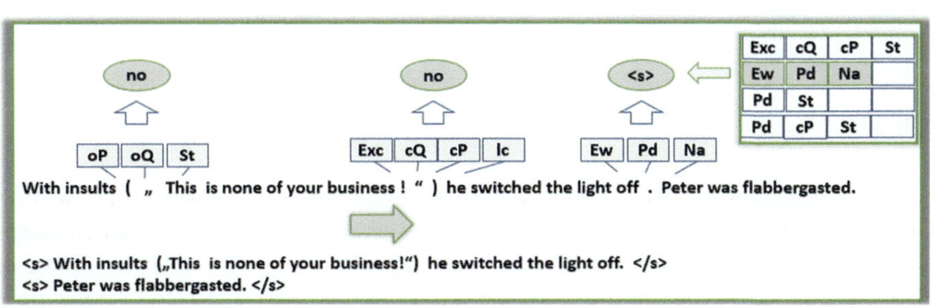

Fig. 4.2 Sentence splitting (LtSSplit)

As mentioned, LtSSplit uses a significant amount of linguistic resources. In addition to the punctuations, also lists of startwords, end words, single and multi-part abbreviations, and names are used. On average these lists are several thousand entries strong.

4.1.2.2 Evaluation

It has been mentioned that sentence splitting is essential for applications like parallel text alignment, Memory-based translation, and Language Processing like Indexing or Named Entity recognition.

1. The first question is how relevant differences in sentence splitting are. A **comparative evaluation** was made using the WACky corpus as reference, cf. (Baroni et al. 2009). The WACky corpus uses the TreeTagger, cf. (Schmid 1994), which is a state-of-the-art tool for sentence splitting and tagging.

The first result is that there are substantial differences in sentence segmentation results between the different tools; with 15–20% of all documents affected. The number of resulting sentences also differs by 15–20%, depending on the segmentation technique selected. While LtSsplit produces more sentences in English, it produces less in German. In cases where sentence segmentation is the basis of alignment, it can be expected that the alignment results would be significantly influenced by the segmentation tools used (Fig. 4.3).

To evaluate the segmentation quality in more detail, about 1000 randomly selected sentences were inspected. The results are given in Fig. 4.4.

The percentage of wrong segmentations is higher for WACky than for LTSSplit; while less clear in English, LtSSplit shows a significant improvement in German. The questionable cases are discussed below, they mainly result from unclear input.

	no. documents	sentences identical	rest text sentences WACky	rest text sentences LtSsplit	difference no. sentences
de	3,900	197,000	101,400	66,600	34,800
en	2,800	134,200	49,500	70,300	20,800

Fig. 4.3 No. sentences affected

	WACky					LtSSPlit				
	total	correct	wrong	%	questionable	total	correct	wrong	%	questionable
de	676	308	172	25.4%	196	424	347	26	6.1%	51
en	450	255	67	14.8%	128	561	364	76	13.5%	121

Fig. 4.4 Sentence evaluation result

Language	English corpus	# sentences	German corpus	# sentences
Corpus	WACKy	2029	WACky	2523
	Europarl	486	Süddeutsche	186
	reuters	570	Europarl	223
			Wikipedia	912
Total sentences		3085		3844
Errors		8		17
Error rate		**0.26**		**0.44**

Fig. 4.5 Corpus evaluation result

2. For an **absolute evaluation** of the LtSSplit quality, a random example about 3000 sentences from different corpora was selected.[6]

Half of the German errors are due to wrong input (paragraphs, missing punctuation etc.), the rest in German and English are due to abbreviation coverage and enumeration problems. As a result, it can be said that LtSSplit is more accurate than usual sentence splitters, esp. for German (Fig. 4.5).

3. There are some problematic cases, however, mostly resulting from text handling problems. This is an example how errors in previous components cause errors in later processing in a pipeline architecture.

- Missing segmentation. Mostly it is a no-split of headings from text, like:

> *Find out more about the Belief and Metaphysics Conference If you haven't done so already,please register with us.*
>
> *Vielfältige Möglichkeiten Die CD-ROM enthält sehr kreative und aufwändige Bearbeitungen vieler Geschichten rund um die Person Jesu.*

Missing information on end-of-heading in HTML downloads is the reason for most of the questionable cases in the data (more than 90% of the questionables above).

- There are cases where a sentence is broken by paragraph marks (<p>). This is sometimes the case in bulleted lists:

> *You need*
>
> - *2 kg of apples*
> - *500 g sugar*
> - *2 eggs*
>
> *to bake this cake*

[6] Each starting from document IDs 70.000 ff.

where Text Handling returns a <p> … </p> for each bullet item. Such structures often return several paragraphs, and a coherent sentence cannot be built anymore as paragraph boundaries are considered to be sentence boundaries as well.

- Scrambled or otherwise **malformed input**, resulting from broken enumerations, tables etc. An example is:

> *Lecturer in Education 05 post School of Social Science and Public Policy/Department of Education and Professional Studies DEPS-Kings College London/Date of entry:29/06/06 Chair of Philosophy of Education School of Educational Foundations and Policy Studies-Institute of Education University of London/Date of entry:29/06/06 Lecturer/Senior Lecturer Philosophy/School of Arts-Roehampton University/Date of entry:27/06/06 Researcher Positions Humanities/Helsinki Collegium for Advanced Studies-University of Helsinki/Date of entry:26/06/06 Lecturer in Philosophy Faculty of Arts Department of Philosophy-University of Leeds/Date of entry:26/06/06 More philosophy jobs...*

  There is not much a sentence splitter component can do.

- A problem are **enumeration** indicators like '*1.2.1*', '*A.*'. They are not part of the textual content but part of the layout, and should be recognised and removed by the text handling component. However they often stay in the text after text handling. While decimal classification headings are rather easy to detect by regular expressions, enumerations using letters are much harder to analyse. They can even be ambiguous:

  In case 1 the 'A.' is an enumeration mark:

  A. *Horner's syndrome causes breathing problems by disrupting the nervous system*
  B. *Hoarseness does so by disrupting the nerves of the larynx.*

  In case 2 it can be part of a person name:

  A. Horner's talk was well received on the conference.

  To solve this a context-sensitive solution must be implemented (something like '*No 'A' without a 'B'*'). This however is missing in LtSSplit, resulting in the ugly situation that such enumerations are detected if followed by a start word, and not if not.

4.1.2.3 LtSSplit for Non-Latin Scripts

As mentioned, non-Latin scripts implement a simple punctuation-based sentence splitting component.

The treatment of Cyrillic script should follow the Latin one as it follows similar rules; however there is an inconsistency in coverage, and Cyrillic script texts do sentence splitting only based on punctuations.

Arabic and Chinese sentence segmentation follow the identification of punctuation characters, without additional sophistication. This is a quite simplified approach; e.g. for Chinese sentence segmentation challenges, cf. (Huang and Sun 2010).

The relevant punctuations in the Arabic script are[7]:

- Sentence-final punctuations: Arabic full stop (\u06D4-), Arabic question mark (\u 061F ؟), end-of-text-mark (\u 061D), (and normal full stop)
- Non-final punctuations: Semicolon (\u 061B ؛) and comma \u (060C ،)

The relevant punctuations in Chinese are:

- Sentence-final punctuations: Ideographic full stop (\u 3002。), fullwidth semicolon (\u FF1B), fullwidth colon (\u FF1a), fullwidth exclamation mark (\u FF1A), fullwidth question mark (\u FF1F)
- Non-final punctuations: ideographic comma (\u 3001 、)
- parentheses: two-byte versions in Chinese: 〈, 《, 『, 【,》 ,〉 ,』 ,】

The relevant language-specific punctuations are collected and annotated in the *ssplit* lists of the respective languages.

4.2 Tokenisation (LtTokeniser)

4.2.1 Rationale

4.2.1.1 Definition: What is a 'Token'?

There is a certain degree of uncertainty as what a tokeniser's task is. The definition in Wikipedia is: '*Tokenization is the process of demarcating and possibly classifying sections of a string of input characters. The resulting tokens are then passed on to some other form of processing*.'[8] Such a description is close to circular, and also holds if strings of five characters would be split away. A bit more concrete is the definition of (Manning and Schütze 1999) that tokenisers ‚*divide the text into units called tokens where each is either a word or something else like a number or a punctuation mark*' (Manning and Schütze 1999:124). Word *or something else*? Others claim that tokenisation means '*segmenting running text into words*' (Jurafsky and Martin 2009: 102). But often enough the units produced by tokenisers are not words.

In a pipeline sequence of tasks, the next step after splitting a text into sentences would be to split the sentences into words. But what is found

[7] https://www.unicode.org/charts/pdf.

[8] https://en.wikipedia.org/wiki/lexical_analysis#tokenization.

between<s>and</s>tags is just a string. However, the step from strings to words is by no means trivial, because the words do not appear in their canonical forms, and sentences contain information other than words, like punctuations etc. And if sentences contain whitespaces, these do not cooccur with word boundaries in many scripts.

Word analysis would be the next step after sentence splitting, i.e. identifying the units in a sentence and assigning them some meaning, mainly by lexicon lookup. This step can either be done in one process (like in LtDeep), or it can be broken down into separated steps like in LtShallow. In the latter approach a tokeniser's task is to prepare lexical analysis by providing its basic units. Tokens are the basic units of lexical analysis. Contentwise they are defined by the lexical analysis component.

The definition of a token therefore depends on the lexical analysis. If there is a lexical entry like '*Yahoo!*' or '*comma,*', then the tokeniser must not split the punctuations. Also, even an obviously wrong tokenisation like de '*Freund(in)*' into '*Freund(in*'+')'[9] may be usable, if the lexical analysis is able to cope with it. In turn, if an analyser is not interested in the recognition of dates, it may let pass a string like '*2019/7/21*' without further notification. But if it is, then a string like the date or like '*300km/h*' would definitely have to be tokenised.

In order to have a clear view of the division of labour of the single components, a distinction should be drawn between the tokeniser which *provides* the basic unit, and the lexical analysis which *interprets* them. As for an example, the string en '*dog's*' would be separated by the tokeniser into '*dog*+ '*s*', and the analyser would attach lexical information like *noun|count|animate* ... to '*dog*', and '*particle|-case=gen...*' to the '*s*.

Misunderstanding the 'upper bound' of tokenisers and trying do segmentation plus full interpretation, there are attempts to burden the tokenisers with the task to recognise Named Entities, like in (Grover et al. 2000), or to combine multiword lexical entries like in (Hassler and Fliedl 2006). Such an approach will scatter the lexical interpretation over quite different components, which is not the best idea. Tokenisers are supposed to deliver the units of analysis; the analysis itself should be performed by the analysers.

Of course there are borderline cases like dates ('*2021/08/07*'), clitics (it '*aumentar-lo*', tr '*ev-iniz*'), multiwords (fr '*parce que*', de '*gang und gäbe*') and others, where it is not completely clear which component should do which task, and it is the combination of tokenisation and lexical analysis which decides on the overall quality of the analysis.

As the tokenisers depend on the capabilities of lexical analysis, it is rather difficult to evaluate the quality of a tokenisation component, as long as it is not clear

9 By stripping the closing parenthesis as resulting from e.g. Juravsky and Martin (2009): 105 and many others.

which lexical analysis intentions should result in which token units. Unfortunately most contributions on tokenisation in the literature do not specify which phenomena their tokenisers should cover. The quality of a tokeniser is defined by the lexical analyser's capability to handle the produced tokens and assign them lexical information.

Most of the research on tokenisers relates to technicalities, like speed, efficient implementation etc.[10] However very seldom the underlying data are considered, and it is a challenge to define how the component can be implemented efficiently before it is clear what it is supposed to do.

4.2.1.2 Requirements

Tokenisers are supposed to prepare the basic units for lexical analysis and further processing. The white-space-focused distinction between single and multitokens[11] is a bit misleading: Single tokens themselves can consist of several tokens, like *500m/h* or *$20.000.-*.

The following requirements define which phenomena the TINA tokenisers should be able to cover; the appearance of whitespace characters is only one of the aspects to consider. Strings of purely digits and purely alphanumeric characters (among which there are e.g. compounds and clitics) would not be further tokenised but left to the lexical analysis proper.

1. General Text Analysis

The recognition of complex token formations is relevant for multilingual indexing, be it because they are object of searches, or be it because they appear in different writing variants (like dates), and it is particularly relevant for several types of Named Entities:

- **digits** containing whitespaces or punctuations: *20 000*, $_{\text{en}}$ *4,000.50* $_{\text{de}}$ *4.500,50.* This is language specific. Only one token should be delivered to lexical analysis.
- **ordinal numbers**: Digit-alpha or digit-punctuation combinations denote ordinal numbers. $_{\text{fr}}$ '*3ième*', 'l*e 16ème siècle*'; $_{\text{en}}$ '3rd', '7th', $_{\text{es/it}}$ '2^{a}' ,'2°', $_{\text{de/tr}}$ '*34.*', '*126.*', $_{\text{ru}}$ '*324-й*', '*36-я*'. The tokeniser should recognise ordinals, it needs a list of ordinal endings as a resource. The result is a textform with the ordinal marker, a POS tag 'ordinal' (*NmOr*), and a lemma just giving the digit.

[10] Examples are (Vijayarani and Janani 2016; Kamali et al. 2022; Diewald 2022; Beesley and Karttunen 2003).

[11] E.g. (Hassler and Fliedl 2006), (Beesley and Karttunen 2003).

- **proper names** containing non-alpha charaters: *AT&T*, *Smith&Wesson*, *O'Leary*. They should come as one token.
- **proper name parts** like (BMW) '*321i*', (Windows) '*NT*', (iphone) '*14 pro max*' are very frequent in texts, and will appear as separate morphemes in most cases. There is nearly no restriction in how these tokens can be formed. Tokens which cannot be assigned to any other category will come out as proper name parts in TINA. Unfortunately they can also occur stand-alone, like in '*The 321i is the fastest car built by the Bavarians*'.
- **dates** not containing alphanumeric month names, like 2022/06/13.[12] These formats are language specific, and different formats can denote the same date. There is a famous difference in date formats of British (*dd-mm-yyyy*) and American English (*mm-dd-yyyy*), which can be resolved, if the Language Identifier delivers the correct result.
- **quote-separated strings:** Quotes have quite different functions and require different treatment.

  a. They can be part of the word, like in *O'Leary* or in. $_{fr}$ *jusqu'à*. Such words should be analysed as just one token.
  b. They can function as morpheme boundary indicator, like in $_{de}$ '*Bach'sche*', $_{tr}$ '*Ankara'daki*'. Such forms should come as one morpheme, and their interpretation should be left to the lexical analysis.
  c. They represent ellipses. In French $_{fr}$ *n'arrive* [pas], *l'exploitation*, *d'un*, *s'installer* the quoted form is an allomorph for '*ne*', '*de*' etc. Similar, in Italian quotes represent allomorphs of determiners (*un'automobile, l'ordine, dell'ottobre*), prepositions (*all'incidente, dell'ottobre, sull'eliminazione, senz'altro, nell'Italia*), pronouns (*quest'anno*) and other constructions (*un po' meglio com'è ovvio*). In English forms like *he'll, we'd, isn't, don't*, the quoted forms would also be allomorphs for their non-quoted counterparts. This also holds for forms like *dog's* which is even a homograph (*dog_is, dog_has, dog's*). In most of these forms the tokeniser should split the strings after the quote and return two tokens. Details are given below.

- **Hyphenated** forms also require more detailed consideration.

  a. They can be part of the word, like in $_{en}$ nice-to-have, stand-off, data-base, $_{it}$ coalizione rosso-bruna, settore lattiero-caseario, $_{fr}$ après-midi, vice-président, procès-verbal, peut-être, $_{de}$ Rauten-Muster, Hans-guck-in-die-Luft. As they are words in toto, the tokeniser should return just one token. However, some forms can have several allomorphs with or without hyphen, like $_{en}$ 'stand-off' versus 'standoff', 'database' versus 'data-base' versus 'data base'. Normalisation will be required in such cases to avoid the creation of different final index terms.

[12] In ISO format.

There are also many cases of ad-hoc lexicalisations, often with prefixes like *pre*, *anti* or *mega*: $_{\text{en}}$ *anti-Iranian, Boston-based,* $_{\text{it}}$ *politica anti-alcol,* $_{\text{de}}$ *Mega-Event*. From the tokenisation point of view they should result in just one token, although lexical analysis would fail in an easy access and would have to analyse the parts separately.

b. Hyphens which separate morphemes and should be removed are for instance $_{\text{fr}}$ *inscrivez-vous, sera-t-il, laissez-moi, elle-même, serait-ce*. These forms should return two morphemes.

c. There are hyphens in German which work as placeholders for compound parts, like in '*Haus- und Grundbesitz*' or '*be- und entladen*'. This is short for '*Haus[besitz] und Grundbesitz*' or '*be[laden] und entladen*'. Such hyphens must stay with the morphems to which they are attached, to indicate a gapped construction.

- **multitokens**: They range from just strings with spaces ($_{\text{fr}}$ '*parce que*', $_{\text{de}}$ '*gang und gäbe*', $_{\text{en}}$ '*to and from*'[13]) to multiword concepts ($_{\text{en}}$ '*nuclear power plant*') and Named Entities ('*Radio Free Europe*'). While some authors, e.g. (Grover et al. 2000) or (Hassler and Fliedl 2006), try also extended Named Entity recognition like person names in their tokeniser, only the first class of multiwords is analysed in LtTokeniser, similar to other systems like (Beesley and Karttunen 2003). The more complicated cases are passed to later components.

Such tokenisation challenges can be found in many texts and can cause considerable confusion if not treated properly. In addition, many tokens must be analysed in more detail in the context of Named Entity Recognition.

2. Named Entity Analysis

The tokeniser should also recognise the units required for Named Entity recognition as these are relevant for multilingual indexing. Among the relevant phenomena are the following:

Communication Means

- **phone numbers**: (0631) 45 60 789. There are different formats in different countries; examples are given in (Manning and Schütze 1999:131). Keywords like '*phone*' or '*mobile*' could help in recognition
- **URLs**: https://fr.wikipedia.org/wiki/Wikip%C3%A9dia:Accueil_principal can contain nearly any character. LtTokeniser looks for keywords like 'https' or 'www' or combinations of period+domain extensions ('*.com*', '*.tr*', '*.biz*' etc.) to recognise '*IBM.com*'. The tokeniser would contain a list of such domain names. It is language independent.

[13] More examples are given by (Beesley and Karttunen 2003).

- **email** addresses:. Emails have a defined syntax,[14] consisting of four parts: a recipient name, an '@' symbol, a domain name, and a top level domain which allows to reuse the URL list.

Units

- **units** are parts of amounts, together with values (digits). They are defined by the *International System of Units (SI)*,[15] maintained by the *General Conference on Weights and Measures*. While the standard states that values and units should be separated by a space this is often neglected in real texts. So there are tokens like *5cm, 17.3°*, 3', *15m/sec*. Percent signs also often are not separated, moreover they can precede or follow the number values (*%24.5, 24.5%*). All the potential units for the tokeniser are collected in a special list. This list is language-independent. In the analysis the units should be treated as (allomorphs of) common nouns, and the amounts together with the digits form complete NPs.
- **currencies** are also a specific class of units. They are identified by currency symbols ('*$*', '*€*') or specific abbreviations (three-letter codes of ISO 4217, like '*USD*' or '*CHF*'). The Symbols can precede or follow the values (*€50* or *50€*) and can have specific forms (decimals like *50.-* or abbreviated forms like *15K* or *6bn*).

Other tokens

- **Bank accounts** are an interesting object for some recognisers. In Europe they follow the regulations of the International Bank Account Number (IBAN, ISO 13616). An example is '*DE76 2300 2100 2393 3999 34*'. The IBAN should be returned as one token by the tokeniser. It is language specific (number of digits).
- **Chemical expressions** like '*2,3,3-Trimethylpentane*' are subject to special retrieval systems for chemical substances. Details are given in (Hawizy et al. 2011) or (Krallinger et al. 2017).
- **mathematical expressions** and formulae are only rudimentarily analysed in LtTokeniser, by splitting digits and mathematical operators (treated as a subcategory of Conjunctions in the Tagset).

As opposed to other tokenisers, LtTokeniser does not segment strings containing only alpha characters. Such strings could be clitic elements in Italian and Spanish, compounds in German, agglutinated forms in Arabic and Turkish. They will be treated in lexical analysis.

---

[14] As defined in RFC5322.

[15] https://en.wikipedia.org/wiki/international_system_of_units.

4.2.2 LtTokeniser

LtTokeniser's input is a string as produced by the sentence splitter, together with the sentence-related annotations (language, topic, global casing, paragraph type). Different analysis strategies are applied depending on script and language.

4.2.2.1 Latin Script Tokenisation

Latin script sentences, as well as Cyrillic script sentences, are analysed by the LtTokeniser in three steps:

1. *Segment* segments the string into basic token classes
2. *Recombine* combines members of these classes into tokens
3. *Annotate* adds annotations to the (possibly newly created) tokens
4. (A forth step, *Normalise*, grew into a separate component outside the tokeniser, cf. Sect. 4.3)

1. **Segmentation** is language-independent and splits the input string into basic elements cf. Fig. 4.6. One of the types defined in Fig. 4.6 is assigned to all input characters.

The members of these classes are defined by tables, including non-standard characters like smart quotes, characters like ‘“’ and ‘”’, fractions like ½, whitespaces like non-joiners, and others.

The segmentation step of the tokeniser is also used in sentence splitting, to identify sentence-final punctuations.

2. The second step is the **recombination** of these basic classes into complete tokens. This is done using a sliding window of length 3. Depending on the patterns formed by the underlying basic token classes the decision is made to split or combine these elements. Additional tests on class patterns can ask for more detailed information; examples like ‘*isOrdinalEnding*’, ‘*isThreePlaceDigit*’, ‘*isLowercase*’, ‘*isUnit*’ and the like. Such tests are collected in a class called ‘*TokenWizard*’. The resulting token element forms the left corner of the next triplet.

- **alpha characters (lower and uppercase)**
- **periods**
- **single quotes**
- **punctuations (all other punctuations, like commata, parentheses, question and exclamation marks)**
- **digits**
- **hyphens**
- double quotes
- symbols (currencies, math operators etc.)
- (mixed. Used during analysis for newly created tokens combining several classes.)

Fig. 4.6 Token classes

token 0			token + 1			token + 2		result token	set feature	example
tclass	test	adj t0+1?	tclass	test	adj t1+2?	tclass	test-t2			
alfa	Currency	y	digit					split1-2		$ 300
alfa		y	punct	OpPar	y	alfa	"in"	combine1+2+3	tclass=mixed, parflag=t	Wirt(in
alfa	No-Abbr	y	period					split1-2		forgotten .
digit		y	period	Dec-Pt	y	digit		combine1+2+3	tclass=mixed	3.6 [en version]
digit		y	alfa	isOrdl				take1		3rd -> 3
digit		-	alfa	isUnit				split-1-2		34 cm
digit		y	hyphen		y	alfa		combine1+2+3	tclass=mixed	24-year
digit		n	digit	is3digit	n	digit	is3digit	combine1+2+3	tclass=mixed	3 000 000
mixed		y	hyphen		y	alfa		combine1+2+3	tclass=mixed	[15-years]-old
mixed	Parflag	y	punct	isCPar				combine1+2	tclass=mixed	[Wirt(in])

Fig. 4.7 Tokenisation example

Whitespaces are not part of this procedure. They are removed from the sliding window and converted into a feature.[16] So there is no need for multiple analyses for '*2022/12/17*' and '*2022 / 12 / 17*' as single token or multitoken. If it is relevant whether or not two basic tokens must be separated by whitespace, then a special test can be used like '*isAdjacent Y/N*'.

Like in other systems, e.g. (Beesley and Karttunen 2003), flags are used to analyse longer-distance phenomena like URLs, opening and closing parentheses, specific numerals and the like.

An example is given in Fig. 4.7. It shows how tokens are built from three items of different token classes, depending on the class itself plus some tests; the result is one or several tokens; examples are given in the rightmost column.

It should be noted that the tokeniser may change the input string, mainly by manipulating inner-token whitespaces into a normalised representation, e.g.'*40 000 000*' –>'*40000000*' or '*$50. –*'–>'*$ 50*'. However the original textform is kept in the Document Model, in case the global IR system wants to use it, e.g. for highlighting such text segments. Also onset and offset refer to the original string.

In turn, multitokens following the pattern '*alfa–alfa*' or '*alfa–alfa–alfa*' are combined into one token, if tests confirm that they are in the multitoken list (e.g. $_{fr}$ '*parce que*').

The tokenisation of quotes and hypens depends on the language.

- In **English**, quotes and hyphens do not split tokens, except for function words like '*he'll*'. Forms like *Boston-based*, *anti-aging* form just one token. Lexical analysis will take care of it.
- The same holds for **German**. Quotes are only used to mark genitive case word Names ending in -s (*Siemens'*) and the incorrect but popular endings in 's (*Gretel's Laden*). This will be taken care of by Lexical analysis
- **French** quotes are only split, if they are pre-vocalic and members of a preclitics list (determiners like *l' d'*, pronouns like *m' t'* etc.); other cases stay in

[16] In the Document Model it is an annotation to the preceding token.

one token (*aujourd'hui*). Hyphens indicate token boundaries only if member of a list mainly of pronouns (*-me*, *-vouz*, *-les-lui*, *-vous-en*).
- **Italian** quotes split tokens, if pre-vocalic and members of a proclitics list (*dall' nell' novecent'* …). Hyphens do not split tokens.
- **Other** languages either do not have this problem (Arabic), or do not split the token (Spanish, Portuguese, Russian). Turkish quotes (*Ankara'da*) also do not split the token and are treated in the Lexical Analysis.

If tokens are split, then the clitic parts (like *l' nell' d'* …) need to be represented in the dictionary for lexical analysis as allomorphs of their non-quoted forms (*la/le, nella, de, novecento*); in turn non-split forms (like *he'll don't*) need also be represented as allomorphs of their respective verb forms. These phenomena clearly indicate that tokenisation needs close interaction with the lexical analysis and its dictionary resources.

3. The recognised token is **annotated** when handed to the Document Model. Annotations are onset and offset, textform and normalised form, casing, and possibly additional features. Details are given below.

4.2.2.2 Chinese Script Tokenisation

There is no officially fixed way to tokenise Chinese.[17] Several phenomena can be observed which allow for different tokenisation:

- compounds: Concepts like '国家公园' ('*National Park*') can be written as one token but can also be split into two ('国家 | 公园') without a change in meaning
- measure nouns are sometimes separated from the numbers, sometimes they are not (1991 | 年 versus 1991年).
- copulae are sometimes given extra tokens, sometimes not: The string '就是' (exactly) can also be split, with an extra token for the copula ('就 | 是'). Splitting certain particles is also an often observed difference in tokenisation (e.g. '– 怒气冲冲地' (angrily) versus '– 怒气冲冲 | 地').

However, even if an input text can be tokenised in different ways, the tokenisation decision is often not so important: Even if texts are differently tokenised texts very often produce identical results, i.e. they show the same meaning, and have identical translations.

More important is the requirement that the tokenisation result matches the dictionary. A missing entry may lead to a segmentation into single characters with misleading content. And if the dictionary contains only a token like '– 怒气冲冲'

[17] (Gao et al. 2005).

(*angry*), then to tokeniser must not return '– 怒气冲冲地' (*angrily*) as one token because it would not be found.

Two lines of technology are generally offered to tokenise Chinese text.

- The character-based line calculates the probability of a Chinese character to occur in word-initial / word-internal / word-final position, and computes the best segmentation on this basis. Corpus data are used to determine these probabilities. An example of such an approach is (Tseng et al. 2005).
- The lexicon-based line uses lexical data. All possible words in an input string are detected, and the best word sequence is computed by a dynamic programming approach and distance measures, e.g. lowest number of words, or cooccurrence probabilities. An example of such an approach is (Chang et al. 2008), or the LtTokeniser described here.

In a multilingual indexing setup such as TINA it is understandable that segmentation should produce tokens usable for indexing, and in addition tokens which can be linked to the global multilingual content net. Such an effect can be achieved by a lexicon-based approach, provided the dictionary can show reasonable coverage. The best results then can be achieved if the segmentation matches the decisions used in the dictionary.

The tokenisation of Chinese texts in LtTokeniser is a modified version of the '*maximum matching*' or MAXLENGTH approach as described in (Jurafsky and Martin 2009: 104ff): From a given point in the input string the longest possible continuation string is searched in the Chinese dictionary. The maximum length is determined by the longest dictionary entry. If not found, the search length is reduced one by one down to a minimum of one. From the last position of the found string the search is continued again with the longest continuation string etc. until the whole sentence is covered.[18]

To determine the maximum token length the dictionary is inspected. The longest entry there shows 37 characters (which is a circumscription of a translation for the Oracle company (美国计算机公司,主要生产Sun系列工作站和网络产品,是Java 语言的创始者)). As it would cost computation to start each search with strings of such a length, the maximum length was set to 10, excluding only about 500 entries (0.1% of the entries) from being searchable. On average the entries were 3.3 characters long.[19]

The MAXLENGTH approach obviously depends on the dictionary resources used. The LtTokeniser uses a token list of 256 K entries, derived from several dictionaries. The resource is designed to cover general-purpose text. Like other such resources it needs to be adapted to the domain which it is supposed to analyse. The challenge is to avoid unknown words because the MAXLENGTH analysis breaks

[18] A method to shorten long Chinese sentences by inserting commata is proposed by (Jin et al. 2004).

[19] (Zhang 2020) reports an average length of 3.86.

Token	Replacement
不等价	
不等价中断	
不等价交换	不等价 交换
不等值	不 等值
不等到	不等 到
不等压头	
不等号	
不等同	不 等同
不等式	
不等式约束	
不等待	不 等待

Fig. 4.8 Example of Chinese list

睡蓮屬的学名*Nymphaea*来自於古希腊语单词 “Νυμφαία”，这个来自於 “Νύμφη”， “Νύμφη”的意思是宁芙。宁芙是希臘神話中一类住在湖泊、森林、河流、巖洞和海邊中的美麗女神，深受眾神和男人的崇敬與愛慕。

Fig. 4.9 Chinese text example

if a word is not found; then the tokeniser can only resume its work at the first known token and leaves the string in between as unknown.

It is known that the MAXLENGTH approach is over-greedy. Therefore the LtTokeniser analysis returns alternative segmentations for a given span whereby a span can be replaced by several shorter tokens for a longer one. The token list stores such alternative segmentations in a second column. If a token candidate has an entry there, then the longer token is replaced by the tokens in the second column. To find the correct tokenisation for such spans, corpus data were inspected manually, and the strings which should be segmented into shorter tokens were marked. About 42 K entries in the token list contain alternative segmentations.

As Chinese texts do not only contain Chinese characters but also digits (both single and double width), punctuations, symbols, Latin text segments etc., a small Finite State analyser is integrated to take care of the different properties of the respective character sets. In the tests using the TED[20] corpus, 0.4% of the tokens were not member of the Chinese script (Figs. 4.8 and 4.9).

[20] www.ted.com. cf. Zeyrek et al. (2020).

4.2.2.3 Arabic Script Tokenisation

The Arabic-Persian script[21] is quite liberal in the use of whitespaces. They seem to follow aesthetic motivations rather than linguistic ones.[22] Many combinations of words (like prepositions and nouns) are not separated by whitespaces, and the insertion of whitespaces into words which otherwise would be hard to read or which look nicer is quite common. An example is پرداخت ها versus پرداختها (*par. dāxt.'hā, payments*); both variants occur in texts. Whitespace tokenisation is an adaptation of Latin-script technology to Arabic scripts.

For this reason the LtTokeniser does not try to interpret or even normalise the whitespaces, but only carries out basic operations. It just separates punctuations from letters. It also cannot rely on different types of whitespaces (joiners/non-joiners)[23] which are sometimes used to combine aesthetic and linguistic requirements, as these are not regularly used in texts, and are normalised away by the Text Handling component.

Specific attention is required for text parts in foreign scripts (mainly Latin script passages). If these text parts are intended to be interpreted, small tokenisers for the respective scripts need to be called during the Arabic script tokenisation, as mentioned in Chinese tokenisation above.

The two Arabic script languages are treated differently in LtTokeniser:

- **Farsi** analysis is based on LtDeep technology. Current shallow approaches like in (Kamali et al. 2022) or (Panahandeh and Ghanbari 2019) try to re-arrange whitespaces as a preprocessing step. In LtDeep for Farsi, tokenisation is included in the Lexical Analysis. Token boundaries are created depending on the morphemes found in the dictionary. Whitespaces are not a priori relevant in this analysis.

  A consequence must be drawn for the POS tagset, however. The analysis will contain morphemes which are classified as bound in other contexts but occur in isolated form here. Such morphemes (like inflections or prefixes) must receive their own POS tag in order to be accessible by morphosyntactic rules, and therefore the tagset must be extended to bound morphemes, like affixes or inflections etc. Details are given in Sect. 5.4

- **Arabic** tokenisation is part of the LtShallow technology. Like in other languages, segmentation of the alpha-character strings is left to the Lexical Analysis component. This relates to combinations of words with conjunctions, prepositions, and clitic pronouns. Unlike proposals like (Attia 2007) the tokeniser itself does only basic token separation, mainly punctuation stripping, but no recombination.

---

[21] A presentation and discussion of the script is given in (Majidi 1996).

[22] More on the aesthetic properties of the script can be found in (Schimmel 1970).

[23] (Megerdoomian and Zajac 2000) use non-joiners in tokenisation decisions. However these characters do not sufficiently reliably occur in everyday Persian texts.

Due to the fact that whitespaces are not stable indicators for token boundaries, TINA has selected a LtDeep application where several tokenisation hypotheses can be followed in parallel (cf. Sect. 5.4).

4.2.2.4 Tokeniser Result

Whichever script was tokenised, all tokenisers are supposed to deliver uniform results and populate the Document Model with their results: Sentences (<s>to</s>are described as a list of tokens (<t/>). Tokens are described by the following feature annotations:

- **ID**: every token has an ID, counted from the beginning of the sentence
- **textform**: the original form of the token as found in the input
- the **onset and offset**, related to the textform
- **normalform**: in case the token has been normalised (like in dates or amounts) the normalised form is given here
- **features**: Some tokeniser actions return information like POS (for ordinals) or more detailed annotations (for parts of dates for instance). This is stored here in feature-value notation.
- **whitespace**: stores the whitespace following the token (if any) in case a reconstruction of the original string is required
- **casing**: values are: *uppercase*, *lowercase*, *capitalised* (for alpha-strings), *digit*, *punctuation*, and *mixed* (two of alpha–digit–punctuation), provided for languages which differentiate casing.

Latin script tokens are truecased. The truecasing modifies the casing of the token in two cases: 1. The first sentence-initial alpha token is lowercased if the token is a startword. 2. In hyphenated tokens the casing value refers to their last element. Tokens like $_{\text{en}}$ '*Boston-based*' or $_{\text{de}}$ '*EU-gefördert*' will therefore have lowercase value.

Whitespace and casing are left empty in non-Latin scripts.

LtTokeniser does not provide alternative tokenisations. If such alternatives are relevant, then an application of LtDeep would have to be chosen where tokenisation and lexical analysis are done in one step and therefore can handle alternative tokenisations. This opportunity is used in the analysis of Farsi, as described in Sect. 5.4.

4.3 Normalisation (LtNormaliser)

Before lexical analysis, the token list undergoes a normalisation step. Its purpose is to reduce spelling alternatives; they would otherwise lead to noise in indexing, and require the query analysis to produce all spelling variants of a given term, in order not to risk coverage losses. In practical terms, the linguistic resources would have to be blown up if spelling variants had to be considered, leading also to an increase in maintenance cost. However, the original form of a textform is stored in the '*textform*' feature, and the normalised form is stored in the '*normalform*' field of the token in the Document Model, so the original form can be reconstructed if required.

4.3.1 String Normalisation

The first step is normalisation on character level. As for **punctuations**, double and single quotes are normalised to just one representation. Smart quotes and language-specific quote characters need not be considered thereafter. This is already done before tokenisation.

Casing is normalised to lowercase, except for proper names and acronyms. This relates to sentence-initial uppercase letters and to casing in English titles. German common nouns are also lowercased as they may occur in compounds in lowercase form. Turkish common nouns, if capitalised, are often proper names (*Yılmaz, Özlem, Sevgi*) and therefore are not lowercased.

Next, **language-specific** phenomena related to writing standards are normalised. This is a reaction to the tendency to colloquialise the writing.

In French, lowercased words may have lost their accents, if converted from uppercased ones as accents need not be written in uppercase forms.

In Italian, grave accents (*città*) can be restored if missing, but words with acute accents may also have legal forms without it (*é* versus *e*), so no correction can be done.

In Spanish, words with acute accents can be ambiguous (*mi* versus *mí*, *donde* versus *dónde*) in other cases, missing accents need to be restored (*educaria –>educaría, limon –>limón*). Also before the last spelling reform characters like *ll* and *ch* were changed to be written as just one character.[24]

In Turkish, the circumflex is not always written anymore. There are both writings (*hâlâ* versus *hala*) representing different words, but in the majority of cases both writings for the same word are possible (*imkan* versus *imkân*). Which form should be the normal form depends on the dictionary to be looked up; the dictionaries in TINA try to follow the regulations of the Turk Dil Kurumu[25] where the circumflexes are written: The normaliser must produce the forms accordingly.

In German, umlauts are often written as e-insertion in older texts (*Haeuser –>Häuser, Zuege –>Züge*); this must be normalised. Also many texts are still produced in old spelling and need to be mapped to new spelling regulations (*greulich –>gräulich, Abfluß –>Abfluss*). This process is not completely regular unfortunately, so the cases must be treated one by one.

[24] Cf. Royal Spanish Academy: Ortografía de la lengua española, published in 2010.

[25] https://sozluk.gov.tr/.

Finally, in English spelling there are two writing systems depending on the locale (British English and US English. *coloured* versus *colored*, *economise* versus *economize*, *rivalling* versus *rivaling* etc.). As indexing should have just one representation of a concept, one form should be mapped to the other one. LtTokeniser allows to select the locale by parameter, and provides a mapping between US and UK English forms; however the decision must be in agreement with the global IR system. For the time being, the UK form has been decided to be used as the standard in TINA.

All these standardisation operations are mainly based on list lookup. The lists are derived from the available dictionary resources, and they contain pairs of<ill-formed, well-formed>strings. They are looked up at runtime, and if an entry is found, it is replaced by the normal form. Given the considerable amount of exceptions in the languages just mentioned, this is the simplest and safest procedure, and it is also the fastest in computation.

Normalisation of **Arabic** texts needs to undergo special treatment. In order to reduce the variance in Arabic writing, several normalisation techniques have been developed,[26] most noticeable the Tashaphyne normalizer.[27] LtNormaliser uses a more detailed analysis, where the normalisation on textform level uses the following criteria:

- remove all honorifics (*\u0610-0614*)
- remove all vowelisation characters (fatha, kasra, damma, sukun) (*\u064E,\u0650,\u064F,\u0652*)
- remove shadda (duplication) and tatweel (underscore) (*\u0651, \u0640*)
- convert initial alef-madda and alef-hamsa to alef (*\u0622,\u0623,\u0625 –>\u0627*)
- convert word-final fathatan to alef (*\u064B–>\u0627*)
- convert Farsi yeh and keheh (یک) to Arabic yeh and kaf (يك) (*\u064A–>\u06CC, \06A9–>\u 0643*)
- convert Arabic-Indic digits to Arabic digits (*\u06F0-06F9–>\u0660-0669*)

Word-*internal* diacritics on alef (*\u0622, \u0623, \u0625*), waw (*\u0624*) and yeh (\u0626) are *not* removed as this would lower the precision. Also, alef maqsura are not normalised, nor is hamsa alone (*\u0621*).

In **Farsi**, only potential misspellings of *yeh* are corrected, and the Arabic digits (\u0660 ff.) are mapped to their Arabic-Indic digits: (\u06F0 ff.).

All normalised forms instantiate the feature '*normalform*' in the document model, its value being the normalised string.

It should be noted that all normalisation actions must be repeated in query processing: If the text normalises the term *itemize* to *itemise*, then the same procedure must be done on the query side, otherwise the term will not be found. The same holds for the dictionaries, esp. for Arabic: When comparing text elements

[26] Examples are (Sallam et al. 2016) or (Al-Shammari 2009).

[27] https://pythonhosted.org/Tashaphyne/Tashaphyne.normalize-module.html.

and dictionary entries then both sides must provide normalised forms as common basis of comparison. This is another reason why normalisation should not require extensive computation and will be restricted to simple list lookup.

4.3.2 Spelling Correction (LtSpeller)

Misspelling is a major source of noise in text analysis, and nearly every system (WinWord, OpenOffice (with Hunspell[28]), Solr, Lucene) provides a spell-checking component. Most of them are restricted to 'non-word' detection ('*tehre*' –> '*there*'), and do not attempt to do malapropism (context-dependent) 'wrong word' detection ('*there*' –> '*three*').[29]

4.3.2.1 Standard Approach

The standard technology for non-word spell-checking is the Minimum Edit Distance, as described in (Jurafsky and Martin 2009:106ff.):

- Every textform is looked up in a 'known word' (full form) list, as provided e.g. by aspell, Hunspell and others.
- If the textform cannot be found, the system tries to find the closest string to the given textform (in terms of number of insertions, deletions, substitutions) from a reference list, based on dynamic programming, and using a distance measure like the Levenshtein distance.
- The n-best findings are correction candidates; they are presented to the authors, which select the final correction.

The quality of such a speller depends on several factors, in particular on the size of the **known-word list**, although sometimes very rare words prevent the speller from finding misspelt words. Practical importance lies in the selection of the reference list, as a dynamic programming comparison of a textform with a known word list of several ten thousand strings is computationally very expensive.

The technology also has its limitations in cases like Turkish or Hungarian where the 'known word' list grows too far due to the morphological richness of these languages. The solution would be to do more complex operations on the 'known word' list before a comparison could be made.

For German spell checking, decomposition of compounds must be tried to reduce the size of the known-words list, and the respective compound parts are tried to be spell-corrected.

---

[28] https://github.com/hunspell/hunspell.

[29] Cf. (Zesch 2012) on detecting malapropisms.

en	es	de
190k	751k	140k
(no en_US forms)		*(no compounds)*

Fig. 4.10 Size of speller lists

4.3.2.2 LtSpeller

A state-of-the-art spell checking system was implemented in TINA for German, English, and Spanish.

The known word list was taken from the TINA dictionaries, enriched by additional word lists from other sources.[30]

Subsets of these word lists were used as a reference list at runtime, consisting of all words starting with the same letter as the textform. Moreover the reference list consists of string-normalised word forms, which further reduces its size as some spelling errors are already corrected by the normalisers (es *tambien* –>*también*, de *Anschluß* –>*Anschluss*). The size of the resources is given in Fig. 4.10.

4.3.2.3 Correction

The **correction** phase of a spell checker differs from interactive correction where human intervention selects the best alternative for the incriminated text form. In text processing contexts like Machine Translation or Multilingual Text Analysis there is no possibility for the manual selection of the best correction proposal. Instead, the decision to pick the best alternative must be taken by the system.

As the only data for this decision consist in distance measures of replacement alternatives, the heuristics applied in LtSpeller are based on the distance between textform and the best replacement, and the distance between the best replacement and the second best. Only if a replacement candidate is sufficiently close to the textform, and significantly better than all other replacement candidates, then an automatic replacement is tried.

Such an approach could need additional evidence which could come from contexts, e.g. triplets could be used to find the best replacement given the left and right context word of the replacement candidate; and the best scoring triplet would be used as additional input for the correction calculus. An additional resource would be needed, consisting of such triplets collected from corpus analysis and domain-adaptation. The difference to Language Models would be that in spell correction only local decisions are required, not the optimal global analysis path through the whole sentence, except if context-dependent 'wrong-word' correction is also intended: Then every word is a replacement candidate.[31]

---

[30] This was possible because no annotations to the entries are required for the speller, only the textform is needed.

[31] An option how to handle this is presented in (Jurafsky and Martin 2009: 201ff.).

There is another option to tackle spelling errors: The errors themselves are not free of regularities. They depend for instance on the proximity of keys on the keyboard, on the sequence of key strokes etc. Such regularities can be exploited. The approach would be to define an additional resource which consists of pairs<*spell error –>correct form*>, and spell checking would be a simple lookup in this list. The resource could be collected from existing index files which contain sufficient undetected spelling errors and can provide material.

Uncorrected misspelt words in indexing cause noise and will not be found, like other out-of-vocabulary terms. They could still be found and corrected in maintenance of the index itself.

4.3.2.4 Tests

The resulting LtSpeller component was tested in a system for the validation of Translation Memories. About 3600 segments of Translation memories of different domains (IT, mechanical engineering, medicine) were inspected.

The findings were that a spell checker is a useful tool as it flags many incorrect textforms. However, to keep the amount of flags to a manageable size, a series of measures needs to be taken:

- recognising and excluding proper names;
- reducing the amount of unknown words overall;
- adaptation of the speller resources to the domain in question (this was particularly relevant for the medical domain);
- recognition of compounds in German.[32]

Without such measures the speller flags too many candidates to be useful, and should perhaps be replaced by a simple 'known errors –>correct forms' list check where the known errors are collected from corpus data.

References

Al-Shammari, E.T. (2009). A Novel Algorithm for Normalizing Noisy Arabic Text, *WRI World Congress on Computer Science and Information Engineering, 2009, pp. 477-482,* https://doi.org/10.1109/CSIE.2009.952.

Attia, M.A. (2007). Arabic Tokenization System. *Proc. of the 5th Workshop on Important Unresolved Matters, ACL Prague, 65–72.*

Baroni, M., Bernardini, S., Ferraresi, A., & Zanchetta, E. (2009). The WaCky wide web: a collection of very large linguistically processed web-crawled corpora. *In: Language Resources and Evaluation 43.3, pp. 209–226.*

Beesley, K. & Karttunen, L. (2003). Finite-State Morphology. *CSLI Center for the Study of Language and Information.*

[32] This is e.g. also part of the Hunspell speller.

Bisang, W. (2016). Chinese Syntax. In Sin-Wai Chan, ed., The Routledge Encyclopedia of the Chinese Language. New York: Routledge.

Chan, S.W., ed. (2020). The Routledge Encyclopedia of the Chinese Language. Routledge.

Chang, P.C., Galley, M., & Manning, Chr (2008). Optimizing Chinese Word Segmentation for Machine Translation Performance. *Proc. Workshop on Statistical Machine Translation (WMT 2008).*

Ching, Y.P., Rimmington, D. (1997). Chinese, An Essential Grammar. Routledge

Damerau, F.J. (1964). A technique for computer detection and correction of spelling errors. Communications of the ACM, 7, 171-176.

Diewald, N., Kupietz, M., & Lüngen, H. (2022). Tokenizing on scale - Preprocessing large text corpora on the lexical and sentence level. *Proc. 20th Conf. of European Association of Lexicography (EURALEX 2022). Mannheim, Germany, pp. 208–221.*

Diewald, N. (2022). Matrix and Double-Array Representations for Efficient Finite State Tokenization. Proc. 10th Workshop on Challenges in the Management of Large Corpora (CMLC-10) at LREC 2022. Marseille, France, pp. 20–26.

Gao, J., Li, M., Wu, A., & Huang, Ch.N. (2005). Chinese Word Segmentation and Named Eintity Recognisiton: A Pragmatic Approach. *Computational Linguistics, 31(4): pp. 531–574.*

Grover, Cl., Matheson, C., Mikheev, A., & Moens, M. (2000). LT TTT – A Flexible Tokenisation Tool. *Proc. 2nd Int. Conf. on Language Resources and Evaluation (LREC 2000), Athens, Greece.*

Hassler, M., Fliedl, G. (2006). Text preparation through extended tokenization. *in: Zanasi, A., Brebbia, C.A., Ebecken, N.F.F, eds.: Data Mining VII: Data, Text And Web Mining And Their Business Applications. WIT press.*

Hawizy, L., Jessop, D., Adams, N., & Murray-Rust, P. (2011). ChemicalTagger: A tool for semantic text-mining in chemistry. *Journal of Cheminformatics 2011, 3:17*

Huang, H.H., Sun, C.T., & Chen, H.H. (2010). Classical Chinese Sentence Segmentation. https://www.researchgate.net/publication/228891221.

Huang, Ch.R., Shi, D. (2016). A Reference Grammar of Chinese. Cambridge University Press

Hunspell Hunspell http://hunspell.github.io/.

Jin, M.X., Kim, M.Y., Kim, D., & Lee, J.H. (2004). Segmentation of Chinese long sentences using commas. Proc. 3rd SIGHAN Workshop on Chinese Language Processing, pp. 1–8, Barcelona, Spain.

Jurafsky, D. & Martin, J.H. (2009). Speech and Language Processing. *Pearson Int. Edition (2nd ed.).*

Kamali, D., Janfada, B., Shenasa, M.E., & Bidgoli, B.M. (2022). Evaluating Persian Tokenizers. https://doi.org/10.1145/1122445.1122456

Karttunen, L., Chanod, J.P., Grefenstette, G., & Schiller, A. (1997). Regular Expressions for Language Engineering. *Natural Language Engineering 1997 Vol. 2 , Issue 4, pp. 305–328.*

Krallinger, M., Rabal, O., Lourenço, A., Oyarzabal, J., & Valencia, A. (2017). Information Retrieval and Text Mining Technologies for Chemistry. *in: Chem. Rev. 2017, 117, 12, pp. 7673–7761.*

Kukich K. (1992). Techniques for automatically correcting words in texts. ACM Computing Surveys, 24, pp. 377-439.

Li, Ch.N., Thompson, S.A. (1989). Mandarin Chinese, A Functional Reference Grammar. Univ. California Press

Majidi, M.-R. (1996). Einführung in die arabisch-persische Schrift. *Buske Hamburg.*

Manning, C.D., & Schütze, H (1999). Foundations of Statistical Natural Language Processing. *MIT Press*

Megerdoomian, K., & Zajac, R. (2000). Processing Persian Text: Tokenization in the Shiraz Project. *Memoranda in Computer and Cognitive Science MCCS-00–322, Computing Research Laboratory New Mexico State University, April 2000.*

Mikheev, A. (2002). Periods, Capitalized Words, etc. *Computational Linguistics 28, 3, 289–318*

Miłkowski, M., & Lipski, J. (2011). Using SRX standard for sentence segmentation in LanguageTool. *in: Vetulani, Z., ed.: LTC 2009, Lecture Notes in Artificial Intelligence 6562 (pp.172–182) Springer.*

Nayak, A.S., Kanive, A.P., Chandavekar, N., & Balasubramani, R., (2016). Survey on Pre-Processing Techniques for Text Mining. Int. Journal Of Engineering And Computer Science IJECS Volume 5 Issue 6 June 2016, .16875–16879.

Panahandeh, M., & Ghanbari, S. (2019). Correction of Spaces in Persian Sentences for Tokenization *5th Conf. on Knowledge Based Engineering and Innovation (KBEI), 2019, pp. 670–674,* https://doi.org/10.1109/KBEI.2019.8734954.

Panda, M. (2018). Developing an Efficient Text Pre-Processing Method with Sparse Generative Naive Bayes for Text Mining. in: Int. Journal Modern Education and Computer Science, 2018, 9, 11–19.

Sallam, R.M., Mousa, H.M., Hussein, M. (2016). Improving Arabic Text Categorization using Normalization and Stemming Techniques. *in: Int. Journal of Computer Applications 135(2):38-43*

Schimmel, A. (1970). Islamic Calligraphy. *Leiden, Brill.*

Schmid, H. (1994). Probabilistic Part-of-Speech Tagging Using Decision Trees. *Proc. of Int. Conf. on New Methods in Language Processing, Manchester, UK.*

Tseng. H.H., Chang, P.C., Andrew, G., Jurafsky, D., & Manning, Chr. (2005). A Conditional Random Field Word Segmenter for Sighan Bakeoff. *Proc. of the 4th SIGHAN Workshop on Chinese Language Processing.*

Verma, T., Renu, & Gaur, D. (2014). Tokenization and Filtering Process in RapidMiner. Int. Journal of Applied Information Systems (IJAIS)–ISSN : 2249–0868. Foundation of Computer Science FCS, New York, USA, Volume 7–No. 2, April 2014–www.ijais.org

Vijayarani, S., & Janani, R., (2016). Text Mining: Open Source Tokenization Tools – An Analysis. *in: Advanced Computational Intelligence: An International Journal (ACII), Vol.3, No.1, January 2016.*

Vijayarani, S., Ilamathi, J., & Nithya (2016). Preprocessing Techniques for Text Mining - An Overview. International Journal of Computer Science & Communication Networks, Vol 5(1),7–16 2016.

Zesch, Th. (2012). Detecting Malapropisms Using Measures of Contextual Fitness. *Traitement Automatique des Langues (TAL). Vol. 53,3, 2012, 11–31.*

Zeyrek, D., Mendes, A., Grishina, Y., Kurfalı, M., Gibbon, S., & Ogrodniczuk, M. (2020). TED Multilingual Discourse Bank (TED-MDB): a parallel corpus annotated in the PDTB style. *Language Resources and Evaluation 54, pp. 587–613 (2020).*

Zhang, H., Miao, J., Liu, Z., Wesson, I.L., Shang, J., (2020). NLPIR-Parser: Making Chinese and English Semantic Analysis Easier and Complete *JADT 2020 : 15es Journées internationales d'Analyse statistique des Données Textuelles.*

Chapter 5
Lexical Analysis

Lexical analysis is the process of turning (tokenised) strings into meaningful lexical entries. Both sides are often called 'words'. However, just talking about 'words' is insufficient in lexical analysis, as 'words' can denote quite different things.

The following chapters use the following terminology:

1. Dictionaries contain **lexical entries**. They are the target of text analysis, and they are used as index terms. They are characterised by a canonical representation, i.e. a lemma, a part of speech, and a semantic description. E.g. '*run*' as noun and as verb are two lexical entries, and so are e.g. '*laker*' (a boat or a fish). Lexical entries are also called **concepts** or **terms** in the following chapters.
2. A **lemma** is the written canonical form of a lexical entry, its graphemic representation, as it would be used in dictionary lookup. Lemmata can occur in different forms in texts: Standing alone as a single word ('*chemistry*'), as part of a multiword ('*traffic light*'), in split form ('*take ... off*'), or in bound form ('*comprar<u>lo</u>*'). The part-of-speech information points to a tagset and selects one of the tags as POS. Semantic information refers to a word sense, to a domain classification, or in the simplest case to a reading number.
3. The occurrence of a lemma in a text is called **textform**. Textforms are realisations of lemmata, with added inflectional or derivational material. Stem variations (like Arabic broken plural, German umlaut, Spanish vowel changes), split forms, agglutinations etc. can also be found in textforms, they are **allomorphs** of the lemma.
4. The task of lexical analysis is to derive the lexical entries from the textforms. It must perform identification (find the lemma from the textforms) and disambiguation (find the right one out of a list of candidates in case the lemma is ambiguous).
5. Both textforms and lemmata have **annotations**. Textform annotations, like case and number, are specific for a given textform (mainly on morphological level), lemma annotations hold for all textforms of a given lemma, like subcategories,

G. Thurmair, *Knowledge-Driven Multilingual Text Analysis and Transparent Information Retrieval*, Cognitive Technologies,
https://doi.org/10.1007/978-3-031-91741-7_5

verb frames, semantic markers. Lexical analysis must interpret the textform annotations and provide the lemma annotations by looking them up in the lemma dictionary.
6. Concepts are the target of **multilingual** analysis, be it in indexing, or Ontology Building, or Machine Translation. As they denote real world objects they form the bridge between the languages in multilingual indexing: They are denoted differently in different languages, i.e. they have different lemmata in different languages, so in multilingual indexing they need to be translated, mainly by using bilingual dictionaries. Named Entities also need to be treated in multilingual indexing, e.g. city names ($_{de}$ '*Mailand*'–$_{it}$ '*Milano*').
7. **Stems** (or roots) are lemmata without derivational affixes. They often have no semantic meaning, cannot be translated (in multilingual indexing), and therefore would not be used as index terms. Examples of stems are $_{en}$ '*integr-*', '*falsif-*'. The term 'stemming' as found in the literature[1] often relates both to lemmatisation and to stemming. But while lemmatisation (linking textforms to lemmata) is a quite necessary step in analysis, stemming is not, and the result of the 'stemming' operation is ambivalent, between helpful and damaging.

Despite these differences all these units are often called **words**. However in the present section the differences are important. And this is the reason why the component is called Lexical Analysis rather than lemmatisation.

5.1 Design

1. The **purpose** of Lexical Analysis is to identify the concepts of the sentence, and assign them all available lexical information.
2. Identifying lemmata in a list of tokens faces some **challenges** which must be overcome by a Lexical Analyser. A simple lexicon lookup is seldom sufficient, and only succeeds in language which do not inflect, like Chinese.[2]

a. First, the textforms (i.e. inflected forms) must be brought to a **lemma** (base form), and the morphological information carried by the textform must be made explicit (i.e. formalised into a feature-value structure). This process is quite different in the different languages supported by TINA, and requires different lexical analysers
b. Productive **derivations** must be recognised, like negation (in Farsi and Turkish), causativation (Turkish), aspect determination (Russian), definiteness (Danish '*hus-et*'). Such a recognition requires affix interpretation. To enable this, affixes must receive some linguistic description, in particular

[1] E.g. used by (Gurusamy and Kannan 2014; Porter 1980).
[2] But also in Chinese a decision must be taken which affixes should be included in the definition of a term, cf. (Gao et al. 2005).

a POS tag. This extends the scope of the tagset from free to bound morphemes, as already stated.

c. A frequent operation to create new concepts is **compounding**. In languages like English compounds are built by adjacency of words, in German they are agglutinated into one word. There are many borderline cases in compound definition ranging from collocations to concepts. As compounding is a productive process the newly created semantic units would not be found in the dictionary, and a step of decomposing is required as part of lexical analysis.

d. There are phenomena where **several morphemes** are involved: (Clitic) pronouns, conjunctions, prepositions are agglutinated into one textform ($_{ar}$ اليمن = *al-yaman* = *Yemen*, $_{tr}$ *hastayım* = *I am sick*); in turn some word parts are located in distance ($_{en}$ *stand ... off*, $_{de}$ *nimmt ... ab*). They must be collected to produce a proper lemma (and concept).

e. Just like multipart tokens exist on the level of tokenisation, **multiword concepts** exist on the level of lexical analysis. They need to be identified by the lexical analyser.

3. The **dictionary** used for lexical analysis is a monolingual resource. It is central to Lexical Analysis as it provides all available information to the words in the text.

Every dictionary needs to provide definitory information for a given lexical entry: The lemma (a canonical representation), a syntactic description (in the form of a POS tag), and a semantic description. More annotations may be provided by the dictionary. The dictionary must also provide a link from the lemma to its derivations (textforms) as they may appear in the text.

LtLemmatiser uses two forms of dictionaries, depending on the analysis strategy:

- **LtShallow** uses a single word full form dictionary (**FWD = full word dictionary**). Each entry is defined by three information items: a full form (the (normalised) textform), a linguistic annotation describing the morphological properties of the textform (in form of a STAG ($_{en}$ *'trees ->NoCo'*) or an XTAG ($_{en}$ *'trees –>NoCo-Np'*), and a lemmatisation instruction.

  The annotations described by the tags in the Full Word Dictionary refer only to those which are specific to a given textform. Annotations which hold for all textforms of the lemma would have to be stored in another dictionary, called the **Basic Lemma Dictionary (BLD)**.
  If the full forms of the FWD contain normalised forms, then these normalisations must match the normalisations used by the LtNormaliser, otherwise mismatches will occur (e.g. text contains *realization* which is normalised to *realisation* but FWD contains *realization*: The form will not be found without extra effort).

- In **LtDeep** the dictionary entries used are **Basic Lemma Dictionaries (BLD)**. They are lists of feature-value pairs, among which the base form, the basic tag (BTAG), and semantic information are obligatory. The relation between lemmata and inflected forms is expressed in the form of inflectional classes describing inflection patterns, supported by different allomorphs (stems) for the respective inflections. Basic Lemma Dictionaries (BLD) in addition contain annotations relevant for all inflections of the lemma, like verb frames, passivation, semantic type, etc.

The relation between the two kinds of dictionaries as shown in Fig. 5.1 is described in the chapter on resources (Sect. 9.2): The LtShallow Full Word Dictionaries (FWD) are created from the Basic Lemma Dictionaries (BLD) also used by LtDeep, by using inflection creators (LtFlexer) as described in the Sect. 9.3.4.

The expansion of a basic tagset (BTAG) into a standard (STAG) or extended (XTAG) tagset is the basis of this operation and is described in the chapter on the tagsets in Sect. 2.3.2. In cases where such an operation is difficult as it would produce too many inflected forms (like in Hungarian or Turkish) a Basic Lemma Dictionary is used and a special kind of lexical analysis based on LtDeep technology must be launched.

4. Often the analysis will return **ambiguous** results. Textforms can point to several lemmata in case of homographs. Lemmata can represent different concepts (semantic units) which would have different syntactic-semantic descriptions (e.g. different verb frames) and would have different reading numbers in traditional dictionaries. Therefore it is necessary to implement components for disambiguation as part of a Lexical Analysis.

Syntactic ambiguities would be disambiguated in LtShallow using homograph analysers, or taggers. Semantic ambiguities would need components from the field of word sense disambiguation. Both components are described in Sects. 7.1 and 7.2.

5. A look into a standard dictionary reveals that about 30% of all entries are not single words but consist of several words, i.e. **multiwords**. Lexical Analysis therefore must be able to cope with this phenomenon.

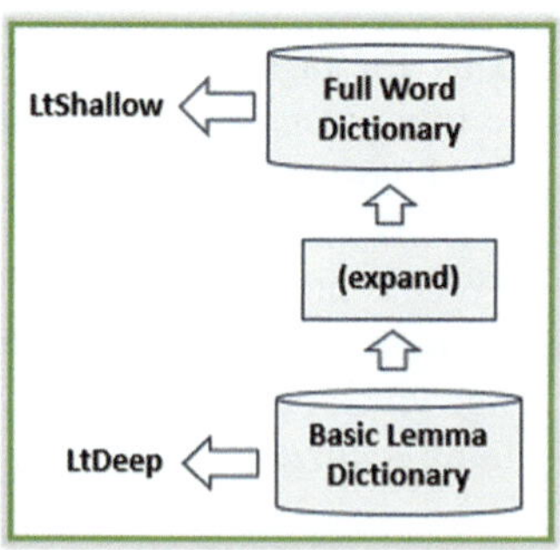

Fig. 5.1 Dictionary resources in lexical analysis

Multiwords require additional lexical annotations: As they are semantic units with an internal syntactic structure this structure must be described in the dictionary. Such a description can then be used to identify them in texts. In TINA, there are options both in LtDeep and in LtShallow to do this; this is described in Sect. 6.1.

6. Independent of the selected analysis strategy its **result** must be returned to the Document Model. The token-based analysis list must be rearranged in cases of multiword tokens and multitoken words, and it must be extended to cover the alternative descriptions.

5.2 Shallow Analysis (LtLemmatiser)

The LtLemmatiser is responsible for the Lexical Analysis in the languages which are supported by TINA. Depending on the language parameter it calls different subcomponents. Each of these handles one particular language and consists of a manager which first calls the resource manager responsible for providing all resources, then a worker component which executes the analysis, and finally hands the output to the LtLemmatiser Manager which populates the Document Model with it (cf. Fig. 1.3 above). The language specific resources are read-only and interact only with the resource manager of the respective language. They are loaded only if this language is required.

The approach to use different technologies for the lexical analysis of different languages is the result of an in-depth analysis of these languages. Many papers on morphology exploit the potential of a new technology by applying it to new languages, and end the description when the harder cases appear (like idiosyncratic plurals, exceptions, derivational phenomena etc.)[3] Others do not talk about morphology at all, e.g. (Yamashita and Matsumoto 2000) or (Toleu et al. 2022). In contrast, the present chapter tries to use the most adequate technology for the analysed languages, including the phenomena described in the respective grammars.

5.2.1 Dictionary Lookup (Chinese)

In Chinese analysis, the tokenisation step also decides on the words in the sentence as Chinese words are not inflected or otherwise modified.[4] Tokenisation and

[3] Examples are given below.

[4] Of course there are many ‚function word ‘ characters. They are needed for syntactic analysis but less relevant for indexing.

Domain	zh > en	zh > de	Annotation
General	854.2K	48K	PKU-tags, semantics, domain, measure words
Technology	-	44.5K	Target Language POS, domain
Economy, Finance	83.1K	36.5K	Target Language POS, domain
Medicine	42.9K	-	Target Language POS, domain
Sports	47.7K	-	Target Language POS, domain
Military	36.9K	10.7K	Target Language POS, domain
Software	18.1K	-	Target Language POS, domain
Names	10.5K	10.5K	PKU-tags, STAG, semantics, domain
Total	1093.4K	150.2K	

Fig. 5.2 Dictionary resources Chinese

lexical analysis therefore are just one step, and the TINA component for Chinese only performs a dictionary lookup.[5]

5.2.1.1 Dictionaries

The dictionaries for Chinese in LtLemmatiser are largely built from bilingual translation dictionaries, translation directions being Chinese->English and Chinese->German. The entries cover different domains and are annotated by domain (using a subset of the domain classification as described in Sect. 3.3).

As translation dictionaries do not contain function words the entries of such word classes have been added for the Chinese general dictionary and for the tokenlist.

The dictionary resources are listed in Fig. 5.2. The general domain dictionaries for Chinese-German contain POS annotations following the PKU tagset.[6] The Chinese-English general dictionary provides also classifier information for a subset of nouns and semantic annotations for a subset of entries, in particular for Named Entities. The Names dictionary also contains PKU tags and a semantic annotation for these names. The other dictionaries contain translations into English/German, domain annotations, and a target language POS annotation.

5.2.1.2 Creating Lexical Entries

A synopsis of the dictionary annotations reveals that a conversion of the tagsets into the TINA tagset faces several challenges although for most of the entries the conversion is straightforward:

[5] More sophisticated and adequate treatments can be found in (Gao et al. 2005; Tseng et al. 2005), or (Packard 2016).

[6] This tagset is discussed in Sect. 2.3.2.

1. The **POS annotations** of the Chinese lemmata are sometimes inconsistent in the used dictionaries. There are hundreds of entries which are categorised as adjectives in one source, nouns and/or verbs in another source. On top, homograph lemmata are presented as one entry categorised as 'homograph' in the PKU tagset while in TINA they form two entries.

The conversion of the PKU tagset into TINA poses additional difficulties in the mapping of POS tags; the PKU tagset and its conversion into the TINA tagset is described in Sect. 2.3.2.

The POS annotations of the domain-specific dictionaries show POS descriptions following the Basic TINA tagset (BTAG), however on the target language side. By experience, the target language POS annotation does not deviate significantly from the source POS (Chinese), in particular for content words. So they would be used also for the Chinese words in TINA.

2. As for the **semantic** description, the domain annotation is the best option available, apart from semantic classification of names. The relevance of topic/domain information has been described in the section on the LtTopicIdentifier (Sect. 3.3), so they serve here as semantic annotations for the Chinese lemmata as well.

As a result, Chinese Lexical Analysis returns lexical entries with a lemma (the string), a POS tag (in terms of the TINA tagset), and a semantic description (semantic information for names, domain information for common nouns). More annotations to the lexical entries would of course be helpful, however they may suffice for the identification of index terms.

5.2.1.3 Evaluation

Tests were made to evaluate the coverage of the Chinese component, i.e. to which extent it covers incoming text. For this purpose, a text corpus was created consisting of the editions of one month of *China Daily* (Jan 1998), containing 2 Mio Chinese characters and about 860 K tokens.

This corpus was tokenised using a large Chinese word list with 140 K entries, and then analysed with a bilingual Chinese-to-German dictionary containing 48 K entries. The result was that 636 K tokens could be translated, 224 K were not found. The coverage was about 74%, and 12.1 K dictionary entries were activated.

Aside from some mismatches resulting from different tokenisation strategies between token list and bilingual dictionary, the main reason for this result are gaps in the dictionary, mainly proper names which however are over-represented in news text.

To overcome such problems, the coverage of a domain-adapted dictionary was evaluated: The bilingual dictionary was extended by about 10 K entries in the drug enforcement and military domain, and about 30 customer-provided documents were tested. The coverage there was between 77 and 92%, with an average of 86%.

So, domain adaptation is a useful strategy for quality improvement also in Chinese analysis, like for other components.

5.2.2 *Lemmatisation (LtLemmatiser): English, German, Italian, Spanish, French, Portuguese, Russian, Arabic*

5.2.2.1 State of the Art

Since the mid-eighties, the leading paradigm in computational morphology is Finite-State Analysis, first presented by (Koskeniemmi 1984), later extended by (Beesley and Karttunen 2003) and many others. The basic idea is that morphological phenomena can be expressed by regular expressions, which then can be compiled into transducers. Later it could be observed that many papers were published showing that such phenomena could be found in other languages as well, which led to the claim that the formalism is suited to both analysis and generation, and that it is even language-independent (because many languages show similar phenomena). And gradually the interest shifted from describing a given morphological system using a specific technology to exploring a given technology using morphological data as illustration. The question whether FS Analysers are the best option if a complete morphological analysis of a language is envisaged is still not answered.

Some observations may help to illustrate the problem:

1. Morphology is not just based on rules but contains a significant amount of **exceptions**. Even the simplest cases are affected: Plural of $_{\text{en}}$ '*dish/whish*' is '*dishes/wishes*' but the plural of '*fish*' is '*fish*'. The automaton needs to be rearranged to cover this case, by writing the rules 'around' the exception. As there are hundreds of such exceptions the whole resource becomes clumsy and hardly maintainable.[7] The solution, to set flags, add postprocess filtering (as proposed in (Beesley 1998)) and other defence measures, is not really encouraging.

Similar cases of irregularities exist in all languages. Turkish nouns show cases which do not follow vowel harmony ($_{\text{tr}}$ '*saat–saati*', '*kalp–kalbi*'). German strong verbs are described in their vowel changes ('ablaut') by (Görz and Paulus 1988) but without mentioning consonant changes involved as well (de '*schneiden–schnitt*' '*ziehen–zog*' etc.). In a paper entitled '*Arabic Morphology Using Only Finite-State Operations*' by (Beesley 1998) there is no mention of the phenomenon of 'broken (inner) plurals' which grammars describe as '*There are hardly any rules about how to form the broken plural from the singular … it is recommended*

[7] „*It is possible to rewrite the regular expression in various ways to eliminate the overgeneration, but this is tedious and dangerous, requiring the making and subsequent parallel maintenance of multiple copies of the noun stems* " (Beesley 1998:53).

that the plural form be learned', so (Abu-Chacra, 2018). If there are no rules, they cannot be described by regular expressions.

2. Another phenomenon is **context-dependency**. It is not just a longer-distance issue of the kind that certain prefixes restrict the possible endings, like $_{\text{de}}$ '*abgebende–abzugebende-*abgegebende*', but also parameters outside the morphological analyser like homograph information which is not yet disambiguated at the stage of morphological analysis: $_{\text{en}}$ '*leaves*' could go to '*leave*' (as verb) but also to '*leaf*' (as noun), and so would $_{\text{de}}$ '*dienst*' with '*dienen*' as verb and '*dienst*' as noun. Some German adjective inflections are six times ambiguous, i.e. would require six readings in the automaton. Ambiguous readings must not just be recognised but also be communicated to the next level of analysis. Another example, in the generation part, would be the creation of spelling variants depending on the text locale (US versus UK English) in consonant doubling ('*levelling*' versus '*leveling*' etc.). It would require some input parameter for the automaton, and its access in the production process.

3. A third observation refers to the **dictionaries**. Nearly no dictionary of serious size does not contain entries which do not belong to the original language and its morphological component. Foreign words, loan words, semi-adapted words, also special terms taken over into the general language, challenge a just rule-based system. In many cases an entry-by-entry approach is required for analysis.

As a result, most of the papers on FS technology explore the potential of the technology but do not intend to write a system component. From the morphology point of view the technology can only cover a part of the morphological analysis of a language. Most papers don't give figures on coverage, i.e. to which part of the words of a corpus their analysis can cover. One reason might be that it needs significant computational resources.[8] But obviously, there are other techniques required which are more adequate to the task at hand.

5.2.2.2 LtLemmatiser

The LtLemmatiser follows a quite different paradigm, which is much simpler and much more efficient. It supports most Indo-European languages and consists of a lexical resource and software which does the lookup and executes the lemmatisation instruction.

[8] The *foma* toolkit (https://code.google.com/archive/p/foma/wikis/MorphologicalAnalysisTutorial.wiki) (Hulden 2009) reports 1.7 kB for a lexicon of 12 entries: This to be used for a standard dictionary of 60 K to 240 K entries in size…

1. Approach

The LtLemmatiser follows a simple *cut&append* paradigm. The lemma is built from a textform by cutting a number of characters from the end of the textform, and appending the appropriate ending. A lemmatisation instruction then consists of an integer determining the cut point, and another integer giving the index to the appendstring table where the string to be appended can be found.

An advantage is that the lemmatisation instruction can be created automatically by a string comparison between lemma and textform to determine the cut point and the appendstring. In turn, the same method can be used to create inflected forms from lemmata (as will be shown in Sect. 9.3.4).

As the textforms need to be linguistically described the lemmatisation resource also contains a linguistic description, usually in the form of a tag (BTAG, STAG, or XTAG).

2. The Full Word Dictionary (FWD)

The core of the lemmatiser is the dictionary. It is a list of textforms, called **FWD** (full word dictionary), as introduced above. This list contains inflected single words.[9] The entries consist of a full form, a linguistic description (a POS tag), and a lemmatisation instruction. In case of homography several entries are given in the FWD, sharing the textform but providing different POS and lemmatisation annotations.

a. The **textforms** are collected from different sources:

- Most entries come from existing dictionaries. Their single word entries are collected, for multiwords all multiword parts are added
- As lemma dictionaries often do not contain function words, such words were added. They often require a more detailed set of annotations on a textform basis than the content words.
- Additional entries are created partly automatically using inflection generators (LtFlexer, cf. Sect. 9.3.4), partly manually whereby inflected forms are created based on descriptions in grammars.
- The resulting list was augmented by corpus analysis procedures in order not to miss relevant and frequent entries and to close lexical gaps; usually tokens with a frequency of>50 were collected from corpora with>1 Mio tokens.

The resulting lists were validated using different heuristics. Their size varies significantly depending on the language in question: It is between 140 K (English)

[9] And the single parts of multiwords.

Language	ar	de	en	es	fa	fr	it	pt	ru
Textforms (K)	161k	841k	143k	330k	110k	201k	461k	454k	1574k
Tagset-size	55	88	82	67	117	77	75	67	63
Appendstrings	9703	3750	403	1236	1218	922	650	1447	1377

Fig. 5.3 Resources for the LtLemmatiser

and 1560 K (Russian),[10] German and Russian being the biggest. Figure 5.3 shows the size of the full form dictionaries used for the lemmatisers, the size of the tagset, and the number of appendstrings.

b. The **linguistic description** consists mainly of a POS tag annotation, and uses one of the tagsets described in Sect. 2.3.2. The annotations were created in a bootstrapping procedure, by running analyses on corpus data, identifying and categorising missing entries and adding them to the textform list as syntactic annotations.

Both the standard (STAG) and the extended tagset (XTAG) are usable for lemmatisation: The standard tagset (STAG) provides parts of speech and syntactic subcategories, containing between 50 and 75 tags per language[11]; the extended tagset (XTAG) contains morphological information (number, person, tense etc.) in addition, and is significantly larger (cf. Fig. 2.19 in Chap. 2). Which tagset to use depends on the intentions for parsing and extended analysis. The current version of TINA uses mainly the standard tagset (STAG).

c. The **lemmatisation instruction** follows the *cut&append* paradigm. The lemma is built by cutting a number of characters from the end of a word, and appending the appropriate ending string.

The appendstrings are collected in a separate list; this way redundancies can be exploited. It is a rather small list overall: Its size is around 1000 on average, exceptions being Arabic (due to the clitic determiner for nouns) and German (due to the infixes in verbs with separable prefixes: $_{\text{de}}$ *ab-zu-gebende ->abgeben*). However, compared to resources needed by other morphological analysers, the footprint in LtShallow is quite small.

The lemmatisation instruction for a given textform then consists of two integers, one specifying how many charaters to cut, and another one pointing to the required line in the appendstring table.

The cut&append approach is quite simple and powerful. It can even replace complete textforms (like $_{\text{en}}$ '*was*' –>cut 3, append '*be*'). However as it analyses words from right to left it is the more efficient the less infixes or prefixes are involved in the lemmatisation process. But even then the only consequence in such

[10] Farsi FWD is not used for analysis but for generation. Details are given in (Sect. 5.4).

[11] The Farsi tagset in Fig. 5.3 is an partly extended tagset (XTAG).

cases will be that the appendlist will grow. This cut&append technique enables the LtLemmatiser to cope with all special cases and subtleties of the language at hand. In addition, the lemmatisation instruction can be created automatically by a string comparison between lemma and inflected form.[12]

Textforms with different annotations are homographs, and have different entries in the Full Word Dictionary. (As a consequence, FWDs for an extended tagset have significantly more entries than FWDs with standard or even basic tagsets).

The FWD has its focus on open word classes (Nouns, Names, Adjectives, Verbs, adverbs). Function words are collected in a special resource, they require much more detailed annotations than content words. These lists comprise several hundred entries per language.

5.2.2.3 Implementation

1. **Input**: Input of the LtLemmatiser is a sentence of the document model, containing the list of (normalised) tokens, with the language attribute as parameter. Some token nodes need not undergo lemmatisation (e.g. digits), some are already lemmatised by the tokeniser (e.g. units). The rest is processed by the LtLemmatiser. Its important input features, provided by the document model, are the normalised token form and the casing information.

2. **Compilation**: At loading time the FWD is compiled into format which collects all POS values and all appendstrings of the dictionary, and assigns a list of readings to each textform (the form may be homograph); each reading consists of a pointer to the linguistic description (POS list) and two integers for the lemmatisation instruction (cut point, index to append list).

3. **Runtime**: The LtLemmatiser only needs two list lookups and two string operations to lemmatise a given textform.

- The first lookup in the *readings* list returns all readings for a given textform.
- For each reading,
  - the POS table lookup fetches the linguistic annotation;
  - the cut integer defines the cut point,
  - the append pointer fetches the appendstring from the append table; the found string is appended at the cut point.

Figure 5.4 exemplifies the lemmatisation. The input list of tokens is looked up in the FWD. For each reading found, a lookup in the POS list returns the linguistic description, and after executing the number of cuts indicated in the lemmatisation

[12] Instead of a lemmatisation instruction the FWD could also just give the resulting lemma form. However there are generalisation possibilities with the instructions, e.g. inflection classes could be built for all forms with the same lemmatisation instructions, cf. Sect. 9.3. This would not be possible if the resulting lemmata are given directly.

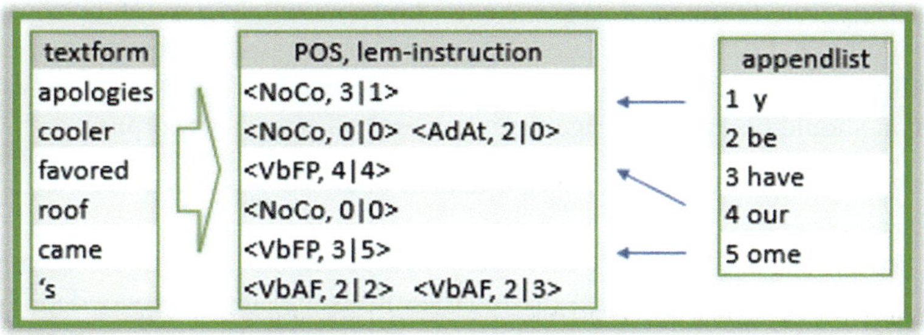

Fig. 5.4 Lemmatiser example

instruction, the specified entry of the append list is appended to the remaining string. The resulting lemma is returned, together with its POS annotation.

4. The **output** of the LtLemmatiser component is a modified record in the Document Model, with an instantiation of the *lemma*, *POS* and *features* annotations.[13]

5.2.2.4 Discussion

It should be noted that this approach depends on the availability of annotated lists of inflected forms. This requirement cannot be satisfied by all languages, however. In Chinese it has been explained that 'inflected forms' do not really exist, there is no difference between inflected forms and lemmata.[14] In Turkish, the number of inflected forms to be produced is close to unlimited. In Farsi there are whitespaces between words and inflections, which would make a (whitespace-based) single word list a bit obsolete. For these languages, different techniques are foreseen in the TINA Lexical Analysis.

5.2.2.5 Evaluation

The kind of lexical Analysis described here crucially depends on the quality of the dictionary. There are two factors which determine the quality of a dictionary:

- *Accuracy*, i.e. correctness of annotations. This is usually evaluated by looking at a subset of the entries, and evaluating its correctness[15].

[13]As a token in the Document Model may allow several annotations (in case of homographs) a temporary extension of the Document Model allows for *lists* of<lemma, POS, annotations>instead of just single values; these lists must later be disambiguated by taggers and semantic interpreters (Chap. 7).

[14]The treatment of word formation has been mentioned above (Packard 2016).

[15]For external delivery of dictionaries, additional criteria are required, cf. the ELRA Validation manual for dictionaries, cf. (Fersøe 2004).

- *Coverage*, i.e. how well does the dictionary match the intended application domain. For instance, a 100 K dictionary of spare parts of agricultural machines does not really help in a medical domain). A 'general purpose' component should be evaluated against a 'general purpose' text corpus.

An evaluation of LtLemmatiser-based lexical analysis was made for German and English.

Accuracy: There is no larger dictionary without errors. A Lexicon lookup component will produce errors when dictionary entries are incorrectly coded. To have an impression on the dictionary error rate, 1000 random entries were extracted from the XTAG German lexicon, and manually inspected.

The error rate was 1.29% for standard tag (STAG) entries (i.e. errors in lemma or POS), and 2.53% for extended tag (XTAG) entries (i.e. half of the errors were due to incorrect morphological annotations). In the experiments presented in the following, only the standard tag annotations were used; so 98.7% of the words resulting from lexical analysis can be assumed to be correct.

Coverage: For German, the test corpus consisted of all tokens of Europarl, EMEA, and JRC corpora, overall 65 Mio. After lexical analysis, 7.795 Mio were left unknown; this is a coverage of 88.1% for the dictionary. The PANACEA control corpus showed similar results, with a coverage of 84% for German.

For English, the test corpus consisted of the WACky (2.3 bn), the Europarl (27 Mio), and the Reuters (233 Mio) corpus, overall 2.56 bn tokens. Of these, 185 Mio were left unknown, which is a coverage of 92% for the lexicon. Of these, Reuters is a bit worse (89% coverage) than the other corpora as the English lexicon contains very few proper names, which however occur more often in news texts.

Both figures for accuracy and coverage are considered to be competitive. Lemmatisation speed is also quite competitive, the data are given in (Thurmair et al. 2012).

5.2.3 Clitic Elements (LtClixer): English, French, Italian, Spanish, Arabic

If the Lexical Analysis for a given token fails, then one of the reasons could be that the token contains additional lexical material.

'Clitics' (for 'clitic particles') are lexical material which is attached to a word, in opening or closing position.[16] The word, together with its clitics, forms one token and does not contain whitespaces. Sometimes special characters like quotes or hyphens are used to mark them ($_{\text{en}}$ *he'll*, $_{fr}$ *l'homme-ci*); this requires harmonisation with the tokeniser treatment of such characters. However there are also clitics without such indicators (e.g. $_{\text{es}}$ *hacerlo*). In addition to a single occurrence

[16] And of course are not punctuation marks.

($_{\text{fr}}$ *chante-le*) clitics can also form sequences ($_{\text{fr}}$ *donne-le-moi, chante-le-lui*). The task of the component called LtClixer is to identify such clitic elements and return (a sequence of) proper lemmata and POS annotations.[17]

The difference to a proper linguistic definition of clitics is the narrow computational perspective to consider only phenomena which are related to tokens/strings, i.e. lexical material which is attached to a word resulting in one token. Generalisations like in Italian where preclitics are used in finite verb forms but enclitics for nonfinite forms and imperatives (Monacchesi 1993) stay outside of this point of view.

5.2.3.1 Specification

The treatment of clitics is language-specific.

English would return a lemma with feature decoration from the lexical analysis for the critical cases (*he'll->will+3.Sg., don't->do+Neg*) as well as the genitive forms (*corner's ->corner+gen*).

German would leave the hyphens indicating a gap (*Haus- und Grundbesitz, be- und entladen*) and wait for a later component which resolve the gapping construction. In both languages no clitic treatment is triggered.

French enclitic forms are identified by hyphens (fr *chante-le*), and are already treated in the tokeniser.

In **Spanish** and **Italian** enclitic elements are mainly pronouns ($_{\text{es}}$ *me, te, se* ..., $_{\text{it}}$ *mi, ti si* ...) or combinations thereof (*-mela, -seles, -oslo*). In both languages there is a small fixed set of clitics which can be suffixed, and there is a contextual constraint: They follow verb forms[18] which are either infinitives or gerunds (*-ando, -endo*) or imperatives,[19] in Italian also participles.[20] Also there is a tendency to lexicalise such constructions, according to (Russi 2008). Unlike in Spanish, the Italian forms can be homograph, e.g. *arma|ti* (verb imperative) versus *armati* (plural noun). Homography is a challenge in cases like *arma|ti* (2 tokens) versus *armati* (1 token) because the TINA document model does not support alternative tokenisations; some ad-hoc heuristics must be applied.

Arabic tokens also use clitic elements. (و, ال). Proclitic forms can be built from conjunctions (و) and particles like (ف), nominal constituents can be preceded by determiners (very frequently, cf. الجزيرة Al-Jazeera) and prepositions.[21] Enclitic

---

[17] The component was developed by Christoph Schwarz and Alois Baumer.

[18] With the exception of ‚*Ecco* ': Eccolo.

[19] In Spanish, in cases of 1.Pl an s-elision (*vamos nos ->vamonos, emos nos ->emonos*), in 2.Pl a d-elision (*meted os ->meteos, decidid os ->decidios* needs to be inserted to find the form. Italian single syllable verbs duplicate the consonants.

[20] Cf. (Monacchesi 1993).

[21] Cf. (Monroe et al. 2014) offering a more complex solution. Cf. also (Sallam et al. 2016).

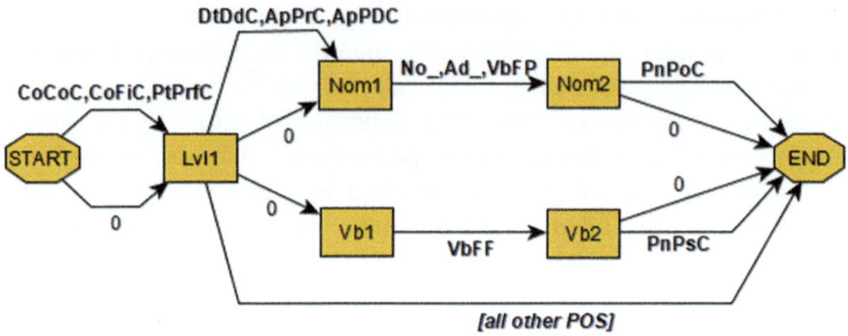

Fig. 5.5 Clitics in Arabic

pronouns can follow nominal and verbal constituents. A small Finite State component processes these phenomena (cf. Fig. 5.5).[22]

Treatment of clitics is not foreseen in other languages handled by LtShallow. In Russian, ли *(li)* acts as a general question marker but is not attached to the word before.[23]

5.2.3.2 LtClixer

The treatment of all these cases in LtClixer is straightforward: As the clitic elements are a fixed set with a limited number of elements (about 40 at most), they can be collected in special lists. Then:

- Every string which has not yet been successfully analysed by the Lexical Analyser is inspected for proclitic and/or enclitic elements stored in the clitic-lists of the respective language.
- If found, the clitic element is removed, and the remaining string[24] is again sent to the Lexical Analysis.
- If it can be analysed, the analysis result is checked for constraints for the clitic, like: '*string must be verb and non-finite*' (STAG: VbFI, VbFG VbFP) or '*string must be nominal*' (No_, Ad_). POS tags are used for this purpose. Also the clitics themselves must be looked up to collect their linguistic descriptions (like number or person).

If the analysis fails, the whole string is marked as '*UNKnown*'. Otherwise the LtClixer returns at least two tokens (the head word and the clitic(s)) with their descriptions (*lemma*, *POS*, *features*), and the token sequence in the document model is adapted.

[22] Treatment of Arabic clitics is also discussed in (Attia et al. 2011).

[23] Portuguese clitics treatment is not yet supported in the current version.

[24] In Italian an -e is attached to the infinitive verb forms.

It should be noted that the treatment of clitics has consequences for the tagset: As not all forms of e.g. possessive pronouns in Italian/Spanish can take a clitic position a special tag is required for the pronouns which do take it (cf. Sect. 2.3.2).

5.2.4 Decomposition (LtDecomposer): German

In languages like German many textforms are not found in the dictionary because they are compounds. Compounding is very frequent and very productive, and it is impossible to store all of them in the dictionary. As a result there is an out-of-vocabulary gap in the linguistic description of an input sequence. However, to continue the linguistic analysis, information about the compound is required, like POS, lemma, feature annotations etc., and as this cannot be provided by the dictionary, it is the task of the decomposer to do this.

1. The goal of the LtDecomposer therefore is to provide the analysis with such information: The lemma would be the non-inflected form of the compound as a whole; the POS would be the POS of the head of the compound, and the semantics are a modification of the semantics of the head. The compound as a whole is the unit to be processed further, not just a part of it (i.e. the head), as can easily be seen in multilingual indexing when the unit as a whole needs to be translated.

2. In an indexing context, the parts and the internal structure of the compound need to be considered: While searching for $_{\text{en}}$ '*hydrodynamic clutch*' with the head ('*clutch*') would return rather good hits, searching with the modifier alone tends to return noise. So the head-modifier relations of the compound parts are relevant for indexing.

3. In general the semantics of the head determines the semantics of the whole compound: '*Unternehmensberater*' ->'*Berater*'. In cases where the head is polysemous the modifier can help in disambiguation: $_{\text{de}}$ '*Mast*' would be disambiguated in cases of $_{\text{de}}$ '*Schweinemast*' ($_{\text{en}}$ '*pig fattening*') versus '*Besanmast*' ($_{\text{en}}$ '*mizen*'). In any case the internal relationship of the compound parts needs to be remembered as this relationship can also be expressed by other syntactic means: '*Schweinemast=Mast von Schweinen*').

4. The decomposition needs to return proper linguistic entities, including correct casing: '*Schweinemast*' ->'*Schwein*', '*Mast*'). This is necessary to use the compound parts in index creation as well as in following operations, like frequency analysis.

5. The current component (LtDecomposer)[25] supports German only, although other languages could also profit from it, e.g. $_{\text{pt}}$ *baratinhas-do-mar>baratinha-do-mar,* $_{\text{en}}$ *database ->data base,* $_{\text{it}}$ *capigruppo ->capogruppo*).

[25] The component was developed mainly by Christoph Schwarz, and revised later on.

5.2.4.1 Linguistic Description: Tagset

Just like a syntactic tagset mirrors the distribution situation in sentences, a decomposition tagset (DTAG) describes the relationships between parts of a word, in this case of a compound.

The tagset starts out from the Basic tags (BTAG) as all words have a POS tag; sometimes the standard tagset (STAG) is used, if DTAG elements refer to subtags (like participles). In addition, specific distributional properties must be annotated in the DTAG, as follows:

- elements which can only be compound-initial (*tele-, sprach-, vernehm-*)
- elements which can only be compound-final (mainly fully inflected forms like *häusern, störtest, herzlichem*)[26]
- elements which can be both compound-initial and compound-final (*geretteten(befragung), gelb(färbung), lese(brille)*)
- elements which cannot undergo compounding at all (final forms like *vertratet, drittes*, but also most function words like *aber, wenn, der* etc.)
- In addition tags are needed for numerals (can only take '*und*': *drei|und|zwanzig*) and infinitives with and without *zu*, nominalised with -s (*abgeben abzugeben lesens(wert)*).

These properties are combined with the Basic Tagset, resulting in a tagset of about 60 tags. An example is given in Fig. 5.6, describing compound options for proper names.

Like with syntactic tagging, all elements in the dictionary must be annotated with DTAGs in addition to the STAG/XTAG annotations. The LtDecomposer dictionary comprises 468 K entries; it also contains compound-specific allomorphs and inflected forms.

5.2.4.2 Decomposition

The decomposition is a two-step process: First the textform is decomposed and its parts are collected in a chart. From a given point in the input word, all possible continuations found in the dictionary are tried. The current DTAG and the DTAG of the continuation candidate are tested by looking up a transition table which indicates if a sequence of these two DTAGs is a legal transition or not. If so, the candidate is added to the chart. It is a breadth-first, all-path analyser.

The transition table is a 60×60 Boolean matrix (transition possible?: $\pm$). It is created by manually inspecting decomposer test output, keeping correct analyses and rejecting incorrect ones. An example (with a simplified tagset) is given in Fig. 5.7.

[26] Note that the input of the decomposer are textforms, not lemmata: The workflow is such that the decomposer is called because the lemmatiser failed, and then the analysis is left with textforms. But of course lemmatised forms can be decomposed as well.

	PROPER NOUNS
NoP-B	proper names as prefixes (Zugspitz-) *precede+-, cpfinal -, alone-*
NoP-F	inflected proper names, except places. Can only be alone. (‚schillers') *preceder -, cpfinal -, alone+ ost-*
NoP-Fc	Proper names, except places. Can be alone or preceders (because of: ‚Adenauerzeit', ‚Deuserband') *preceder +, cpfinal -, alone+ ost-*
NoPl-F	inflected place names. Like other proper names, but can be prefixed by cardinal points (‚ostzyperns'). *preceder-, cpfinal -, alone +, ost+*
NoPl-Fc	compoundable place names. Can be prefixed by cardinal points (‚ostzypern'), and used in compounds (‚ostzypernreise').. *preceder +, cpfinal -, alone +, ost+*

Fig. 5.6 Decomposer tagset (DTAG) example: proper nouns

from\to	$A	Ad	Av	CoCo	Nm	NoC	PtNg	PtVb	Ptzu	VbF	VbI	VbP	$E
$A	-	+	+	+	+	+	+	+	-	+	+	+	+
Ad	-	+	+	+	+	+	+	+	-	-	-	+	+
Av	-	+	+	+	+	+	+	+	-	-	-	+	+
CoCo	-	+	+	+	+	+	+	-	-	-	-	+	+
Nm	-	+	+	+	+	+	+	-	-	-	-	+	+
NoC	-	+	+	+	+	+	+	+	-	-	-	+	+
PtNg	-	+	+	-	-	+	-	-	-	+	+	+	+
PtVb	-	-	-	-	-	+	-	+	+	+	+	+	+
Ptzu	-	-	-	-	-	-	-	-	-	-	-	-	-
VbF	-	-	-	-	-	-	-	-	-	-	-	-	+
VbI	-	-	-	-	-	-	-	-	-	-	-	-	+
VbP	-	+	+	-	-	+	+	+	-	-	-	-	+
$E	-	-	-	-	-	-	-	-	-	-	-	-	-

Fig. 5.7 Decomposer transitions (simplified)

In a final step, in a reverse search of the chart all possible compounds found in the input are collected.

5.2.4.3 Filtering

Even with the filter of the transition table, there are still many decomposition results after an analysis, most of which are useless. So the challenge is to identify the correct decomposition(s) from a list of candidates. To do this, a set of rules has been developed which identifies the best decomposition in a candidate list.

Examples of such rules are given in Fig. 5.8:[27]

[27] These rules were written by Chr. Schwarz, cf. (Hjelm and Schwarz 2005).

0	**catchIllegal**	Deletes all senseless decompositions where finite forms are involved.
1	**filterOnlyVadjVinfl**	if the last part is an inflected verb, the preceeding parts must all be verbal particles (Pt-Vb).
2	**filterLowCap**	If the word is not capitalized, this rule prefers the non-nouns.
3	**filterVadjVinfl**	If the word is capitalized any decomposition where the last part has a Vb-category, the first part is a verb particle (Pt-Vb) loses.
4	**filterNoVadjPart**	If some constituent is a participle and the preceding part is not a verb particle, participle reading wins over infinitive or inflected verb.
5	**filterNLast**	This rule prefers compounds with nominal head category over other compounds.
6	**filterNMax**	For nominal-compounds, the largest number of noun parts wins.
7	**filterNSeq**	Chooses the reading with the largest number of noun constituents in sequence, counting backwards from the last constituent.
8	**filterLongR**	Recursively tries to find the longest constituent from the right. Compares noun parts with noun parts, adjective parts with adjective parts and so on.
9	**filterDefault**	If still decompositions are left, they are filtered according to a hierarchy of constituents tags involved. It uses the following hierarchy: Noun > adjective > Pt-Vb > Noun-min > Verb > Adj-min > Verb-min >nction word.

Fig. 5.8 Decomposer filter rules (example)

Such a rule-set is much more elaborate than the usually used heuristics (like: smallest number of parts, longest rightmost part).

The implementation of these rules uses a scoring system. The scoring computes a local score for each decomposition element, and computes a global score for the whole word. Local scoring is applied during chart building for each decomposition element. Local scores consider the tag, the length of the textform, and its position within the decomposition. The global scores compare the different tags of the whole decomposition in addition. The decomposition with the best overall score is returned.

The filtering intends to match the intuition:

- good lexical POS at the end of the word are favoured (noun wins over other alternatives)
- longer strings towards the end of the word win (a kind of right-side-longest-match heuristic)
- minor categories lose (as they are scored down) against better competitors
- decompositions with less elements win against decompositions with more elements.

5.2.4.4 Output Production

The resulting structures undergo several final steps:

Exceptions: There are still compounds which would be analysed incorrectly. There is an option in LtDecomposer to correct such cases, by using an exception list which for a given compound provides the correct decomposition. This list is some thousand entries strong.

Complete analysis: Decomposition was intended to help in lexical analysis for non-found words. To complete the analysis, the dictionary lookup is repeated just for the head of compound (the rightmost part). The lookup is supposed to return the lemmatisation instruction, the POS information (STAG or XTAG) and the feature annotations for the head lemma. In case of alternatives, the compound often helps in disambiguation ($_{de}$ *'strecke'* may be noun and verb but *'weg|strecke'* is only noun).

This analysis is also done for the non-head compound parts: All of them should be returned with proper POS and lemma.

Truecasing: Non-initial compound parts are always lowercased. For proper entry creation it is necessary to deliver proper truecasing. The initial element also needs to be truecased: Nouns like *'Beicht|vater', 'Bügel|eisen'* have compound-initial verbal parts which should be lowercased: *'beichten|Vater', 'bügeln|Eisen'*.

Output generation: If the LtDecomposer is used stand-alone, not as a component in LtShallow, several options for output presentation are available, e.g. one to support the creation of training files for Statistical MT (Moses, cf. (Koehn 2010)) whereby the compound is just replaced by the lemmata of its parts; or another one providing POS and lemma for collecting linguistic data (Fig. 5.9).

If the decomposer is part of LtShallow, then the component is supposed to return a proper record as required by the LtDocument model:

- textform, casing, and normalform do not change, they are provided by the tokeniser/normaliser
- for POS the POS of the compound head is taken (in BTAG)

Textform	Lemma	Decomposition
Legeröhre	Legeröhre No	[legen(Vb) + Röhre(No)]
Raufrost	Raufrost No	[rau(Ad) + Frost(No)]
Flachseile	Flachseil No	[flach(Ad) + Seil(No)]
herausnehmbar	herausnehmbar Ad	[heraus(Fw) + nehmbar(Ad)]
dunkelgrünen	dunkelgrün Ad	[dunkel(Ad) + grün(Ad)]
Gegendrehmoment	Gegendrehmoment No	[gegen(Fw) + drehen(Vb) + Moment(No)]
herausgebrochen	herausbrechen Vb	[heraus(Fw) + brechen(Vb)]
Wegfahrsperren	Wegfahrsperre No	[weg(Fw) + fahren(Vb) + Sperre(No)]
Hausaufgabe	Hausaufgabe No	[Haus(No) + Aufgabe(No)]

Fig. 5.9 Examples of decomposition output

- for lemma the compound lemma is taken (lemmatised head plus all non-lemmatised compound parts)
- the decomposition results are returned in the features section, in a format similar to the multiword description (head, lemmalist, POSlist). The compound structure is given in the features *head*=... and *modifier*=....[28] This enables the search system to give higher scores to heads than to modifiers.

However, even after decomposition, the Lexical Analysis faces tokens which are unknown. They will be treated by the LtDefaulter described below.

5.2.4.5 Evaluation

To test the LtDecomposer performance, a corpus of 3.1 Mio tokens was collected from several corpora. All types which were returned as UNKnown by the Lexical Analysis were collected, which resulted in a file of 199 K different textforms. This file was sent to the decomposer and evaluated for accuracy and coverage.

Accuracy: As for the types analysed by the decomposer, the output was evaluated manually; the result is given in Fig. 5.10.

It can be seen that correct cases (a+b+y+z+o) were found for 112.014 textforms, which is about 94.46%. Incorrect cases (x+e+u) were 6.569, 5.54%. These cases also contain material that should not have been sent to the decomposer at all (e+u, proper names and rubbish). If those are left aside, then the percentage of really wrong decompositions (x) is 0.01%.[29]

mark	explanation	size	example
a	correct	107.502	Eliten|Bildung Ost|Tirol
b	correct, alternatives possible	190	Trommel-$_{\text{No}}$|Fell-$_{\text{No}}$ vs. trommeln-$_{\text{Vb}}$|Fell-$_{\text{No}}$
y	correct but more or less parts also possible	458	Mittel|meer|gebiet vs. Mittelmeer|gebiet
z	correct but strange lemma (participles)	1.517	ersatzgeschächt -> ersatz/schwächen
o	correct but naming of POS to be improved	2.347	agro, mini, öko
x	wrong decompositions	1.494	zurück|Verlangen-$_{\text{No}}$ Wasser|Scheu-$_{\text{No}}$
e	decomposed proper names	3.788	Bauernfeind, Althof, Liebknecht
u	wrong rest (typos, acronyms, foreign word	1.287	G-GmbH, hoffendlich

Fig. 5.10 Decomposer evaluation

[28] Not shown in the example Fig. 5.9.

[29] Another test set was evaluated by (Hjelm and Schwarz 2005), comparing the LtDecomposer with GERTWOL, where the decomposer performed better.

Coverage: The decomposer was able to handle 118.5 K forms, which is 59.5% of all types remaining after lexical analysis. Most remaining non-analysed forms were foreign language words, proper names, acronyms, and gaps in the decomposer dictionary.

5.2.4.6 Static Decomposer

A special form of output is called 'Static Decomposer': In cases where a fixed set of text material, not too voluminous, needs to be analysed, a static decomposer resource file can be created, consisting of a textform and the decomposition of this textform. At runtime a simple lookup can provide the decomposition result, instead of starting the complete LtDecomposer component. Such a tool has been prepared for the Multilingual Thesaurus of the Assistive Domain in the EASTIN system. This is described Sect. 8.3.2. The technique is suitable for query analysis in retrieval.

5.2.5 Local Defaulting (LtDefaulter)

Out-of-vocabulary words have significant impact on the overall analysis quality. Therefore attempts have been made to reduce the negative effects of such words, mainly by assigning POS tags to them, and creating a default lemma.

There are several options how this could be done for a given unknown word form: Assigning equal probabilities to all tags, or assigning probabilities in relation to other hapax legomena[30] are easy options. More sophisticated solutions take capitalisation into account, or use frequent suffixes for POS detection. (Brants 2000) reports good results for the recognition of unknown adjectives based on endings like *-able*. Context dependent solutions have been proposed by (Toutanova et al. 2003): They look for company name suffixes like *Inc.* or *Co.* within a 3-word window.

5.2.5.1 Analysis

A closer look at out-of-vocabulary words shows that nearly all of them fall into one of the following categories:

- Spelling Errors. Such errors are very frequent, esp. in HTML texts
- Foreign language words, mainly English words in non-English text corpora. As an example, the analysis of the 1000 most frequent out-of-vocabulary words in the analysis of the Turkish Wikipedia shows that one third of them is foreign words, mostly English.
- Proper names

---

[30] Details can be found in (Jurafsky and Martin 2009:192).

- Acronyms and other mixed-character classes, like *HTTPS, HU3i, GSL, 328i* etc.
- Unknown regular words and proper names.

The LtDefaulter uses different heuristics to cover these phenomena. It is called 'local' as it takes just single tokens as input, and does not look at contextual information (like agreement patterns or POS context).

1. As for the **spelling errors**, the normalisation component described in Sect. 4.3.2 should already have taken care of it. The corrected form is already part of the Document Model and stored in the '*normalform*' feature.

2. **Acronyms** and mixed character words are nearly exclusively nouns. The defaulter assigns both a common noun (*NoCo*) and a proper noun (*NoPr*) tag to them; they must be disambiguated in later analysis steps: In '*BMW is the only company that ...*' the *NoPr* reading is activated whereas '*People love to tune their BMW ...*' would show the common noun reading.

3. **Foreign words** are a frequent source of ooV words. To integrate such words into the dictionary complicates the dictionary process as such words do not follow the language-specific systematics on which a dictionary is based, in terms of annotations like gender, inflection class, subcategorisations and the like. Therefore they are marked as special cases in the dictionary, or are kept in a special list.

This list is based on the resource already used for language identification (described in Sect. 3.2). It could be enriched by POS information taken from the dictionaries of the other languages, however it is questionable if they show the same distributional properties as in their native language. This holds in particular for function words being part of expressions like '*out of the box*'. In a German sentence like '*Er kauft die Textilien nicht out of the box*' POS assignment (Av) would hold for the whole English expression, not for its parts. For the time being, foreign language words are assigned a common noun POS (NoCo).

4. For the other out-of-vocabulary forms, i.e. **unknown regular** words, a special component was built which does the defaulting. It tries to assign information to unknown strings based only on the shape of such strings. For example, it makes use of the fact that strings ending in '*-ersson*' are person names, or strings ending in '*-isations*' are plural common nouns. But it goes beyond the use of a few suffixes, which often are ambiguous in themselves (like-*ity*).

5.2.5.2 Approach

The component, called LtDefaulter, uses a discrimination tree based on the ending of words.[31] The discrimination tree is built on the basis of the full word dictionary of a language. All endstrings of length n are compared. If all of them show the

[31] It was invented by (Ring 1978).

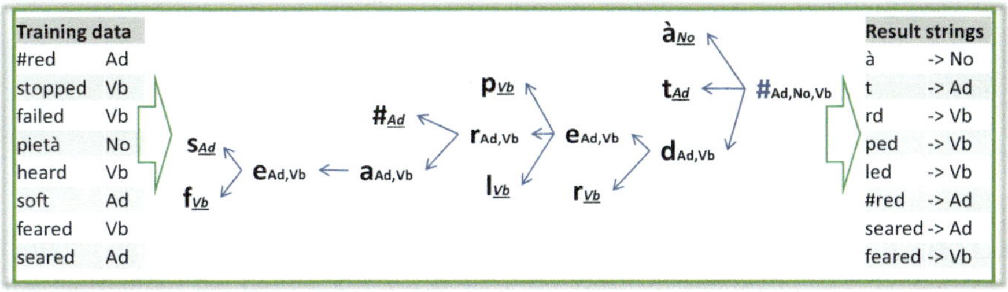

Fig. 5.11 Defaulter

same tag, the combination of<endstring, POS tag>is saved. Otherwise the next character (length n−1) is inspected. This process is repeated until the endings are disambiguated into unique POS tags. (Of course there are special cases like systematic homographies, the analysis ends with a homograph string then).

In the example Fig. 5.11 it can be seen that all words ending in '*à* ' are defaulted as nouns, words ending in '*t* ' as adjectives whereas words ending in '*d* ' stay ambiguous. Next, while words ending in '*rd*' will be recognised as verbs, '*ed*' endings stay homograph, and are only disambiguated after some more steps. The results are collected in an endstring table. Of course all this depends on the dictionary used for training.[32]

5.2.5.3 Implementation

The LtDefaulter consists of a training part and a runtime part.

The **training** takes pairs of<textform, POS>as input. This information is taken from the textform dictionaries (FWD). Only open word classes are defaulted, without function words and proper names.

It uses the discrimination tree technology just described, and outputs, for every produced endstring, a POS tag and a probability (in relation to other POS tags ending in this endstring).

Figure 5.12 gives the size of the training data, and the size of the resulting strings. The more variance a language has the more ending strings are required for defaulting.

The **runtime** part first executes some 'easy' tasks:

- Textforms containing **hyphens** (like '*data-base*') are split at the hyphen position, and the right element is resent to the Lexical Analysis
- **URLs** are identified and marked as proper nouns (NoPr)
- **Acronyms** (i.e. textforms with uppercase letters in non-first position) are tagged as both common (NoCo) and proper nouns (NoPr).[33]

---

[32] Originally it was used to develop a lexicon-free lemmatisation component, cf. (Fischer 1982).

[33] Non-noun capitalised strings like '*Great Wall*' are usually not unknowns, and covered by lexicon lookup.

Language	de	en	es	fr	it	pt	ru
training size (k)	819	170	337	207	487	521	1560
result strings (k)	174	93	84	62	162	150	150

Fig. 5.12 Defaulter result strings

- **Foreign words** are detected by using a special list of frequent foreign language words, as explained above. They are assigned a noun POS tag (NoCo) as long as no better option is available.
- In German, **capitalisation** is also used as filter after defaulting: All POS candidates except nominals can be removed.
- the remaining textforms are looked up in the defaulter endstring table. The POS of the longest endstring found will be returned.

As the result of lexical analysis, *all* input tokens in the Document Model have a lemma and a defined part-of-speech tag, delivered as a lattice of readings per token in case of homography; so they can be analysed by taggers etc. with some chance of success.

5.2.5.4 Evaluation

The evaluation strategy was defined such that if *one* of the POS tags returned by the LtDefaulter was correct, the assignment was considered valid, as the following tagger has the chance to find the correct POS in the returned POS set. If this was not the case, the entry was counted as an error. Also, tokeniser errors were counted as errors here (as the *final* assignment is incorrect).

The first test was about quality. Assuming the LtDefaulter would have to work in a dictionary-free environment the component would show correct POS assignments in 94.7% of the entries. This is similar to human assignments. So regular word formation is well covered.

The second test was an in-praxi setup. Tests with the 'leftovers' of the Europarl/EMEA/JRC data analysis were carried out. The test file contained all entries which were not in the dictionary, and/or could not be decomposed or otherwise POS-described. 21.000 entries of this test file were manually evaluated.

The results for German show that.

- if only *German* or German-looking words are considered, i.e. if the defaulter works with 'in-language' material, the error rate is 10.62%
- if *all* words are considered, it drops to 20.5%, the reason being that there is a high percentage of foreign language words in this test set.

A control set collected by the PANACEA project, cf. (Bel 2010), was also evaluated, the error rate was 20,2%, in line with the results of the large test set.

It could also be noticed that the defaulting quality depends on the parts of speech to be assigned: Some parts of speech are defaulted more reliably than others, esp.

nouns have rather low error rates, and so have participles. Verbs, esp. finite ones, have higher error rates in the corpus data, which may indicate that most verbs are already in the dictionary, so the defaulter produces mainly incorrect hypotheses.

As a result, textforms not identified by lexical analysis and/or decomposition, after dictionary lookup and decomposition, would be assigned a correct part of speech in 80–90% of the remaining material. This is quite superior to approaches where *all* possible POS values are assigned to an unknown token, with or without relative probabilities.

5.2.6 Large Scale Test: English, German

For the evaluation of languages processed with the shallow LtLemmatiser tool, a test was carried out to evaluate their usability in large-scale industrial settings where indexing of massive text material is required.

It has been mentioned previously that errors in a pipeline architecture tend to accumulate. So even if the single components show acceptable results, it may be the case that the outcome is not acceptable once the whole pipeline is considered. Moreover the input to a component changes as compared to stand-alone tests: For instance, a defaulter which is fed only with the leftovers of lexical analysis and decomposition will produce results which differ from the ones achieved when processing normal text. To have a better insight into such problems a large-scale test was carried out.

The evaluation followed the following criteria:

- Software stability: This is a prerequisite in industrial contexts.
- Speed: The analysis should run in acceptable speed. The tests showed that lexical analysis needed 31 K words per second, decomposition 11 K words, and defaulting 30 K words per second. Overall about 10 K entries per second could be achieved on a standard PC as available in 2012.
- Quality: The analysis should produce as few incorrect analysis results as possible
- Coverage: The analysis should be able to analyse as many input words as possible.

5.2.6.1 Test Data

The following corpora were used for test and evaluation:

- For German: dpa news, Süddeutsche Zeitung, WACKy (Baroni et al. 2009), Wikipedia, Europarl (Koehn, 2005), JRC/Acquis (Steinberger et al. 2006), emea (Tiedemann 2009). The overall size of the corpus material is more than 2.5 bn tokens.
- For English: Reuters news, WACKy, Europarl, JRC/Acquis, emea, overall resulting in also more than 2.5 bn tokens.

In addition, a control corpus was created from the PANACEA corpus ((Pecina et al. 2011), (Toral et al. 2011)) for the topics environment (ENV) and labour (LAB); 50 documents were taken from each set, containing 160 K (ENV) and 50 K (LAB) tokens respectively.

5.2.6.2 Workflow

The workflow of lexical analysis was configured for the large scale tests as shown in Fig. 5.13.

The tokenised strings were first sent to Lexical Analysis. If found, the information of the dictionary was returned. If not found, then the German decomposer was called. If found there, the annotations of the decomposed POS and lemma would be returned. If the tokens could not be decomposed either, they would undergo a treatment by the defaulter, and the defaulted annotations (POS, lemma) would be returned.

As the result of lexical analysis, *all* input tokens would have a part-of-speech tag which is understood by the following components, delivered as a lattice of readings per token in the Document Model; so that they can be analysed by taggers etc. with some chance of success.

5.2.6.3 Component Evaluation

Component Evaluation has been given above for the Lexical Analysis as well as for Decomposer and Defaulter. Coverage of the Lexical Analysis of English and German is reported between 84 and 88%. For the German Decomposer error rates between 3 and 4% have been found. The German and English Defaulter results lie in the nineties for normal texts as input, and in the eighties for 'leftover' textforms.

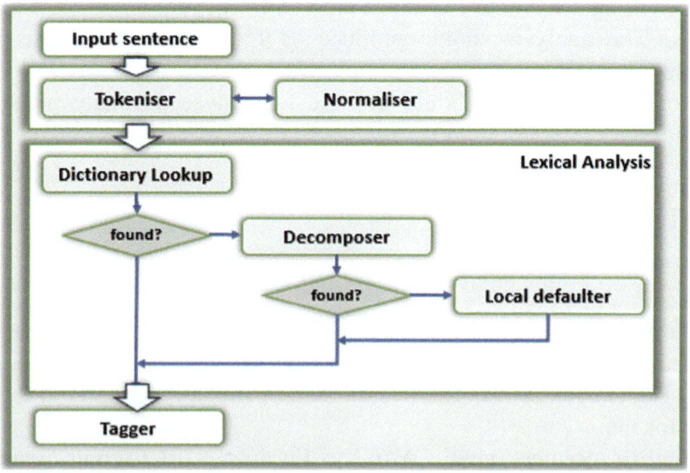

Fig. 5.13 Large scale evaluation

5.2.6.4 Workflow Evaluation

If the different components of lexical analysis are combined in a workflow, then the single components may have to cope with 'non-standard' input, and the error rates need to be adapted: Usually errors in a workflow accumulate, and the different components, fed with incorrect input, tend to produce more errors than in single tool evaluation.

For instance, the decomposer called in the overall workflow must also process tokens which are not compounds (as it is given *all* tokens that did not pass the lexical analysis). Its performance in the workflow therefore differs from its performance as a stand-alone tool (where only compound candidates are analysed). Similarly, the defaulter in the process flow has significantly more incorrect or odd tokens than in normal text input. Different input leads to different evaluation results.

For the workflow analysis, the whole *chain* of tools needs to be inspected, whereby the respective next component operates on the output of the previous one. So the test procedure is as follows:

- All tokens of the corpus are sent to Lexicon Lookup. The relation of processed to non-processed tokens gives the coverage. The processed tokens are inspected to find wrong analyses, this gives the error rate.
- (Only) the non-processed tokens are handed to the decomposer. Again, the coverage is determined by the decomposed tokens as compared to the decomposer input tokens. The decomposer results themselves are again inspected to determine the error rate.
- The tokens which could not be analysed by the decomposer are sent to the defaulter. As the defaulter always returns a POS tag the coverage is no issue, but the error rate is of interest.

As for the overall coverage, although the coverage is 100% as the defaulter always assigns a POS tag, it is still interesting to see which component contributes how much to the overall analysis. The accuracy can be evaluated by comparing how many errors each of the participating components will produce, and how the error rate of the whole workflow is.

The workflow test was carried out for German (as English has no decomposer); with both the large test set and the small PANACEA control test set.

The results are collected in Fig. 5.14.

Lexicon Lookup is the first component. Even considering a huge corpus the analysis could cover about 95% of it (a bit less for the control corpus). The quality is dependent on the quality of the dictionary, which had an error rate of 1.29%. for 1000 randomly selected entries and the standard tagset (STAG); the extended tagset (XTAG) doubles the error rate to about 3%.

The **Decomposer** had as input all missing forms left by the Lexical Analysis (about 7.8 Mio tokens). Decomposition quality / *accuracy* was evaluated by manually inspecting a random subset (14.300 entries) of decompositions. Of these, 2.2% were found to be incorrect. For the PANACEA control corpus, *all*

LARGE SET	k tokens	k analysed	remaining	coverage %	k evaluated	# errors	accuracy %
lemmatiser	65568	62380	7796	95.1	1000	13	98.7
decomposer	7796	5814	1982	74.5	14300	312	97.8
defaulter	1982	1982	0	100	2000	397	80.1
total accuracy							**98.1**
PANACEA	**k tokens**	**k analysed**	**remaining**	**coverage %**	**k evaluated**	**# errors**	**accuracy %**
lemmatiser	206	173	33	83.9	3500	49	98.6
decomposer	33	22	11	66.6	22000	567	97.4
defaulter	11	11	0	100	11000	2539	76.9
total accuracy							**97.1**

Fig. 5.14 Workflow test

compounds in the set were evaluated (20.000), the error rate is slightly higher (3,14%).

Main error types were proper names identified as compounds (esp. location names like ‚*nieder+au* ‘, ‚*klein+schön+ach* ‘, ‚*renn+steig* ‘), lexicon gaps (e.g. ‚*günstigkeit* ‘ in ‚*günstigkeits+prinzip* ‘ or ‚*vernässung* ‘ in ‚*wieder+vernässung* ‘ were not in the lexicon), and foreign language words (like ‚*attribut+ion* ‘).

To have an impression of the recall, all tokens of the test file remaining unknown after decomposition were inspected, to find out cases which *should* have been decomposed but in fact were not. It turned out that 98.2% of all compounds in the corpus were in fact detected by the decomposer; so it missed only very few compounds.

The **defaulter** took as input all tokens left unanalysed by either Lexicon Analysis or Decomposition. If the defaulter is evaluated in a framework of the open word classes of a language (i.e. ‘in-language-domain’), the error rate would be about 10.6% as mentioned above. But in the workflow tests, where the defaulter must cope with the ‘left-over’ entries of the corpus, the error rate doubles to 20%. This is due to proper names, foreign language words and all kinds of exceptions. The PANACEA control corpus shows a similar behaviour. About half of the errors in the defaulter component are in fact due to errors of the tokeniser, which are propagated through the analysis chain.

Overall, the test shows that by far the most tokens in the corpus can be handled by the TINA components, with an error rate of about 2–3% on average. The index terms provided by the TINA Lexical Analysis can contribute to the indexing quality by providing linguistically clean terms.

5.3 Lexical Analysis with Segmentation: Turkish

Turkish morphological analysis cannot rely on techniques applied to languages presented so far. These were based on the possibility to create lexical resources which collect (more or less) all inflected forms of this language, with defined exceptions (clitic determiners and prepositions in Arabic, clitic pronouns in Spanish and Italian, compounds in German).

However such an approach cannot be applied to Turkish as there is no possibility to create such a dictionary resource. Turkish is an agglutinating language which allows for a quasi-unlimited creation of word forms. These forms include elements of several parts of speech and their transitions. They can even contain whole sentences ($_{tr}$ '*öğretmeninizim*' ->'*I am your teacher*'). Although words like'*muvaffakiyetsizleştiricileştiriveremeyebileceklerimizdenmişsiniz*' do not occur in real world texts they show the potential of this language, even if the average word length of Turkish texts does not differ significantly from other languages.[34] There is no point in trying to generate a dictionary of inflected forms for Turkish.

1. As a consequence, Turkish analysis in TINA is based on **morphological segmentation** and the interpretation of these segments. Such an approach requires a lexical resource which is quite different from the ones of the languages analysed so far: It is not full-form oriented but presents base forms, i.e. lemmata, and information about their combinability.[35] As a consequence, the usual tagsets, like 'inflected participle', 'attributive adjective' etc. are not usable in a baseform-driven analysis without modifications. As a result, Turkish uses a modified STAG tagset, as explained in the chapter on the tagsets (Sect. 2.3.2).[36]

2. The **dictionary** organisation in Turkish differs significantly from the dictionaries used so far: It is not a full word dictionary (FWD) but a Basic Lemma Dictionary (BLD): The definition of an entry by<lemma, syntactic description, semantic-information>still holds. But the BLD collects base forms, not inflected forms; its part of speech information follows the Basic Tagset (BTAG), not a syntactically more complex tagset (STAG, XTAG). Just its semantic information, if available, would stay unchanged.

In addition the dictionary contains a description how a given base form can appear in texts: This is usually done by assigning the lemma an inflection class and a set of **allomorphs**. And it must state under which constraints such allomorphs are used in the word analysis.

The Turkish Basic Lemma Dictionary contains about 91 k entries.

3. Analysis also requires lexical entries which hitherto were not considered: The dictionary must contain **affixes**. They must be provided to be able to analyse inflected forms. They need constraints on their combinability, and features describing their contribution to the description of the inflected form (number, person etc.). All this requires an extension of the tagset to bound morphemes, as will be shown below.

---

[34] German has 11.66, English 8.23, Turkish 11.01 as average word length; according to (Güngör 2003).

[35] Of course morphological segmentation can be used in German, Italian etc. as well, and has been, cf. (Thurmair 1984), but the approach taken in TINA is more efficient and performant for those languages.

[36] Similar to (Öztürel et al. 2019).

5.3.1 Existing Approaches

Early attempts to Turkish analysers used techniques like Augmented Transition Networks as proposed by (Güngör and Kuru 1993) whereby the states of analysis were modelled by nodes, and the morphophonemic challenges were placed in procedures which fired when a given transition was tried.

The most important and influential progress was the contribution of Oflazer.[37] It defines the analysis as a finite state network, covering nominal and verbal classes, and the possibilities of interaction between the two. On top, it defines rules based on the two-level morphology approach to handle morphophonemic variations like vowel harmony.

While the two-level morphology, as invented by (Koskenniemi 1984), allows for a concise scientific definition of the Turkish morphophonemic rules, it loses its simplicity and clarity when real data come into play: It is developed for rules, not for exceptions. However, Turkish lexical entries provide lots of idiosyncrasies: Many entries do not follow the wordfinal innervocalic weakening rule for *t* | *k* | *p* |ç ->*d* | *ğ* | *b* | *c*; many entries do not follow vowel elision; there are exceptions to aorist formation (*alır görür olur*), there are exceptions to vowel harmony (*saat*|*i rol*|*ü*), and the like. While it would be hard to integrate these cases into a description framework relying only on rules, the only way to safely analyse and form proper word forms is to store such phenomena in the dictionary, as part of the lexical entries. As a consequence no two-level component is used in the TINA morphological analysis.

The work of (Çöltekin 2010) follows the approach presented by Oflazer, but makes the analyser publicly available as TrMorph.[38] It also uses a Finite State machine and a two-level morphological component. For segmentation the component uses a spellchecker-derived lexicon; augmented by some entries to cover irregularities. As opposed to Oflazer, Çöltekin presents evaluation data based on corpus analysis: For the Wikipedia corpus, it reports a coverage of 83%. The tool was further developed in terms of tagset, software basis, and g2p integration, cf. (Çöltekin 2014).

Different approaches have also been tried: (Eryiğit and Adalı 2004) propose an analysis which is based on suffix stripping: Inflectional and derivational suffixes are stripped from a word form until the stem is left. The approach does not use a lexicon but works with the information which suffixes can precede other ones. The remaining stem undergoes morphophonemic treatment (like *kitab*|*im* ->*kitap*) to produce the most plausible stem. However, a lexicon-free approach faces the challenge that it is not clear whether a string looking like a suffix is really a suffix or rather a part of the stem: Strings like *dolar, keler, kiler, ekler* are (singular) stems in themselves, and the 'plural' morpheme (*ler*) must not be removed. Also, concepts like *dinleyiciler* (= auditory) form concepts in itself, including all the suffixes.

---

[37] Among others: (Oflazer et al. 1994; Oflazer 1994).

[38] Available at Github: https://github.com/coltekin/TRmorph.

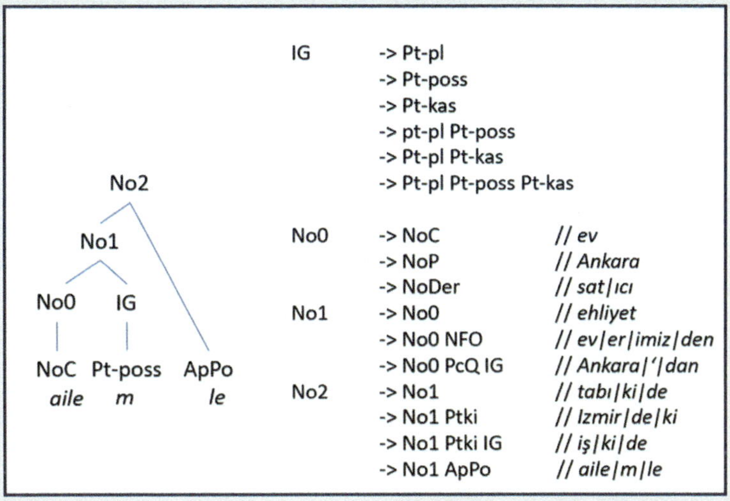

Fig. 5.15 Turkish analysis with LtDeep

There has also been an attempt to use Deep Processing (LtDeep) for Turkish analysis.[39] It used a dictionary like the one proposed here, a chart-based analyser, and a set of context-free rules for analysis. The rule set uses inflection groups (IGs) an nonterminal elements, recursively attaches affixes to the IGs, and collects the feature information in this context. Corpus tests showed a preliminary coverage of about 60%, however the project could not be continued (Fig. 5.15).

5.3.2 Lexical Analyser Turkish

5.3.2.1 Lexical Material

The lexical material follows the conventional part-of-speech classification as its first subdivision. Most lemmata of Turkish belong to open-word classes (nouns, adjectives etc.) and are classified along the Basic Tagset (BTAG).

The dictionary also provides information on the morphological potential of the entries. The most important information is their allomorph, this is how the lemma appears in texts, and provides the 'handle' for word analysis. Endings are often weakened when inflected (*kitap sözlük kağıt amaç ->kitab̲la sözlüğ*|*ü kağıd̲*|*la amac̲*|*la*) or they elide vowels (*şehir–şehr*|*i, burun–burn*|*u*). The dictionary stores such forms as **allomorphs** of the lemma. Unlike the current opinion (and many finite state analysers) these morphophonemic changes are not completely regular;

[39] It was mainly developed by Vera Aleksić, cf. (Aleksić and Thurmair 2007).

there are sufficiently many forms that stay unchanged (*elbet faik gömgök haşlak harap*[40]). The only safe option in such a situation is to mark this behaviour in the dictionary.[41]

In addition to allomorphs entries are also marked for offset (strong–weak–vocalic), verbs need to provide the number of syllables of which they consist (as it influences aorist inflection) and some other properties. Such feature information must declare to which allomorph it refers.

In addition to entries coded with a BTAG there are lexical entries which hitherto were not required: The dictionary must contain **affixes**. They are required in the analysis of inflected forms. Like any other entry, these affixes need at least a lemma, a POS tag, and some 'semantic' description (although the 'content' of many of them are just grammatical functions).

Given the huge morphophonemic variability, the **lemma** of the affixes uses the convention of using uppercase letters as variables.[42] So '*DIK*' represents *dik duğ tik tüğ* and others, 16 allomorphs in total.

The affix entries need also a **POS** tag. Therefore the tagset must be extended. Affixes would be classified as particles (BTAG = Pt), next according to their position (prefix–infix–suffix–inflection), next according to the POS they can be attached to (nouns, verbs etc.), and finally according to their specific content.

Like other entries, affixes are annotated with feature-value pairs, like tense, person, mood, number etc. These features are used in analysis to describe the textform in question (Fig. 5.16).

5.3.2.2 Transitions

The analyser proper defines how the lexeme classes in the dictionary can be combined in word formation. It models their sequence in an (inflected) word form. It should define all possible transitions through the word form, and block all incorrect ones: So '*ev* | *imiz* | *den*' would be accepted by the analyser while *'*ev* | *tik* | *dan'* would be rejected.

A simple representation of nouns like *ev-ler-imiz-de-ki* or *Ankara'da-ki* has been given in Fig. 5.17. The system accepts common nouns as well as names, which in Turkish must be followed by a quote. Next there can be plurals, followed by possessive pronouns, and a case particle. Finally, there can be a copula (like in '*doktor-um* '*I am a doctor*') but also a suffix like *-ki* which turns the noun into an adjective ('*Ankara'-da-ki profesör* ' '*the professor from Ankara*'), but not both.

---

[40] Following the official Türk Dil Kurumu Sözlükleri: https://sozluk.gov.tr/

[41] Similar phenomena can be obseverd with vowel harmony: There is a whole number of exceptions, like *saat–saati*, *kalp–kalbi*, *rol–rolü* and many loan words, cf. (Clements and Sezer 1982; Bacanlı et al. 2020).

[42] Following (Ersen-Rasch 2004) but different from (Oflazer et al. 1994).

Tag	Head	Meaning	Example
PtFNC	*nominal*	case marker	*DEn, E*
PtFNP	*nominal*	possessive pronouns	*Im, sIn*
PtFV	*verbal*	verbal inflections	*Im, In, Inlz*
PtSNC	*nominal*	copula	*yIm, Inlz*
PtSNN	*nominal*	number (better PtFNN)	*lEr*
PtSVC	*verbal*	conditional	*sE*
PtSVD	*verbal*	derivatives (causat., passiv, reflex.)	*tIr, Il*
PtSVG	*verbal*	gerunds/converbs	*dIkçE, Ip, mEktEn*
PtSVI	*verbal*	infinitives	*mEk, mE*
PtSVM	*verbal*	modals	*Ebil, mEli*
PtSVN	*verbal*	negations	*m, mE*
PtSVP	*verbal*	participles	*en, DIK*
PtSVT	*verbal*	tenses	*Iyor, DI, Iyormuş*

Fig. 5.16 POS tags for Turkish affixes

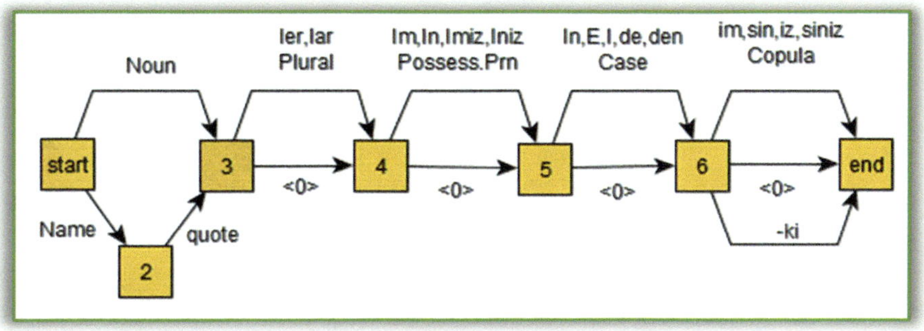

Fig. 5.17 Graphic definition of a simple Turkish noun analyser

Zero transitions allow to analyse simpler forms (like '*profesör* ' or '*kedim* ' '*my cat*'), but add linguistic information to the analysis: the<0>between states 3 and 4 annotates '*number=singular*', between 5 and 6 it gives '*case=nominative*'.

Of course, the above example is far from being sufficient.

- Nouns ending in vowels have different possessive and copula particles from the ones ending in consonants (*güzel-im* '*I am nice*' versus *hasta-yim* '*I am sick*'),
- Some lemmata have allomorphs when followed by a vowel (*kitap*: *kitap-lar* '*books*' versus *kitab-ım* '*my book*');
- nouns ending in strong consonants use other particles than ones ending in weak consonants (ablative *kitap-tan* '*from books*' versus *amiral-dan* '*from the admiral*').

If these issues are not considered, the automaton would accept incorrect inputs like **kitab-lar-tan*.

As a result, the graphic representation grew into a size and complexity which made it less usable for Turkish than for other languages. As a result, the automaton was designed not as a graph but as a table as shown in Fig. 5.18, where the row

isFinal	from\to	N_0v	N_0s	N_0w	N_1v	N_1s	N_1w	N_2v	N_2s	N_2w	N_Pl	N_Posv	N_Posc	N_Kasv	N_Kasc
--	<start>	0v	0s	0w	1v	1s	1w	2v	2s	2w	--	--	--	--	--
+	N_0v	qu,Dll,Dcl,DcEslnE	Dllk,Dclk	Dslz	--	--	--	--	--	DllG,DclG	lEr	Psl,PlErl	Pm,Pn,Pmlz,Pnlz	KyE,Kyl,KdE	Knln,KdEn
+	N_0s	Dll,DCl,DCEslnE	qu,Dllk,DClk	Dslz	--	--	--	--	--	DllG,DClG	lEr	Pl,PlErl	Plm,Pln,Plmlz,Plnlz	KE,Kl,KtE	Kln,KtEn
+	N_0w	Dll,Dcl,DcEslnE	Dllk,Dclk	qu,Dslz	--	--	--	--	--	DllG,DclG	lEr	Pl,PlErl	Plm,Pln,Plmlz,Plnlz	KE,Kl,KdE	Kln,KdEn
+	N_1v	Dll,Dcl,DcEslnE	Dllk,Dclk	Dslz	--	--	--	--	--	DllG,DclG	lEr	Psl,PlErl	Pm,Pn,Pmlz,Pnlz	KdE	KdEn
+	N_1s	Dll,DCl,DCEslnE	Dllk,DClk	Dslz	--	--	--	--	--	DllG,DClG	lEr	PlErl	--	KtE	KtEn
+	N_1w	Dll,Dcl,DcEslnE	Dllk,Dclk	Dslz	--	--	--	--	--	DllG,DclG	lEr	PlErl	--	KdE	KdEn
--	N_2v	--	--	--	--	--	--	--	--	--	--	--	--	KE,Kl	Kln
--	N_2s	--	--	--	--	--	--	--	--	--	--	Pl	Plm,Pln,Plmlz,Plnlz	KE,Kl	Kln
--	N_2w	--	--	--	--	--	--	--	--	--	--	Pl	Plm,Pln,Plmlz,Plnlz	KE,Kl	Kln
+	N_Pl	--	--	--	--	--	--	--	--	--	--	Pl,PlErl	Plm,Pln,Plmlz,Plnlz	KE,Kl,KdE	Kln,KdEn
+	N_Posv	--	--	--	--	--	--	--	--	--	_lEr	qu	lEr	KnE,Knl,KndE	Knln,KndEn
+	N_Posc	--	--	--	--	--	--	--	--	--	_lEr	--	lEr	KE,Kl,KdE	Kln,KdEn

Fig. 5.18 FSA Turkish example

and columns give the source and target state, and the cell contains the (list of) transition conditions which release it, written as lemma of the respective affixes.[43] The matrix is 90×90 big.

5.3.2.3 Grammar Coverage

The FSA codes a grammatical description of Turkish morpheme sequences. This is the heart of the analyser. It is created by linguists and provides the following coverage:

Basic noun structures are given in Fig. 5.17. They cover number and case, clitic possessive pronouns, some derivational phenomena (nouns ending in *-siz, -li, -ki*), and nouns with copula suffixes (i.e. copula sentences). Adjectives follow the noun analysis.

Verb analysis covers derivational features like frequent causatives, reflexives, and passivations. Finite forms cover person and number; tenses both simple (like *-iyor*, *-ecek*) and composite (temporal extensions with *-idi* and *-imiş*), imperatives and conditionals; nonfinite forms cover infinitives (both short and long forms), participles (ending in *-en* and *-dik*) and coverbs / gerunds[44] (ending in *-erek*, *-ip*, *-ince* etc.).

For numerals, cardinals, ordinals (*ikinci*), and distributives (*ikişer*) are given. Postpositions can take possessives and cases (*üst-ümüz-de*). Other parts of speech (pronouns, conjunctions) are covered by lexical entries as full forms.

Overall, the coverage follows the grammatical description given in most Turkish grammars.[45]

5.3.3 *Evaluation*

Two types of evaluation were carried out for the Turkish component: The first interest was how good the grammatical analysis is (both in accuracy and in coverage), and the second interest was how good it is in analysing corpus data (coverage).

5.3.3.1 Grammatical Tests

To test the grammar coverage, a special test suite containing word forms as used as examples in grammars was collected. Word forms were collected from text

[43] It can be seen that the matrix distinguishes vocalic (v) from string (s) and weak (w) consonant endings as they influence the following affixes.

[44] The literature uses both gerund and coverb for these verb forms.

[45] Cf. (Ersen-Rasch 2004; Lewis 2000; Moser-Weitmann 2001; van Schaaik 2020; Göksel and Kerslake 2011).

Grammar coverage	total	correct	incorrect	coverage %
total words	5.417	5.355	62	98.86
Grammar accuracy	inspected	correct	incorrect	accuracy %
total inspected	270	261	9	96.6

Fig. 5.19 Turkish grammatical tests

books and grammars as well as from special tables as provided by (Landmann 2009) and (Seçkin 2008). They cover all sections of the grammars which deal with word formation, and are grouped into nouns (including names) and adjectives, finite verbs, non-finite verbs, and function words (numerals, adpositions etc.). The total amount of forms is 5400 overall.

Figure 5.19 presents the results. The word forms of the test data were nearly completely analysed; the non-found types were mainly input errors.

For accuracy evaluation, every 20th analysed word was manually inspected, and wrong analyses were identified.

As a result it can be said that the analyser is highly accurate both in accuracy and in coverage.

5.3.3.2 Corpus Tests

An important measure for corpus analysis is coverage. It answers the question which percentage of a running text can be covered by the Turkish analyser, i.e. how good the component 'fits' the text which it is supposed to analyse. This can also be determined on the basis of coverage and accuracy. Coverage tells how many word forms can be analysed, accuracy tells which of them present a linguistically meaningful reading. Of course, a form can have several analyses but all these forms should be linguistically meaningful.

An important factor in such tests is the selection of the corpus. It is a fact that corpus data show quite different behaviour in linguistic analysers: Legal texts, medical texts, newspaper texts etc. differ already in their top 100 most frequent tokens. To avoid such a biased evaluation, a Wikipedia corpus was selected: Wikipedia claims to cover all relevant knowledge domains of a language. Although this corpus is biased as well (e.g. it has much more proper names than other text types), it can be hoped that all relevant domains are somehow covered.

So, a Wikipedia dump of July 2010 was taken as a reference. The texts were deformatted, cleaned up (e.g. by eliminating all text portions with Chinese, Arabic, Cyrillic etc. scripts), and tokenised (using a simple tokeniser which basically removes punctuations from tokens). All wordforms with a frequency > 1 were kept, resulting in 640 K types, representing 27.6 Mio tokens.

Corpus	Wikipedia	METU-Sabancı
tokens	27.625k	67.8k
tokens analysed	23.038k	25.6k
tokens not found	4.587k	2.2k
token coverage %	83.45	96.89

Fig. 5.20 Turkish corpus tests: coverage

In addition to Wikipedia, the METU-Sabancı corpus[46] was used for comparison purposes. This corpus is in the focus of research in the CoNLL activities to build a treebank for Turkish, cf. (Kayadelen et al. 2020). It would be linguistically 'cleaner' than the Wikipedia corpus. It has about and 19 K types representing 67 K tokens.

Figure 5.20 shows the results of the analysis in terms of **coverage**. The results for METU-Sabancı are quite satisfactory, the Wikipedia is also in the range of other tests: (Çöltekin 2010) reports similar results.

To have a closer look to the non-analysed Wikipedia wordforms, every 20th of the misses was collected, and the top frequent 1000 tokens were inspected. The result was that one third (360 tokens) was in a language other than Turkish (mainly English) (like *never, human, my, all, to*), another third were non-Turkish proper nouns (like *Alfonso, Stalingrad, Johnston, Benny*), and only 213 tokens (20%) would have been relevant for Turkish morphological analysis (still including many Turkish proper nouns like *Bahçeköy, Osmangazi, Sivasspor*). The high number of proper names is due to the fact that the corpus was taken from Wikipedia.

The METU-Sabancı results are quite satisfactory, and as the Wikipedia contains much more non-Turkish types and more proper names than the METU corpus, its figures are lower. The coverage for Wikipedia can easily be improved by adding proper names to the dictionary, a measure which is required for domain adaptation of a Named Entity recogniser anyway.

The result shows, by the way, how dependent of the selected corpus the reports of linguistic results are.

Next, the question whether the found analyses (i.e. precision) were correct was investigated. Figure 5.21 presents the results of the accuracy tests. Every 20th type of the two corpora was manually inspected, and wrong analyses were identified. In all cases, precision is in the high 90'ies, which means that the Lexical Analyser Turkish is highly accurate.

If the Turkish analysis is part of larger systems (tagging, syntactic analysis), then the number of analyses per type is of interest. The number of segmentations per word is also given in Fig. 5.21, in the grammar there were 2.09 readings per type on average. It shows that the component does not overgenerate. Some types show as many as 15 possible analyses, however the majority (about 70%) has

[46] (Say et al. 2002; Oflazer et al. 2003; Öztürk et al. 2014).

Corpus	Wikipedia	METU-Sabancı
inspected types	850	956
incorrect analyses	16	16
accuracy	98.1	98.3
no. analyses per type	2.11	2.14

Fig. 5.21 Turkish corpus tests: accuracy

just one or two results. The average is about 2.1 analyses per type. This is less than reported elsewhere[47] and reduces the workload for the following components (esp. tagger, chunker, syntactic analyser): Many parsers have an upper bound of hypotheses which they can handle, and combinatorial effects and chances of wrong analyses are significantly reduced, if only few and correct readings need to be considered.

5.3.3.3 Observations

A closer look to the output of the analyser reveals that the majority of analysis results per type consists of more than one reading: While 46% have only one result, 26% have 2, 12% have 3, and 16% more than 3 analysis results. There are several reasons for this phenomenon:

- Ambiguity of affixes and morphological patterns

  Typical endings represent different morphological configurations: Possessive-3sg and accusative in cases like *kuzen|i*, Possessive-2sg and genitive in *kitab|in*, Possessiv-2sg and possessive-3sg in cases like *rol|ün|ü* versus *rol|ü|nü*. Also, if copula suffixes come into play, forms with *lar/ler* are systematically ambiguous: *sporcu|lar* can be plural (*sportspersons*) or a copula clause (*they are sportspersons*). In cases with possessives, three options are possible: *dönem|ler|i* (plural, accusative and possessive-3sg) versus *dönem|leri* (possessive-3pl).

  Similar ambiguities can be found in verb inflections: verbs in future tense, (*geç|ecek*), aorist (*ak|ar*), perfect (*yaralan|mış*) etc. are systematically ambiguous between finite and participle forms.

  All these cases are systematic, must be produced by the analyser, and wait for disambiguation by the following analysis components (parser).

[47] Öztürel et al. (2019) report 12 analyses on average. A 5-word sentence would then have close to 250 k reading possibilities, and the parser would explode quite quickly.

- Lexical ambiguities

  A second source of ambiguity is lexical: Such forms have different lexemes as origin. One reason is that a lemma is a homograph with different parts of speech: *soğuk* is represented as both noun and adjective, so it has two readings in the output; similar for *sade* (adjective and adverb).

  Another reason is that some derivations (like passivation) are also lexicalised: A form like *denilir* is a passive aorist (*den|il|ir* -> *denmek*) but the passive also already exists in the dictionary (*denil|ir* -> *denilmek*); a form like *karıştırdı* originates from *karmak*: *kar|ış|tır|dı* (*mix, mingle*) but also from the reflexive *karışmak: karış|tır|dı* (*mix*), and finally from the causative *karıştırmak: karıştır|dı* (*cause to mix*). This phenomenon also holds for nouns: *filler* originates from two lemmata, *fil* (*fil|ler*) and the plural *filler*. Such phenomena are quite frequent in the dictionary. The richer forms may have been added to the dictionary because they show different translations in singular and plural form (here: *fil* -> *elephant* versus *filler* -> *species of trunked animals*). For such reasons the Turkish component selects the longest segmentation in such cases, i.e. the most specific lexicalisation.

While some of these ambiguous readings can be handled locally (e.g. by selecting the longest-match alternative), others can only be disambiguated in context, i.e. by parsing or tagging. Disambiguation is analysed in serveral papers like (Hakkani-Tür and Oflazer 2002), (Yuret and Türe 2006).

5.3.3.4 Result

Like all other language components of the Lexical Analysis, the Turkish analysis returns a set of records for the Document Model: *Textform*, *casing*, *normalform* stay unmodified. They are followed by the number of readings, and for each reading, the lemma and part-of-speech as provided in the lexicon are given; in the additional feature section, the usual feature-value information (number, person, tense, case etc.) are returned. The feature information would be sufficient to compute additional STAG or even XTAG tags, if required. The fact that these results are created with a completely different technology than other languages is completely transparent to the outside.[48]

5.4 Deep Lexical Analysis: Persian, Pashto

Deep Analysis (LtDeep) is used in TINA for Persian and Pashto. Both are Indo-European languages, Persian belonging to the West-Iranian languages, Pashto to the East-Iranian ones. Both languages show significant varieties: Kandahar and

[48] (Çetinoğlu and Oflazer 2006).

Peshawar are centres for East and West Pashto varieties, Teheran and Kabul denote the Persian varieties, also known as Farsi (Teheran) and Dari (Kabul). Differences relate to phonetic and lexical issues. Lexical entries can be marked for their affiliation to Farsi and/or Dari.[49] Grammatical differences between the varieties are less significant, and justify a common analysis component each for Farsi/Dari and East/West-Pashto.

Of course both languages (Persian and Pashto) are different in many respects. Pashto shows more complexity in phonetics (e.g. the retroflex consonants), it has more characters for text analysis, and is morphologically richer, e.g. nouns have gender (masculine/feminine) and case (rectus/oblique). For the purpose of analysis, however, there are significant similarities: Both languages use the Arabic script, which shows similar challenges for the text analysis. Both languages show proclitics (conjunctions, prepositions) and enclitics (pronouns), like the genitive د (de) in Pashto, so the analysis challenges are similar.

The following section presents the analysis with LtDeep for Persian. The analysis for Pashto follows the same lines, but of course the content of the resources (dictionaries, grammars etc.) is different.

5.4.1 Context

5.4.1.1 Rationale

Unlike the other languages, the lexical analysis of Persian uses the Deep Analysis (LtDeep). The main reason for this is the fact that in the Arabic script whitespaces are not congruent with morpheme boundaries:

Like in Arabic or Turkish, a textform can contain several morphemes. In Persian there are proclitic conjunctions (و, *waw*) and prepositions (بی, *be*), and also enclitics can be found like pronouns (بزنش *bezan-aş, beat him)* or the affix for the direct object (را,rā) which is attached to the last element of a noun phrase in clitic position.

In turn, there are single morphemes which are separated by whitespace, esp. if they contain he (ه) (e.g. هم به, linking). Also noun inflections or comparative affixes are often separated from their nouns by whitespace (ها وحشی, *barbarians*, ها صبح, *in the morning*). Such cases are very frequent. Farsi also contains infixes in morphemes, esp. negation in verbs.

These writing conventions are not stable, and often both options ('آموزشهای ' (āmuzesh-hāje, educations) as well as 'های آموزش ') can be found in texts. This would lead to duplicate efforts in analysis (once with, once without white space analysis).

The consequence of these facts is that the standard (shallow) approach by which a tokeniser delivers just one string of morphemes must be extended such

[49] A contrastive dictionary of Farsi versus Dari lexemes is given in (Newid and Thurmair 2023).

that the segmentation of a sentence string can return not just one but also alternative segmentations. As a consequence, the Lexical Analysis of Farsi just ignores whitespaces, considers the whole input sentence at once, interprets it as a sequences of morphemes (segmentation), with several possible alternatives, and writes a grammar which identifies the lexical elements in such a sequence (interpretation). This approach uses LtDeep technology and combines tokenisation with lexical analysis in one component.

Once LtDeep is used for Lexical Analysis, some of the following components which also use LtDeep can easily be added, like syntactic analysis, multiword analysis (important due to the high number of support verb constructions in Persian, as presented by (Family 2006), or Named Entity recognition.

5.4.1.2 Approaches

Most of the approaches towards Persian morphology follow the approach of shallow processing and arrange the data such that it fits to a line of processing as given by languages using Roman script. The challenge is to add whitespaces in cases of clitics, and to remove them in cases of inflections (like 'hā'), infixes (like 'mi') and such cases.

Attempts to use special whitespace characters like non-breaking spaces, as proposed by (Megerdoomian and Zajac 2000), do not really help as there are many texts which do not follow such conventions, and therefore could not be analysed[50].

One alternative would be to try to normalise the whitespace setting similar to the Roman script and then try to do a 'standard' processing similar to other Roman-script languages. Several toolsets of this kind for Persian were presented,[51] a comparison of tokenisers is given in (Kamali et al. 2022). Such whitespace handling is error-prone and does no really contribute to the understanding of the language. It is just an attempt to apply a known technology to an object to which it is extrinsic. However it would be better to accept that the Persian script is different here, and look for a more adequate solution for a language which cares for whitespaces more under aspects of calligraphy[52] than text analysis.

5.4.2 Lexical Analysis Persian and Pashto

In contrast the TINA Lexical Analyser Farsi and Pashto ignores whitespaces and analyses the whole sentence string as a input with LtDeep technology. It does not consider whitespaces at all but just sequences of morphemes, and applies knowledge to them using dictionaries and (morphosyntactic) rules.

[50] Therefore TINA normalises such kinds of spaces to usual whitespaces in texthandling.

[51] E.g. (Shamsfard et al. 2010; Seraji et al. 2012).

[52] A described by (Qadi 1959) or (Schimmel 1970).

The Persian analyser consists of two main components: Segmentation splits the input sentence into morphemes, and analysis groups them to words and interprets them.

5.4.2.1 Morphological Segmentation

The segmentation component splits the input sentence into morphemes and collects them in a chart. This could be seen as a kind of tokenisation process; the difference to LtShallow tokenisation is, however, that segmentation returns *all* possible segmentation options for a given input sentence, i.e. also alternative segmentations, and not just one linear sequence of tokens, as the whitespace tokeniser does (Fig. 5.22).
It uses two resources for this purpose:

- the morpheme list
- the segmentation automaton.

a. The **morpheme list** contains all possible morphemes of Persian, including all affixes and all forms of content words (incl. broken plurals).[53] It serves only the purpose to produce legal sequences of morphemes. It is derived from the Basic Lemma Dictionary (BLD). The difference to a full word dictionary (FWD) as used in the shallow approach of LtLemmatiser is that it does not contain whole word forms but just morphemes.

Each entry of this dictionary has two main information elements: the morpheme and a POS annotation. The morpheme is used to identify the string in the input; the POS is used for the morphological filter. The tagset used is the standard tagset (STAG) because in some transitions more expressive power than a BTAG is required. The morpheme list also contains all the affixes with their respective POS information.

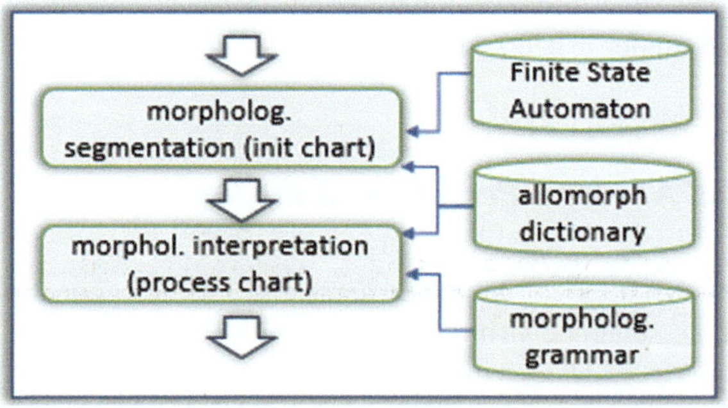

Fig. 5.22 Persian analysis

[53] In case of dictionary entries of multiwords the allomorph dictionary provides all allomorphs of the multiword parts, with the non-head constituents tagged 'MWB' (multiword part).

Of course the morphemes can be homograph, i.e. point to different lemmas. However this is irrelevant for the morphological filter as long as the same POS tags are involved: They will be disambiguated later in the analysis phase. Only if different POS tags are involved, the morpheme list will contain more than one entry.

b. The **segmentation automaton** describes the legal transitions between parts of speech of morphemes. Of course the chart could be filled with all possible morphemes at any position in the sentence. As this would create significant overhead in parsing, the automaton is used as a filter, basically excluding impossible morpheme sequences but without further analysing the possible ones.

The transitions can basically be described as follows:

- Textforms can be surrounded by punctuations
- Textforms can show proclitic conjunctions
- Textform analysis depends on the POS of the main allomorph:
  - nominal textforms can contain prepositions, inflections, possessive pronouns and determiners (rā, yeh)
  - verbal textforms can be negated, they can contain prefixes, inflections, and personal pronouns
  - adjectives can contain comparative suffixes and determiners
  - numbers can be simple or complex and can contain numeral suffixes
  - all other parts of speech can be single words, also inflections with or without determiners (های) can be found in non-bound position
- unknown textforms are marked as ‚unknown'.

Like in Turkish the automaton can be developed using a graphical editor, and will be compiled into an appropriate data structure by the software (Fig. 5.23).

It should be mentioned that the automaton slightly overgenerates; however such analysis results would be eliminated by the following morphological analysis.

All segmentations which pass the filter are collected in a chart, which is the basis for the morphological analysis.

5.4.2.2 Morphological Analysis

The task of the analysis is to build words with feature annotations as given in the segments of the sentence. It uses two resources for this purpose: a monolingual dictionary and a morphological grammar.

a. The **monolingual dictionary** contains all entries known to the system. It is a Basic Lemma Dictionary (BLD). Each entry is defined by<lemma, POS, semantics[54]>as in the other TINA languages, and contains additional linguistic annotations related to phonological, morphological, syntactic (subcategories, transitivity), semantic or pragmatic (*dialect=farsi | dari;*) information.

[54] Semantics can be a reading number, a domain classification, or an explicit *sem* feature value.

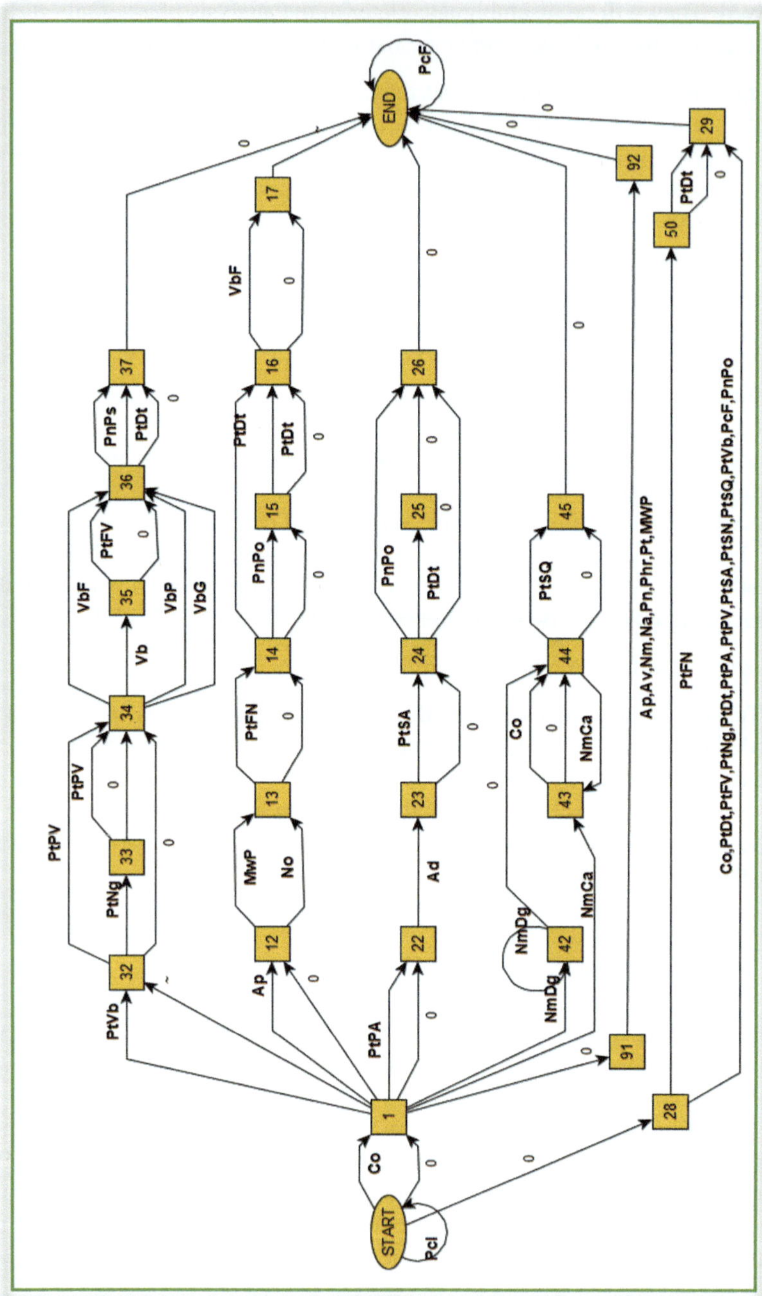

Fig. 5.23 Persian segmentation

For morphological analysis the dictionary section on morphology is of interest. This section stores the lemma together with all its allomorphs.[55] Each lemma has one or two allomorphs (only a few have more): For verbs it is the present and past stem, for nouns it is the regular stem and a plural stem (in cases of broken plurals or special inflection classes).

Each allomorph has annotations describing it. Allomorph-specific information is.

- morphophonemic properties (like*: onset|offset* being *vocalic|consonantal*),
- type of stem (*for verbs: present|past; for noun: is singulare tantum; is singular but lemma can pluralise; for adjectives: is gradable*).

This information is required to prevent incorrect analyses like combination of a present time inflection with a past tense verb stem.

In addition to allomorphs and their information annotations, the lemma contains more morphological information like number (like *plurale tantum*) and in particular the inflection class.

Unlike Turkish where plural is always formed with *ler/lar* Persian nouns and verbs can take different inflections. Therefore the noun entries contain an **inflection class**, specifying which inflections it can take. Correspondingly the inflectional affixes are annotated with the inflection classes which they support. Noun and affix inflection class information must be unified in analysis to produce a proper inflected form. For nouns, several inflection classes are used (cf. Figure 5.24[56]), for verbs just *regular/irregular* inflection is stored. Verb inflections follow the paradigms described in (Amin-Madani and Lutz 1972).

Affix entries carry their 'classical' features like number, person, tense, negation, short versus long infinitive etc.; in addition they show annotations regarding their combination potential: position and morphophonemics of the head, separability etc. The STAG gives the kind of particle (prefix, suffix, inflection) and the POS of the head.

The monolingual Basic Lemma Dictionary comprises about 59 K entries overall.

b. The design of the **morphological grammar** adapts the X-bar scheme for syntactic analysis, cf. (Jackendoff 1977). As this scheme does not include morphology, an adaptation is required 'downwards' such that the terminal node in the X-bar scheme is already a projection of morphology rules: An X0 node would be built from rules like '*N0 ->NStem+NFlex*'. The terminal elements of the grammar would not be (inflected) words as in the standard X-bar scheme but morphemes (stems, flexes).

The morphological rules combine the morphemes forming a word, unify their features and percolate them to the higher nodes. Once they are interpreted the affix nodes can be removed from the analysis tree by pruning transformations, and only the lexemes are kept. For all this the LtDeep formalism as described in Sect. 2.2.1 is used.

---

[55] Multiwords give the allomorph of their head.

[56] First developed in (Thurmair and Newid 2007), also used in (Newid and Mumm 2007).

head	class	flexive	cmt
No	1	hā	Plural in –hā
	2	ān	Plural in –vn
	3	yān	Plural in –yān (allomorph of 2)
	4	gān	Plural in –gān (allomorph of 2)
	5	āt	Plural in –āt
	6	yāt	Plural in –yvt (allomorph of 5)
	7	in	Plural in –in
	8	yun	Plural in –yun
	9	br	broken Plural (arabic plural)
	10	jāt	Plural in –jāt (arabic plural)
	11	t-āt	Plural in –āt (split t)t
	12	h-āt	Plural in –āt (split h)
Vb	reg		regular stem
	irr		irregular stem
PtFV	1		paradigm 1
	2		paradigm 2
	3		paradigm 3

Fig. 5.24 Persian inflection classes

The morphological rules used expand the following rule schema, cf. Figure 5.25:[57]

The bulk of work needs to be done by feature handling: Unifying the inflection classes and morphophonemics, propagating the features of the affixes to the next level etc. (Fig. 5.26).

The **noun** rule in Fig. 5.27 builds a No0 node from the dictionary entries No پدر (pedar, father) and the inflection ان (-ān). The tests exclude impossible inflection classes and unifies the ones between stem and inflection. The example would build an No node for پدر ان (pedar-ān, fathers) (or پدران, without whitespace).

The No0 node are built in analogy to the No0 nodes of other languages where they are the result of a simple dictionary lookup. Here the STAG value is set to NoCo, and the node must have number annotation percolated to the head from the affix.

The Farsi **verb analysis** does not build an INFL node as can often be seen in X-bar theory but uses the traditional way of attaching the inflection information to the Vb node.

Figure 5.27 is a rule to analyse complex finite verb forms like فرو نمیرود (*'foru-ne-mi-rav-ad'*: *does not sink*). Potential whitespaces can be ignored.

Finite verb forms need to cope with the following affixes:

- inflection affixes (prefixes and suffixes) for finite and nonfinite forms. Affix and lemma features must unify. The lemma is formed from the (long) infinitive form of the verb, the inflections are removed (only the lemma is given in the Vb0 node, everything else is featurised.

[57] There are some more, e.g. for verbal nouns, for adjectives created from participles, and the like.

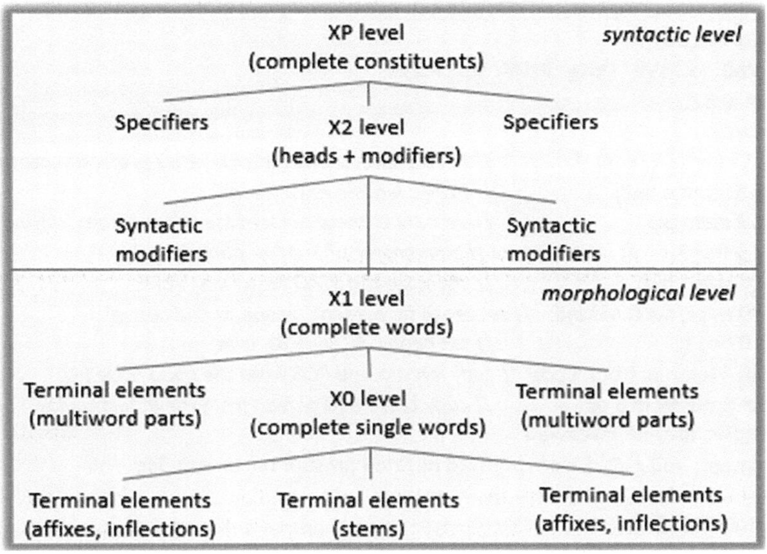

Fig. 5.25 Grammar design

No0	::=	No		*// common nouns*
No0	::=	Na		*// proper nouns*
No0	::=	No	PtFn	*// unify inflection class ...*
Ad0	::=	Ad		*// adjectives do not inflect*
Ad0	::=	Ad	PtSA	*// comparative*
Vb0	::=	VbF | VbP | VbG		*// finite forms may occur in dictionary*
Vb0	::=	PtVb? PtNg? PtPV? Vb PtFN?		*// the verb analysis scheme*

Fig. 5.26 X0 level rule schema

- negations (infix). Negation is turned into a feature (*neg* = *true;*), the infix is also removed.
- verbal separable particles are treated as follows: The lemma is built from the stem plus the particle (e.g.: بازکردن (*bāskardan, to open*). The allomorph is only the stem (کردن) but with a feature requiring the particle (*reqprf* = باز*; bās*). The analysis is only valid, if the required particle is really found in the input (it would not need not be adjacent) and its lemma باز (*bās*) can be unified with the value of the *reqprf* feature.

```
ID: VbFin-121
PS: Vb0  -> PtVb  PtNg  PtPV  Vb  PtFV
Prob: 0.87
Features:
  int 1 lemma 4 reqpref;          // PtVb's lemma must match verb's prefix-annotation
  req 3 lemma 'mi';               // PtPV's lemma must be 'mi'
  req 4 stem pr;                  // verb must show present stem ('rav' from 'raftan')
  req 5 flex 3;   // inflection must be member of 3rd flex paradigm
  set 0 stag VbFF;                // then: set the STAG feature to VbFF
  set 0 tn pr; set 0 md ind;  // set tense to 'present', mood to 'indicative'
  set 0 neg t;                    // set negative value to 'true'
  from 4 lemma; from 4 pos; // copy lemma and POS from the head node (4)
  from 5 nu; from 5 ps;           // copy tense and person from the inflection (5)
Example: foru-ne-mi-rav-ad
Comment: Vb0 / VbFF from prefixed negated present tense verb 3sg
------------------------------------------------------
ID: No0-021
PS: No0  ->  No  PtFN
Prob: 0.64
Features:
  nrq 1 flex br;                  // noun flexclass must not be 'broken':
                              // those don't take inflections
  int 1 flex 2 flex;              // both need same flexclass (here 'ān')
  set 0 STAG NoCo;            // Then:  set the STAG
  from 1 lemma; from 1 pos;  //  copy head features
  from 2 nu;                      //  copy number. Neither gender nor case are marked
Example: pedar-ān
Comment: No0 node from No
```

Fig. 5.27 Examples of X0 rules for No and Vb

The interaction of the components of lexical analysis is explained using the string وباباپانرا (*va|bābā|yān|rā*: *and the fathers, accusative*). In Fig. 5.28[58], segmentation first produces a list of morphemes (conjunction و before noun بابا before inflection یان before determiner را). Analysis combines them to words or sentence parts and interprets them: number from the plural morpheme, case and definiteness from the determiner suffix. It also shows that the conjunction is a separate constituent although agglutinated to the noun in preclitic position.

[58] The tree is to be read from right to left.

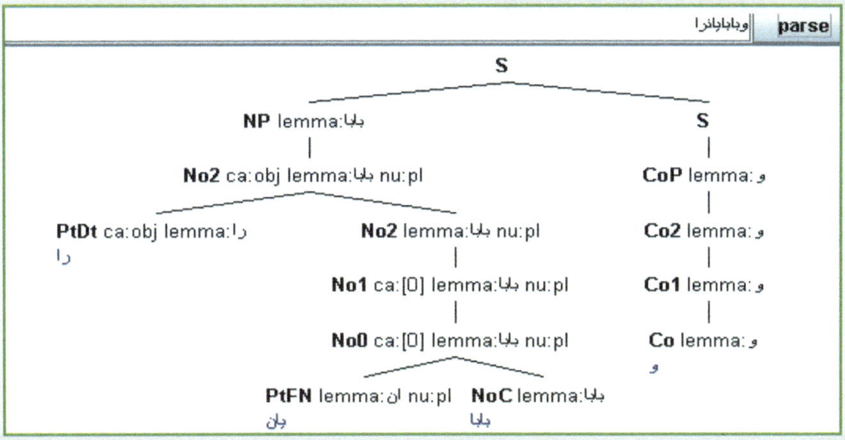

Fig. 5.28 Parse for وبابایانرا

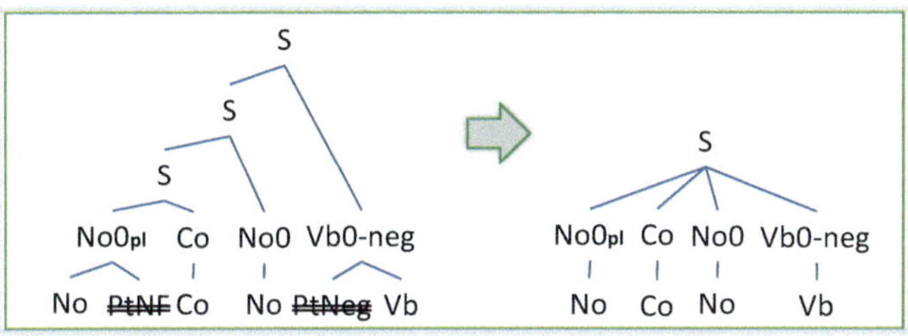

Fig. 5.29 Wordlist creation

5.4.2.3 Creation of a Word List from Sentences

In order to be compatible to the results of a shallow analysis the morphological analysis not only has to return nodes with the same annotations as the nodes of LtShallow (lemma, STAG, XTAG feature annotations) but also must return a list of words as used in the Document Model (Fig. 5.29).

For the collection of the list of words for the Document Model, the S rules must be modified such that an S rule collects all lemma nodes in the sentence built by the grammar. This is done by recursively applying S rules of the kind *S->S No0, S->S Co, S->S Ad0, S->S Vb0* etc., starting from the left corner. If the intermediate S nodes are pruned away, a proper list of lemmata can be returned by just collecting the nodes below the S node.

From there the lemmata can be handed back to the Document Model.

5.4.3 Evaluation of Persian and Pashto

The lexical analysis components were tested for both languages in similar ways. Corpus data from news domains were collected: Aryanet, Hamshahri, BBC and other sources were used for Farsi, mainly BBC, Azadiradio and China Broadcast for Pashto. Two sets of document were built for Farsi, set 1 containing 133 documents, set 2 containing 352 documents (3.84 Mbyte). The Pashto set had 177 documents (1.42 Mbyte).

All sets were prepared in the same way:

- The documents were split into sentences
- The sentences were whitespace-tokenised
- The tokens were cleaned: All tokens with non-Arabic scripts were removed, all digits and punctuations were removed.
- The tokens were frequency-sorted.

Next the single whitespace tokens were sent to the Lexical Analyser, and the results were evaluated. This is documented in Fig. 5.30.

Although the results for both Farsi and Pashto are quite acceptable (with about 90%) it should be remembered that the tests were made with white-spaced tokens. Some of these are not translated (mostly due to clitics) and returned as unknowns, and some tokens are in fact parts of words (like inflections) and are analysed, however not as a single whitespace token. To eliminate these aspects a manual inspection was carried out, based on 14 documents with about 4850 tokens. Of these, only 130 were really found to be unknowns, the rest fell into the whitespace issue. So the coverage would increase to 97.3% for Farsi.

For Pashto it should be mentioned that the orthography is not really standardised, and a normalisation component could help to improve the results.

To streamline the tests with the other TINA languages, a coverage tests for Persian was carried out on the basis of the Hamshahri corpus, cf. (AleAhmad et al. 2009), similar to Turkish tests with the METU-Sabancı corpus. The result is shown in Fig. 5.31.

	Set 1 Farsi	Set 2 Farsi	Set Pashto	Manual 1 Farsi
no. documents	133	352	177	14
no. types	22,000	14,200	8,200	
not analysed	5,200	5,910	3,830	
coverage in %	76,36%	58,38%	53,29%	
no tokens	161,000	401,600	84,308	4,849
not analysed	18,990	37,780	8,327	130
coverage in %	88.2%	90.6%	90.1%	97.3%

Fig. 5.30 Tests Farsi and Pashto

	no tokens	coverage	no types
Total	63,131k		440.1k
frq 1			255.8k
frq > 99	61,174k	96,89	18.7k
frq > 49	61,710k	97,75	26.4k

Fig. 5.31 Tests with Hamshahri corpus

Hamshahri shows a comparable coverage: Persian analysis could cover 97,75% of all tokens there, which activated 26.500 entries of the dictionary.

Overall the tests show that the Lexical Analysis for Farsi and Pashto are usable in practical settings.

5.5 Stemming (LtStemmer)

5.5.1 Rationale

The task of a stemmer is to reduce a lemma to its stem, i.e. its un-affixed kernel, in order to broaden the search space by adding related terms which share the same stem. This is intended to increase the recall. The challenge is to prevent a loss of precision, if a given search term is expanded by a stemmer.

5.5.2 Definition

It should be noted that not every cutoff at a word ending is stemming, cf. (Harman 1991). Stemming is an operation relating a *lemma* to its (morphological) stem, by removing *derivational* affixes. It must not be confused with lemmatisation (stripping the *inflectional* affixes from *textforms*). Stemming rather assumes that lemmatisation is already done.[59] Stemming has lemmata as input, not inflected forms, and not compounds. A stem is defined by morphology to be an elementary meaning-bearing morpheme, and its relation to the lemma is a derivational one: Lemmata are stems plus their (derivational) affixes.[60] This distinction between lemmatisation and stemming is fundamental as lemmatisation is a required component in multilingual indexing while stemming is not: Lemmata represent concepts and therefore can be translated but stems cannot. Many papers[61] and even software neglect this difference.

---

[59] And also that a lemma has one stem, i.e. that a word has been decomposed before if necessary; otherwise defective stems like ‚*behindertenintegr* | *ation*‘ would be returned.

[60] Or other variations like ablaut, broken plural etc.

[61] E.g. (Gurusamy and Kannan 2014).

As far as the processing chain is concerned, stemming assumes that the input forms have already been lemmatised and decomposed. The task is to identify some 'semantic kernel' of the lemma. The purpose is to find *semantically* similar lemmata for query expansion by means of *morphological* operations. These are different levels of description, and they do not match at all times. This is the reason why the increase in recall is often parallelled with a decrease in precision. A good stemmer does the one but avoids the other.

5.5.3 Approaches

5.5.3.1 Only Suffix Stripping

The first and simplest stemmers simply look for suffixes and strip them away. The most famous example is the Porter stemmer, cf. (Porter 1980).[62] Many of such stemmers do not differentiate between lemmatisation and stemming, and just do suffix stripping. Such stemmers have been built for many languages, e.g. for Persian, cf. (Jadidinejad et al. 2009).

An example of such an approach is a first version of the LtStemmer of TINA.[63] It assumes lemmatised words as input, and consists of lists of suffixes and how they should be treated. The example in Fig. 5.32 shows such a list of suffixes and combinations thereof, sorted by length, and how they should be treated in terms of *cut&append* manner to create the stem: A word like '*soluble*' would identify the suffix '*luble*', then cut 4 and append a '*v*' to result in '*solv*' as stem.

As such simple attempts produce many incorrect results, there are exception lists for many of the suffixes, which are consulted to cover nonstandard cases.

suffix	***c&a***	***exc***
ful	0	-
ism	3	+
alism	5	+
ionism	6	+
ium	0	+
um	0	+
ion	3	+
lion	0	+
ification	7	+
luble	4v	+

ium:
labium 2
medium 2

Fig. 5.32 Stemming by suffix stripping

[62] cf. (Jurafsky and Martin 2009: 102).

[63] This version was built by Alois Baumer.

So '*medium*' will be stemmed to '*medi*' rather than to '*med*', to distinguish it e.g. from '*medical*', and to avoid to mingle them into one cluster. For each such suffix there is a mark whether or not an exception list exists, cf. Fig. 5.32.

The component was developed for German and English; the suffix lists comprise several hundred entries, the exceptions list several thousand exceptions.

When this approach was evaluated some fundamental weaknesses were detected.

1. The technique leads to stem clusters which combine quite **semantically divergent words**: A common cluster of words around the stem '*med*', for both '*med | ical*' and '*med | ium*', would not improve search results but rather create ballast. The same for '*stat*' in '*stat | istics*', '*stat | ion*', '*stat | ionery*', '*stat | e*', '*stat | ist*', similar in German for '*reg | ieren*' '*reg | al*' '*reg | en*'. Such cases will increase the recall in retrieval but lower the precision.

   A solution could be to abandon the morphological foundation of stemming and allow for better fitting 'stems', like '*med*' for '*medical*', '*medicine*' etc. '*medi*' for '*medium*', '*media*', and '*mediat*' for '*mediation*', '*mediator*' clusters. Another solution could be to only use affixes which do not change semantics but only change the part of speech ('*flight*'–'*flight | y*').

2. The technique leads to **corrupted stem candidates,** if the suffix stripping results in non-words. While the results of suffix stripping are often meaningful bound lexical morphemes (like '*integr |ation* versus *|ieren*') this cannot be said in many other cases: Examples would be '*b | atist*' '*st | atist*' '*def | eatist*'. German examples would be '*bemust | erung*', '*beze | chen*', '*sportn | ation*'. Such 'stems' would never be of help in search.

   Both types of errors indicate that it is necessary to gain control of the word parts which are found *before* the suffixes, and to make sure that they contain meaningful stems.

5.5.3.2 Use of Dictionary Information

Approaches which follow this consideration would be lexicon-based. Incoming words are split into a lexically motivated stem and its suffixes. The suffix list can be taken from the previous approaches; the stems need to be built by inspecting the word clusters, manually grouping the semantically similar words into clusters, and assigning a common 'stem' to them. If such a stem represents more than one cluster, it could be extended e.g. with an index ('*cut1*', '*cut2*' etc.). The important thing is not the naming of the stem, i.e. the label of the cluster, but the semantic similarity of the words that are its members.

The advantage of such an approach is that it creates much better stems as stemming implies linguistic knowledge and experience, and can create semantically well-formed word clusters. However, the creation of a lexical resource for a stemmer requires manual effort, and faces some tricky design decisions.

This approach is followed by the second version of the LtStemmer.[64] It uses a dictionary of stems, augmented by lexicon-free procedures for cases where the input string is unknown.

It is implemented in two flavours: The first one is rule-based and develops rules how a stem can be created from a given word (stripping the suffix, changing the lemma etc.). Such rules can also be applied to unknown input words, i.e. where no lexical entry is available; this would be identical to the simple suffix-stripping version described above. The second option is to create a static resource for a mainly stable vocabulary, and use a simple and fast lookup technique at runtime.

5.5.4 LtStemmer

The LtStemmer uses three resources:

- the **list of stems**. It contains about 23 K entries. Each entry consists of the stem, a part-of-speech annotation, stem-building instruction (in terms of *cut&append*, i.e. change), and a hint on further stemming alternatives.
  - The part of speech can of course only be given for stems which are words in themselves (free morphemes), like '*dog*', '*music*' or '*fire*'; for bound morphemes like '*refus*', '*regulat*' no part of speech can be given, they are marked as '*bound morpheme*'.
  - The stem building instruction is required for cases like e-elision (de '*handel–handl* | *ung*', '*zeichen* versus '*zeichn* | *ung*'), derivational umlaut ('*tat*'–'*tät* | *ig*', '*hart*'–'*härt* | *en*'), verb ablaut ('*nahm*' versus '*nehm*'), alternations ('*absenz*' versus '*absent* | *ieren*')
  - Stemming alternatives are related to prefix issues. Strict morphological approaches would split prefixes from stems (de '*nehm*' versus '*ab* | *nehm*', '*an* | *nehm*', '*auf* | *nehm*'; en '*set*', versus '*on* | *set*', '*off* | *set*'), however such prefixed verbs would be semantically quite different from each other; so LtStemmer decided that prefixes stay attached to the real stem, so the stems would be '*abnehm*' '*annehm*' '*aufnehm*' instead of just '*nehm*'. However the stem list contains also the information on prefixes and 'real' stems in the third column. There are of course prefixes which never part of a stem, like '*ab-zu-nehmen*', '*ab-ge-nommen*'; however they are already treated by the lemmatiser and would not show up in the stemmer input list.

  The list of stems was created the following way: Starting point was the decomposer dictionary: This is the only resource that is lemmatised *and* contains only single words. This resulted in a list of 84K entries. Then irrelevant entries (proper names etc.) were removed, all entries with no suffixes (i.e. free morphemes) were marked, and all other entries were analysed by

[64] While the first version was developed for German and English, the second version covers German only.

applying the suffix-based version of the stemmer described above to them, and the results were reviewed manually. The resulting stem list contains 24K entries.

- the **list of suffixes**. It contains the affixes and all their combinations,[65] and again a stem-building instruction, e.g. '*-lich*' would indicate an umlaut in the stem. This instruction is not a redundancy, however, as there may be words in the input which are not covered by the stem list, and the stem of those words would undergo this stem building instruction. The German affix list has about 1900 entries.
- the **list of exceptions**. Words are stored in the exception list in cases where the regular strategies would lead to incorrect results. For instance, most cases keep the suffix '*-e*' for nouns as cutting it off would often lead to semantically irregular results, like '*wüst*' versus '*wüste*', '*streb | en*' versus '*strebe*', '*taub*' versus '*taube*', '*west | en*' versus '*weste*' etc. The exception list comprises about 500 entries.

With these resources the stemmer processes an input word. The input string must be lemmatised, it must be decomposed and it must not be a proper name.

5.5.5 Stemming Results

Then the stemmer tries to split the input word into a stem part and a suffix part. For this purpose it looks for the longest match between input word and stem list. The assumption is that longer stems are mostly more specific than shorter ones, so for '*logistik*' the version '*logist | ik*' would win over '*log | ist | ik*'. Also if a stem does not find a matching suffix (like in '*brummi | *g*'), the longer stem may be more expressive.

The result of the splitting attempt is categorised as follows:

1. The input word has no suffix. it is a free morpheme, identical to the stem ('*haus*', '*baum*'). The stem is identical to the word.
2. The input word can be split into a stem and a legal suffix cluster: '*invest | ment*', '*fris | eur*'. The stem is the first part of the split, in cases of ambiguity the longest stem is taken.
3. No stem is found but the input word ends in a legal suffix combination. The result is similar to the suffix-based stemmers à la Porter.
4. A stem is found but it cannot be completed with any affix.
5. Neither a stem nor a suffix is found.

[65] The alternative would be to write a morphological grammar for suffix combinations, as proposed in (Thurmair 1984). It would make the whole component more complex however.

While the cases (1) and (2) return useful stems, (4) and (5) definitely fail in stemming. Case (3) is a borderline case, it could contain proper stems (which just are missing in the stem list), so it needs more detailed inspection.

The output of the stemmer shows, for each input word: the input word, the split result (1–5), the stem, and the suffix combination used.

5.5.6 Evaluation

To evaluate the stemming quality a series of translation memories was used, and their German part was extracted. From about 8 Mio sentences, a word list of 87 k types was created, representing 1.52 Mio tokens. The types were normalised, lemmatised, but not decomposed, and all entries with a frequency of>50 were selected, resulting in a list of 58 k lemmata representing 355 k tokens. This list was analysed with the decomposer, and the final single word test corpus contained 12.9 k entries. This list was stemmed.

The result was as follows:

1. Case (1): Free morphemes comprised 110 entries
2. Case (2): Correct stemming was found in 10.300 entries
3. Case (3): Stems based only on suffix combinations were found in 1600 cases, many of them incorrect, so they could not count as properly stemmed.
4. Case (4) (only a stem found) was found for 110 entries, showing some decomposer errors
5. Case (5) (not analysable) was found in 550 entries, like acronyms etc.

If only cases (1) and (2) are counted, the stemmer accuracy is about 81.7% of the single word forms. But, as the input list contains also many compounds the compounds in the test corpus were decomposed and their parts were stemmed, and evaluated according to the 5 cases. The final results are given in Fig. 5.33.

This result for the high-frequency words seems to be acceptable given the fact that the test corpus was a different domain, and that the errors of the decomposer in this domain also lead to errors in the stemmer.

	Lemmata / types	in %	tokens	in %
Total words	58.400		39.52 m	
correct analysis	51.400	88,00%	36,47 m	91,64%
incorrect analysis	7.030	12,00%	3.05 m	8,36%

Fig. 5.33 Stemmer evaluation

5.5.7 Clustering

Once a corpus has been stemmed it is a plausible continuation to group all words sharing the same stem into clusters, and create a static resource which can accessed at query expansion time by a simple look-up. For a given word, the resource would return all words which are related by the same stem.

To create such a resource a wordlist was used created from Wikipedia as this is the most domain-independent resource at present. The Wiki dump contained 497 Mio tokens and 10 Mio types. All types with frequency > 14 were kept, normalised and lemmatised. The resulting list was compared to the stemmer lexicon, and 33 K tokens were retained.

The resulting list was processed with the stemmer, the correct stemmings (class 1 (free morphemes) and 2 (correct stemmings)) were kept. Clusters were built around the identified stems. From those clusters, about 9500 had only one member, they were eliminated as they would not have an effect in expanding a search query. 8100 stems formed clusters with more than one term.

A manual inspection of all clusters with 5 or more members was performed in order to identify semantic incompatibilities in such clusters. Clusters with incompatible members of were split, and two/three stems were assigned to them. Examples are: '*pilot*' versus '*pilot* | *ieren*', '*harmon* | *ie*' + '*harmon* | *isch*' versus '*harmon* | *ium*', '*konserv* | *e*' + '*konserv* | *ieren*' versus '*konserv* | *ativ*', '*konservat* | *iv*' + '*konservat* | *ivismus*' versus '*konservat* | *orium*'. As for the prefixes, the ones which did not change the semantics of the base form too much were added ('*lachen*' versus '*aus* | *lachen*', '*tippen*' versus '*ab* | *tippen*).[66]

As a result, the number of stems increased ('*pilot1*'–'*pilot2*' etc.) but the semantic consistency of the clusters grew substantially. In turn, the resulting clusters are semantically homogeneous, and can safely be used for query expansion in retrieval contexts. The result can be used as a static resource, for the 33 K words to be enriched by semantically related terms. An example is given in Fig. 5.34.

In order to evaluate the quality of the stemmer results, proper retrieval tests would be required, with measures of recall and precision of (query-expanded) search requests over a defined document set. However this would go beyond a stemmer component. But looking at the results of the stem-based clusters of search terms a certain confidence that the results would improve would be justified.

As already mentioned, stemming is an attempt to create semantic correspondences between terms based on morphological similarities. It could be shown that such an attempt can be successful to a certain extent.

However there are obvious limitations, beyond the difficulties to filter the non-matching terms from the set of terms containing the same stem. One such limitation relates to polysemic words like $_{\text{en}}$ '*cell*' (monastery) versus '*cell*' (biology)

---

[66] Of course there is a area of uncertainty in these semantic differentiations.

stem	lemmata
fall	fall, fallen, fallweise, rückfall, rückfällig
fäll	fällen, fällung
fällig	allfällig, fällig, fälligkeit
falsifiz	falsifikat, falsifikation, falsifizierbar, falsifizierbarkeit, falsifizieren, falsifizierung
falt	faltbar, falte, falten, faltig, faltung, fälteln
falz	falz, falzen
famili	familie, familiär
famul	famulatur, famulus
fanat	fan, fanatiker, fanatisch, fanatisieren, fanatismus
fang	fang, fangen, fänger
fantas	fantasia, fantasie
fantast	fantast, fantastisch
farb	farbe, farben, farbig, farbige, farbigkeit, farblich, farblos, farblosigkeit, färbbar, färben, färber, färberei, färbung
färb	anfärbbar, anfärben, anfärbung, durchfärben
farm	farm, farmer, farmerin
fasan	fasan, fasanerie
fasch	fasche, faschine
faschis	faschismus, faschist, faschistin, faschistisch, faschistoid, fascho
fasel	faseln, gefasel
faser	faser, faserig, fasern, faserung
fass	fass, fässchen
fassen	fassbar, fassen
fassung	fassung, fassungslos, fassungslosigkeit
fasz	faszie, faszikel
faszin	faszination, faszinieren, faszinosum
fatal	fatal, fatalerweise, fatalismus, fatalist, fatalistisch, fatalität

Fig. 5.34 Example of stemmer clusters

versus '*cell*' (battery), or $_{\text{de}}$ '*Kiefer*' (tree) versus '*Kiefer*' (body part), de '*Steuer*' (tax) versus '*Steuer*' (steering wheel). Different techniques, like word sense disambiguation, will be required in such cases, they are presented in Sect. 7.2.

In a context of *multilingual* text analysis, stemming is a side-line component. Stems cannot be translated, only words can be, so target language equivalents for source language stem are not available. However they could be used in query expansion, linking related words (and their translations) to a given search term via a common stem.

References

Abu-Chacra, F. (2018). Arabic, An Essential Grammar. *Routledge*

AleAhmad, A., Amiri, H., Darrudi, E., Rahgozar, M., & Oroumchian, F. (2009). Hamshahri: A Standard Persian Text Collection. *Knowledge Based Systems 22,5, 2009*

Aleksić, V., & Thurmair, Gr. (2007). TSTD Projektbericht Türkische Morphologie. *unpublished.*

Amin-Madani, S., & Lutz, D. (1972). Persische Grammatik. *Heidelberg (Julius Groos).*

Arabsorkhi, M., & Shamsfard, M. (2006). Unsupervised Discovery of Persian Morphemes. Proc. 11th Conf. of European Chapter of the Association for Computational Linguistics (EACL), Trento, Italy.

Attia, M., Pecina, P., Tounsi, L., Toral, A. & van Genabith, J. (2011). Lexical Profiling for Arabic. *Proc. 2nd Conf. of Electronic Lexicography in the 21st century (eLex) 2011, pp. 23–33.*

Bacanlı, E., Flynn, D., & Pounder, A. (2020). Vowel Harmony and Other Morphological Processes in Turkish. *in: Proc. of the Workshop on Turkic and Languages in Contact with Turkic 5. 26–40.* https://doi.org/10.3765/ptu.v5i1.4782

Baroni, M., Bernardini, S., Ferraresi, A., & Zanchetta, E. (2009). The WaCky wide web: a collection of very large linguistically processed web-crawled corpora. *In: Language Resources and Evaluation 43.3, pp. 209–226.*

Beesley, K. & Karttunen, L. (2003). Finite-State Morphology. *CSLI Center for the Study of Language and Information.*

Beesley, K.R. (1998). Arabic Morphology Using Only Finite-State Operations. Computational Approaches to Semitic Languages.

Bel, N. (2010). Platform for Automatic, Normalized Annotation and Cost-Effective Acquisition of Language Resources for Human Language Technologies: PANACEA. *Proc. Intern. Conf. of the Spanish Society For Natural Language Processing (SEPLN), Valencia.*

Bergenholtz, H., & Mugdan, J. (1979). Einführung in die Morphologie. Stuttgart (Kohlhammer).

Brants, Th. (2000). TnT - A Statistical Part-of-Speech Tagger. *Proc. of the sixth Applied natural language processing Conf., ANLP-2000, Seattle, USA, pp. 224–231.*

Brockelmann, C. (1992). Arabische Grammatik. Langenscheidt.

Çetinoğlu, Ö., & Oflazer, K. (2006). Morphology-Syntax Interface for Turkish LFG. *Proc. 21st Int. Conf. on Computational Linguistics (COLING) and 44th Annual Meeting of the Association for Computational Linguistics (ACL), Sydney, Australia, pp. 153–160.*

Clements, G.N., & Sezer, E. (1982). Vowel and Consonant Disharmony in Turkish. *In: H. van der Hulst and Norval Smith (eds.) The structure of phonological representations, II. Dordrecht: Foris.*

Çöltekin, Ç. (2010). A Freely Available Morphological Analyzer for Turkish. *Proc. LREC Malta.*

Çöltekin, Ç. (2014). A set of open source tools for turkish natural language processing. *Proc. 9th Int. Conf. on Language Resources and Evaluation (LREC 2014), , Reykjavik, Iceland, pp. 1079–1086.*

Csató, E., & Nathan, D. (2003). The Turkish Suffix Dictionary. http://www.dnathan.com/language/turkish/tsd/

Eisenberg, P. (1986). Grundriss der deutschen Grammatik. Stuttgart (Metzler).

Eisenberg, P. (2004). Grundriss der deutschen Grammatik vol. 2: Der Satz. Stuttgart (Metzler).

Engel, U. (1988). Deutsche Grammatik. Heidelberg (Groos).

Ersen-Rasch, M.I. (2004). Türkische Grammatik für Anfänger und Fortgeschrittene. *München (2. Aufl.) (Hueber).*

Eryiğit, G., & Adalı, E. (2004). An affix stripping Morphological Analyser for Turkish. *Proc. Int. Association of Science and Technology for Development (IASTED) Intern. Conf. Artificial Intelligence and Applications, Innsbruck.*

Eryiğit, G., & Oflazer, K. (2006). Statistical Dependency Parsing of Turkish, Proc. 11th Conf. of the European Chapter of the Association for Computational Linguistics (EACL), Trento, Italy.

Family, N. (2006). Explorations of Semantic Space: The Case of Light Verb Constructions in Persian. *Diss. Paris*

Feili, H., & Ghassem-Sani, Gh. (2004). An Application of Lexicalized Grammars in English-Persian Translation. Proc. 16th Conf. of European Coordinating Committee for Artificial Intelligence (ECCAI), Valencia, Spain.

Fersøe, H., (2004). Validation Manual for Lexica. *ELRA*.

Fischer, H.G., ed. (1982). Information Retrieval und natürliche Sprache. Integrierte Verarbeitung von Daten und Texten im Modell CONDOR. *K.G.Saur*.

Fischer, W. (2002). Grammatik des Klassischen Arabisch. Harrassowitz.

Forbes, D. (1861). A Grammar of the Persian Language. New Delhi (Cosmo).

Gao, J., Li, M., Wu, A., & Huang, Ch.N. (2005). Chinese Word Segmentation and Named Eintity Recognisiton: A Pragmatic Approach. *Computational Linguistics, 31(4): pp. 531–574.*

Göksel, A., & Kerslake, C. (2011). Turkish, An Essential Grammar. *Routledge*

Görz, G., & Paulus, D. (1988). A Finite State Approach to German Verb Morphology. *Proc. 12th Int. Conf. on Computational Linguistics (COLING), Budapest 1988*

Grebe, P. (1973). Duden Grammatik der deutschen Gegenwartssprache. Dudenverlag, Mannheim

Güngör, T., & Kuru, S. (1993). Representation of Turkish morphology in ATN. *Proc. of Second Symposium on Artificial Intelligence and Artificial Neural Networks, 1993,6, 92–104.*

Güngör, T. (2003). Lexical And Morphological Statistics For Turkish. *Int. XII. Turkish Symposium on Artificial Intelligence and Neural Networks - TAINN 2003.*

Gurusamy, V., & Kannan, S. (2014). Preprocessing Techniques for Text Mining. https://www.researchgate.net/publication/273127322_Preprocessing_Techniques_for_Text_Mining

Hakkani-Tür, D., & Oflazer, K. (2002). Statistical Morphological Disambiguation for Agglutinative Languages. *Computers and the Humanities 36.*

Harder, E., & Schimmel, A. (2007). Arabische Sprachlehre. Julius Groos, Tübingen.

Harman, D. (1991). How effective is suffixing? *Journal of the American Society for Information Science, 42, 7-15*

Haywood, J.A., & Nahmad H.M. (1965). A new Arabic Grammar of the written language. Lund Humphries.

Heidolph, K.E., Flämig, W., & Motsch, W., eds. (1984). Grundzüge einer deutschen Grammatik. Berlin (Akademie-Verlag).

Hjelm, H., & Schwarz, Chr. (2005). LiSa-morphological analysis for information retrieval. *Proc. Nordic Conf. on Computational Linguistics (NODALIDA) 2005, Joensuu, Finland, pp. 65–70.*

Hulden, M. (2009). Foma: a Finite-State Compiler and Library. *In Proc. of the Demonstrations Session at EACL 2009, pages 29–32, Athens, Greece.*

Jackendoff, R. (1977). X-bar Syntax: A Study of Phrase Structure. *MIT Press, Cambridge (MA)*

Jadidinejad, A.H., Mahmoudi, F., & Dehdari, J. (2009). Evaluation of Perstem: A Simple and Efficient Stemming Algorithm for Persian. *Proc. 8th Workshop of the Cross-Language Evaluation Forum (CLEF) 2009.*

Jurafsky, D. & Martin, J.H. (2009). Speech and Language Processing. *Pearson Int. Edition (2nd ed.).*

Kamali, D., Janfada, B., Shenasa, M.E., & Bidgoli, B.M. (2022). Evaluating Persian Tokenizers. https://doi.org/10.1145/1122445.1122456

Kayadelen, T., Ozturel, A., & Bohnet, B. (2020). A Gold Standard Dependency Treebank for Turkish. *Proc. 12th Int. Conf. on Language Resources and Evaluation (LREC), Marseille, France, pp. 5156–5163.*

Koehn, Ph. (2010). Statistical Machine Translation. *Cambridge University Press.*

Koehn, Ph. (2005). Europarl: A parallel corpus for statistical machine translation. *Proc. Machine Translation Summit X, Phuket, Thailand.*

Koskenniemi, K. (1984). A General Computational Model for Word-Form Recognition and Production. *Proc. 10th Int. Conf. on Computational Linguistics (COLING) and 22nd Annual Meeting of the Association for Computational Linguistics (ACL), Stanford, California, USA, pp. 178–181.*

Koskenniemi, K. (2018). Guessing lexicon entries using finite-state methods. Proc. 4th Int. Workshop for Computational Linguistics for Uralic Languages (IWCLUL 2018), Helsinki, Finland, pp. 59–77.

Lambton, A.K.S. (1953). Persian Grammar. Cambridge Univ. Press

Landmann, A. (2009). Türkisch, Tabellen zur Deklination und Konjugation. *Wiesbaden (Harrassowitz).*

Lewis, G. (2000). Turkish Grammar. *Oxford (2nd ed.).*

Mace, J. (2003). Persian Grammar. RoutledgeCurzon.

Madani, M.H. (1993). Elementarbuch der persischen Sprache. Ebermannstadt.

Megerdoomian, K., & Zajac, R. (2000). Processing Persian Text: Tokenization in the Shiraz Project. *Memoranda in Computer and Cognitive Science MCCS-00–322, Computing Research Laboratory New Mexico State University, April 2000.*

Megerdoomian, K. (2000a). A Computational Analysis of the Persian Noun Phrase. MCCS-00–321, Univ. New Mexico.

Megerdoomian, K. (2000b). Unification-Based Persian Morphology. Proc. 3rd Int. Conf. on Computational Linguistics and Intelligent Text Processing (CICLing), Mexico.

Megerdoomian, K. (2004). Finite-State Morphological Analysis of Persian. Proc. 20th Int. Conf. on Computational linguistics (COLING '04).

Moïnfar, D. (1978). Grammaire du Persan. Documents de Linguistique Quantitative Nr. 34. ed. Jean-Favard.

Momenipour, F., Keyvanpour, M.M. (2015). Comparative Study of Various Persian Stemmers in the Field of Information Retrieval Journal of Information Proessing Systems.

Monacchesi, P. (1993). Object clitics and clitic climbing in Italian HPSG. *Proc. of the 6th Conf. on European chapter of the Association for Computational Linguistics (EACL), pp. 437 - 442.*

Monroe, W., Green, S., & Manning, Chr. (2014). Word Segmentation of Informal Arabic with Domain Adaptation. *Proc. 52nd Annual Meeting of the Association for Computational Linguistics (ACL), Volume 2, pp. 206–211, Baltimore, Maryland.*

Moser-Weithmann, B. (2001). Türkische Grammatik. *Hamburg (Buske).*

Müller, N., & Riemer, B. (1998). Generative Syntax der romanischen Sprachen. Tübingen (Staufenberg).

Newid, M.A., Mumm, P.A. (2007). Persisches Lesebuch. *Wiesbaden (Reichert)*

Newid, M.A., Thurmair, Gr. (2023). Wörterbuch Farsi – Dari, Dari – Farsi. *Buske.*

Oflazer, K. (1994). Two-level description of Turkish Morphology. *Literary and Linguistic Computing, 9(2): 137–148.*

Oflazer, K., Göçmen, E., & and Bozşahin, C. (1994). An outline of Turkish morphology. *Report to NATO Science.*

Oflazer, K., Say, B., Dilek, Z., Hakkani-Tür, D.Z., & Tür, G. (2003). Building a Turkish Treebank. *in: Abeille, A. ed. (2003). Building and Exploiting Syntactically-annotated Corpora, , Kluwer Academic Publishers.*

Öztürel, A., Kayadelen, T. & Demirşahin, I. (2019). A syntactically expressive morphological analyzer for Turkish. *In Proc. of the 14th Int. Conf. on Finite-State Methods and Natural Language Processing (pp. 65–75).*

Öztürk, S., Sankur, B., Güngör, T., Yilmaz, M.B., Köroğlu, B., Ağın, O., İşbilen, M., Ulaş, Ç, & Ahat, M. (2014). Türkçe Etiketli Metin Derlemi - Turkish labeled text corpus. https://doi.org/10.1109/SIU.2014.6830499

Packard, J.L. (2016). Chinese Morphology *in: Chan, S.W., ed. (2016). The Routledge Encyclopedia of the Chinese Language. Routledge, pp. 215-226*

Pecina, P., Toral, A., Way, A., Papavassiliou, V., Prokopidis, Pr., & Giagkou, M. (2011). Towards Using Web-Crawled Data for Domain Adaptation in Statistical Machine Translation. *Proc. 15th Conf. of the European Associtation for Machine Translation (EACL), pp 297–304, Leuven, Belgium.*

Penzl, H. (1955). A Grammar of Pashto. A descriptive Study of the Dialect of Kandahar, Afghanistan. American Council of Learned Societies, Washington.

Porter, M.E. (1980). An algorithm for suffix stripping. *Program, 14(3), 130-137.*

Qadi A. (1959). Calligraphers and Painters, transl. V. Minorsky, *Washington.*

QasemiZadeh, B., & Rahimi, S. (2006). Persian in Multext-East Framework. In FinTAL, p. 541--551.

Raverty, H.C. (1880). Language of the Afghans (The Pushto Manual). Star Publications New Delhi.

Reumuth, W., & Winkelmann, O. (2001). Praktische Grammatik der italienischen Sprache. Mannheim (Egert).

Reumuth, W., & Winkelmann, O. (2003). Praktische Grammatik der spanischen Sprache. Mannheim (Egert).

Reumuth, W., & Winkelmann, O. (2005). Praktische Grammatik der französischen Sprache. Mannheim (Egert).

Ring, H. (1978). PELIKAN - Ein Lernsystem für linguistische Klassifikationsalgorithmen. *Nachr. f. Dokumentation 29,6, pp. 224-226*

Russi, C. (2008) Italian Clitics: An Empirical Study *Walter de Gruyter.*

Sallam, R.M., Mousa, H.M., Hussein, M. (2016). Improving Arabic Text Categorization using Normalization and Stemming Techniques. *in: Int. Journal of Computer Applications 135(2):38-43*

Say, B., Zeyrek, D., Oflazer, K., & Özge, U. (2002). Development of a corpus and a treebank for present-day written Turkish. *Proc. 11th Int. Conf. of Turkish linguistics (ICTL), 183–192.*

Schimmel, A. (1970). Islamic Calligraphy. *Leiden, Brill.*

Seçkin, Ş. (2008). Verbtabellen Türkisch. *Stuttgart (Klett).*

Seraji, M., Megyesi, B., & Nivre, J. (2012). A Basic Language Resource Kit for Persian. *Proc. 8th Int. Conf. on Language Resources and Evaluation (LREC'12), Istanbul, Turkiye.*

Shamsfard, M., Jafari, H.S., Ilbeygi, M. (2010). STeP-1: A Set of Fundamental Tools for Persian Text Processing. *Proc. 7th Int. Conf. on Language Resources and Evaluation (LREC'10), Valletta, Malta.*

Steinberger, R., Pouliquen, B., Widiger, A., Ignat, C., Erjavec, T., Tufiş, D., & Varga, D. (2006). The JRC-Acquis: A multilingual aligned parallel corpus with 20+ languages. *Proc. 5th Int. Conf. on Language Resources and Evaluation (LREC'2006), Genova, Italy.*

Tegey, H., & Robson, B. (1996). A Reference Grammar of Pashto. Center for Applied Linguistics, Washington DC

Thurmair, Gr., and Newid, M. (2007). Term Substitution Dari-Deutsch *TSPD Project Report, unpublished.*

Thurmair, Gr., Aleksić, V., & Schwarz, Chr., (2012). Large-scale lexical analysis. Proc. 8th Int. Conf. on Language Resources and Evaluation (LREC'12), Istanbul, Turkey.

Thurmair, Gr. (1984). Linguistic Problems in Multilingual Morphological Decomposition. *Proc. 10th Int. Conf. on Computational Linguistics and 22nd Annual Meeting of the Association for Computational Linguistics, pp. 174–177, Stanford, California, USA.*

Tiedemann, J. (2009). News from OPUS – A collection of Multilingual Parallel Corpora with Tools and Interfaces. *in: Nicolov, N., Bontcheva, K., Angelova, G., & Mitkov, R., eds. (2009). Recent Advances in Natural Language Processing (vol V) (Benjamins).*

Toleu, A., Tolegen, G., & Mussabayev, R. (2022). Language-Independent Approach for Morphological Disambiguation. *Proc. of the 29th Int. Conf. on Computational Linguistics (COLING), Gyeongju, Republic of Korea, pp. 5288–5297.*

Toral, A., Pecina, P., Way, A., & Poch, M. (2011). Towards a user-friendly webservice architecture for statistical machine translation in the PANACEA project. *Proc 11th European Association for Machine Translation (EAMT), Leuven, Belgium.*

Toutanova, Kr., Klein, D., Manning, Chr., Singer, Y. (2003). Feature-Rich Part-of-Speech Tagging with a Cyclic Dependency Network. *Proc. 2003 Human Language Technology Conference of the North American Chapter of the Association for Computational Linguistics NAACL), pp. 252–259.*

Tseng. H.H., Chang, P.C., Andrew, G., Jurafsky, D., & Manning, Chr. (2005). A Conditional Random Field Word Segmenter for Sighan Bakeoff. *Proc. of the 4th SIGHAN Workshop on Chinese Language Processing.*

van Schaaik, G. (2020). The Oxford Turkish Grammar. *Oxford University Press*
Voigt, W., ed., (1989). Leitfaden der russischen Grammatik. Leipzig /VEB-Verlag.
Weinrich, H. (1993). Textgrammatik der deutschen Sprache. Mannheim (Duden-Verlag) .
Yamashita, T., & Matsumoto, Y. (2000). Language Independent Morphological Analysis. *Proc. of the 6th Conf. on Applied natural language processing (ANLC '00), pp. 232–238.*
Yuret, D., & Türe, F. (2006). Learning Morphological Disambiguation Rules for Turkish. *Proc. of the Human Language Technology Conf. of the North American Chapter of the ACL, pp. 328–334, New York, June 2006.*

Chapter 6
Special Entries

In the context of index term production, lemmatisation as described in the previous chapter is a key element. However there are two types of lemmata which deserve special attention. This is the treatment of **multiword entries** because they are not covered by the techniques applied so far, and this is the treatment of (proper) **names** as they cannot completely lexicalised by definition.

6.1 Multiword Recognition

Multiword expressions are semantic units built from constituents with internal morphosyntactic structure. Calzolari et al. (2002) define them as '*Expressions whose linguistic behaviour is not predictable from the behaviour of its component words when they occur in isolation*'.

As a multilingual indexing component deals with concepts, i.e. semantic units, it is essential to include them into the analysis chain and treat multiwords the same way as other semantic units (single words): Why should the multiwords ‚*planetary gear set*' or ‚*Salt Lake City*' differ in text analysis contexts from single words like ‚*Planetengetriebe*' ‚*diacid*' or ‚*Chattanooga*'?

Multiword treatment was always considered a '*pain in the neck*'[1] for text analysis. TINA was one of the first projects to analyse multiwords in Information retrieval (cf. Schwarz 1985). Meanwhile there is significant literature on multiword extraction, and even toolkits[2] and also Web Service offers are available.

[1] (Sag et al. 2002).

[2] For Python, see (De Araujo et al. 2011), and for Java, (Kulkarni and Finlayson 2011). For web services, cf. (Quocchi et al. 2012).

G. Thurmair, *Knowledge-Driven Multilingual Text Analysis and Transparent Information Retrieval*, Cognitive Technologies,
https://doi.org/10.1007/978-3-031-91741-7_6

The standard procedure is to identify candidates which show certain features, and apply filters to them to increase precision. Such features can be statistical as in (Tu and Roth 2011), contextual as in (Fahramand and Martins 2014), or linguistic (e.g. patterns of part-of-speech sequences). Additional features are also proposed, for instance, use dependency treebanks, cf. (Martens and Vandeghinste 2010). And of course combinations of features are also tried.

Applications of multiword recognition can be found in information retrieval,[3] sentiment analysis, cf. (Moreno-Ortiz et al. 2013), or translation, as in (Thurmair 2018).

However, the recognition of multiwords is widely neglected in text analysis workflows.[4] The reason is that the internal morphosyntactic structure of multiwords needs to be recognised in order to identify it as semantic unit. From the point of view of linguistic descriptions, multiwords belong to quite different classes,[5] which sometimes require specific processing, cf. (Vincze et al. 2011). Examples are compound terms, light verb constructions, split verbs, or named entities. The TINA analysis draws special attention to multiword analysis and their integration into a text analysis system.

6.1.1 Rationale

A closer look into resources which are usually used for multilingual indexing reveals that multiword terms take a significant part there.

a. An essential resource in indexing are **thesauri**. Many applications use thesauri, like Food Science Technology Abstracts (FSTA),[6] MeSH (Medical Subject Headings)[7] which is a controlled vocabulary thesaurus used for indexing articles for PubMedMedical, ISO9999 which is a thesaurus for assistive domain technology,[8] or Eurovoc[9] which is a thesaurus of the European Commission. Such thesauri consist to an overwhelming part of multiwords. This is illustrated in Fig. 6.1.

b. Another important source of information in indexing are **multilingual dictionaries**. They are standard means to map the denotation of terms in different languages. Such dictionaries also use multiwords to a significant extent. In commercially used dictionaries, multiwords contribute to the overall number of entries to about 33%; details are given in Fig. 6.2.

[3] Examples are (Schwarz 1985) or (Acosta et al. 2011).

[4] An example would be the Shiraz system for Persian analysis (cf. (Amtrup et al., 2000)) or (Sagot and Walther 2010)).

[5] They reach from compound structures to complex verb constructions, cf. (Sag et al. 2002), (Hurskainen 2008).

[6] FSTA: https://www.ifis.org/fsta.

[7] MeSH: https://www.nlm.nih.gov/mesh.

[8] iso9999: https://www.iso.org/standard/72464.html.

[9] Eurovoc: https://eur-lex.europa.eu/browse/eurovoc.html.

Eurovoc	es	de	el	en	ro	pt	it	fr	hu
total terms	6798	6798	6798	6798	6798	6798	6798	6798	6798
single word terms	1600	4288	1679	1606	1544	1646	1623	1680	3072
multiword terms	5198	2510	5119	5192	5254	5152	5175	5118	3726
% single word terms	23,5	63,1	24,7	23,6	22,7	24,2	23,9	24,7	45,2
% multiword terms	76,5	36,9	75,3	76,4	77,3	75,8	76,1	75,3	54,8

Fig. 6.1 Multiword terms in the Eurovoc thesaurus

dictionary	ar	de	en	es	fa	fr	it	ps	pt	tr	zh
total entries (k)	52.4	175.2	268.6	140.1	55.8	122.8	95.2	47.8	76.9	63.9	--
singlewords	23.2	158.7	111.4	71.5	21.7	64.4	65.7	26.4	52.0	44.1	
multiwords	29.2	45.4	157.1	68.6	34.1	58.3	29.4	21.3	24.9	19.7	
% multiwords	55.7	9.4	58.5	49	61.1	47.5	31	44.7	32.4	31	--

Fig. 6.2 Multiword terms in bilingual dictionaries

The table shows that the amount of multiwords is more than one third in all languages except German. German shows rather few multiwords as its specific compounding mechanism puts many compounds into the singleword section.

While the recognition of multiwords is essential even in monolingual text understanding, it is indispensable in multilingual indexing. As soon as translations are involved, single word terms and multiword terms interact quite freely in translations, as the examples in Fig. 6.3 show.

c. In the area of multilingual **ontologies** and multilingual wordnets like EuroWordNet,[10] terms are mapped to abstract representations (like concept IDs). Even there a mix of single and multiword terms is required to cover all nodes of the respective ontology.

d. As a result, it can be said that text analysis systems which do not provide multiword analysis have a reduced coverage of at least 30% percent right from the beginning, and index files which do not contain multiwords (like nearly all of them) are inadequate in multilingual contexts.

It should be noted that the task at hand is not to find hitherto unknown new multiwords in texts: Unknown multilingual term extraction is a heavily discussed topic, e.g. (Bonin et al. 2010). In contrary, the focus here is on identifying multiwords which are already known, and stored in resources like dictionaries.

6.1.2 Description

The **definition of a multiword** as a semantic units is a difficult task not only in computational environments but also in ordinary lexicography or translation. There

[10] EuroWordNet (Vossen 2004; Bond et al. 2020). These representations are discussed in Chap. 8.

single -> multiword		
de-fr	ermuntern	remonter le moral
it-en	allacciare	hook up
en-es	lawsuit	acción legal
fa-en	(emšab) امشب	today evening
multi -> singleword		
it-en	bicchiere colmo	bumper
de-fr	prozentualer Anteil	pourcentage
en-es	admonish strongly	urgir
fa-en	(osul-e hesabdari) اصول حسابداری	accounting
en-de	nuclear power plant	Kernkraftwerk
en-de	planetary gear set	Planetengetriebe
single -> singleword		
it-en	immatricolazione	enrolment
de-fr	Ermüdung	fatigue
en-es	cupidity	envidia
fa-en	(tanāb) طناب	rope
multi -> multiword		
it-en	squadra avversaria	opposing team
en-es	spreadsheet element	elemento de hoja de cálculo
fa-en	(ehtiāt kardan) احتیاط کردن	be careful

Fig. 6.3 Mix of single and multiword terms in translation

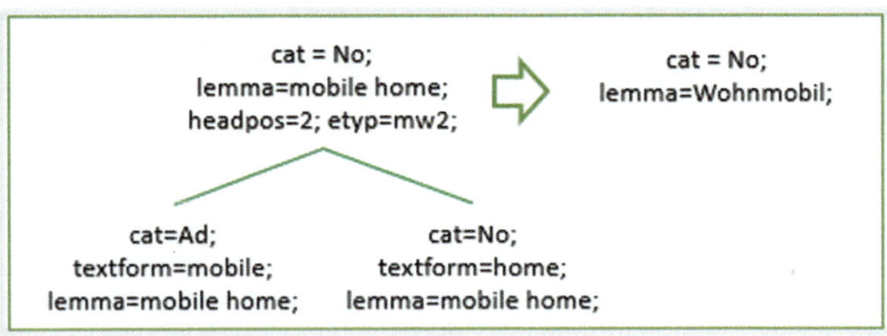

Fig. 6.4 multiword translation example

is no clear borderline between collocations (composing its semantics from its parts) and multiwords (with unique semantics). A pragmatic means to identify multiwords is to rely on previous work as documented in bilingual dictionaries: Different translations for a multiword, not predictable by its parts, indicate a multiword semantic unit. Of course this is a very coarse approach, however viable; it would take the multiword entries as given, and accept the distinctions which were defined by previous lexicographers: So, treating a multiword entry as a semantic unit finds its justification in a bilingual dictionary entry which provides a non-compositional translation of a source term. In the example in Fig. 6.4, treating $_{en}$ '*mobile home*' as a multiword may be justified by a specific German translation ($_{de}$ '*Wohnmobil*').

6.1.2.1 Properties of MWs

The following observations define the scope of analysis which multiwords should undergo:

- All nodes involved in the multiword structure need a **POS** tag. This holds for the entry as a whole but also for all of its parts.
  Most multiwords are nouns. Multiword verbs are also rather frequent, esp. in languages with a significant share of support verbs like Persian, cf. (Family 2006). Multiword adjectives and adverbs are less frequent but more difficult to describe, and the POS of the head of the entry often differs from the POS of the whole multiword (e.g. adverbials of the type $_{en}$ '*in addition*' $_{es}$ '*en realidad*[11]') which is categorised as adverb(ial) although the head is a preposition; to mention also $_{de}$ '*letzten Endes*', '*mittleren Alters*' which are adverbials although their head is a noun).[12]
- The multiword and all its parts receive a common **lemma**. E.g. the token '*into*' denoting '*take into account*' will have '*into*' as its textform but '*take into account*' as its lemma, differing from a textform '*into*' with the lemma '*into*' to be used as lemma for other cases. This way multiword entries can be unified by their lemma feature in lexical analysis.
- Multiwords can be inflected. **Inflection** is attached to the head element of the multiword. Therefore the analysis must know what the head of the multiword is. The multiword children behave differently in respect to inflection: As a rule of thumb, tokens before a preposition may inflect and need to be lemmatised while tokens after it stay constant and can be identified on a textform basis. Entries like $_{es}$ '*tasa crítica de rentabilidad*' or $_{de}$ '*ungeschütztes Speichern pro Einheit*', '*volle Puffergröße für Datenpakete*' etc. would have to lemmatise the adjectives but could succeed with a simple comparison of the textforms of the PP elements following the multiword head. Some TINA analysers therefore assign a special part of speech MWP (multiword part) to such invariant post-preposition tokens.
- Except for the verbs, most multiword parts are in **adjacent** position. Just support verb elements, cf. (Deksne et al. 2008), or separable verb particles are set into split position, while for nouns, adjectives and adverbs adjacency is the standard.[13]
- Some languages use special means to **mark** (nominal) multiwords. Turkish compounds attach clitic possessive pronouns to the head as a multiword marker: $_{tr}$ '*cankurtaran arabası*' ($_{en}$ *ambulance*), '*okul servisi*' ($_{en}$ *school bus*). This also holds for larger compounds, like $_{tr}$ '*demiryolu taşımacılığı derneği*'

---

[11] Translated into German by the adverb ‚*eigentlich* ‘.

[12] Sometimes the head must be chosen arbitrarily, like in fr ‚*parce que*‘.

[13] Of course there are exceptions, e.g. ‚$_{fr}$'*centre de recherche*' allows for adjective insertion, e.g. '*centre Belge de recherche*'.

($_{en}$ *'rail transport association'*, $_{de}$ *'Eisenbahnverkehrsverband'*). In Farsi, compound elements are linked by the insertion of an '*-e-*'[14] called ezafe, e.g.' سلیمان تخت ('*Taht-e-Suleiman*').[15] The formation of the lemma of such a multiword needs to *generate* such elements if they cannot be found in textforms.

- Lemmatisation of multiwords sometimes includes **generation** aspects, mainly in cases where the non-head parts of multiwords are inflected, like $_{de}$ *'regulärer Ausdruck'* ($_{en}$ *'regular expression'*) from the lemmata *'regulär'* and *'Ausdruck'*, $_{es}$ *'país asociado'* from *'país'* and *'asociar'*. This holds mainly for adjectives and participles: If the lemmata should be built, such elements need to be inflected.
- As all multiword parts have a **POS assignment**, the multiword as a whole can be described by a **POSlist** giving the POS tags of its elements, like *'Ad,No'* as BTAG sequence for *'mobile home'*, *'Ad,No,No'* for *'nuclear power plant'*. Such POS patterns are far more complex than the well-investigated *'N de N'* and *'N à N'* descriptions, in e.g. (Daille 1994): The TINA dictionaries show hundreds of such patterns for a given language. This is discussed below. The size of the patterns of course depends on the tagset used: Descriptions with a basic tagset (BTAG) have much less patterns than those built from standard tagsets (STAG).[16]

Phenomena like the ones just mentioned must be taken into account when a component for multiword analysis is to be designed.

6.1.2.2 Lexical Representation of MWs

Multiwords need to follow the design specifications of lexical entries in the dictionary, but on the other hand they require additional information.

1. Standard Annotations

The lexical representation of multiwords should be in line with the one of other entries. So, like all other dictionary entries, also multiword entries are defined by a lemma, a part-of-speech tag, and a semantic description. They also need morphosyntactic annotations, mainly inherited from the head node, and they need syntactic and semantic annotations, which are specific for the multiword as a whole, especially selection restrictions and verb frames. For instance, a verb like $_{en}$ *'take'* does not take a *that*-clause whereas a multiword like *'take into account'* does.

---

[14] *-i-* in Dari.

[15] Ezafe is only written in case of tokens ending in a vowel, otherwise it is just spoken. For ezafe, cf. (Kahnemuyipour 2014).

[16] Inflected forms are also participles, which causes a difficulty as they act like adjectives in multiword constructions but their BTag is Vb.

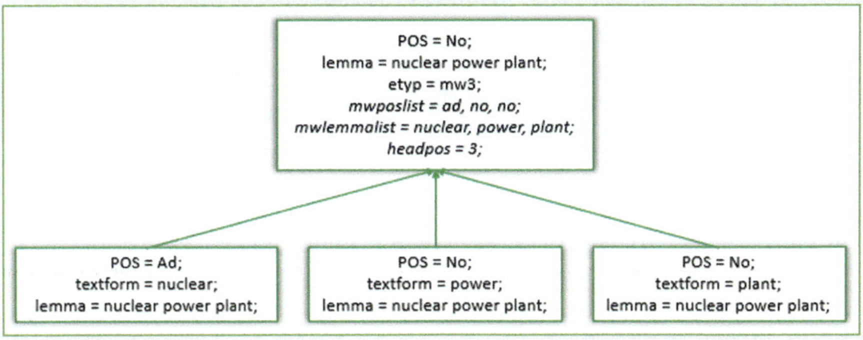

Fig. 6.5 Multiword term representation

2. Extension

Compared to single words, multiwords need additional annotations. Unlike single words they have an internal syntactic structure, with a node representing the multiword, dominating the nodes for its parts (cf. Fig. 6.5). The children nodes will be needed to identify the multiwords in text flows. They need to be annotated for this purpose, and also their internal structure must be described.

An option is to include tree structures, however not in the form of full syntactic trees,[17] which would lead to a representation beyond the basic typed feature structure used in TINA dictionaries. The alternative is to consider only what is absolutely necessary to identify the multiword parts: This is:

- their POS and their lemma: The list of POS/lemma defines the tree nodes. The tree layout is always a flat list of nodes dominated by the multiword node (cf. Fig. 6.5). The sequence of nodes needs not be described as it is invariant. So the multiword tree can be described by the POS and lemma of the child nodes.
- Beyond this, only the position of the head must be given as it is not always derivable from the tree structure. (cf. below).

So the minimal information for a multiword is the head position and the POS and lemma of its parts. This information can be defined in the feature value setting as follows, using list-valued features:

- *Head position*. This is simply an integer pointing to the head position. For a MWE like '*take into account*', this would be '*1*'.[18] The head of the MWE must be known, as it inflects.

[17]As proposed e.g. by (Grégoire 2009).

[18]*Position* was preferred to *lemma* as identification means, as lemmata can be duplicated, e.g. in many Turkish multiwords like '*koşa koşa*', cf. (Oflazer et al. 2004).

- *List of parts of speech* of the multiword (*mwposlist*), based on the BTag tagset: '*take into account*' would have the sequence '*Vb,Ap,No*', '*Salt Lake City*' would be coded as '*No,No,No*'.
- *List of the lemmata* forming the multiword (*mwlemmalist*): '*take into account*' would have the sequence '*take,into,account*'

There is also the *Entry type* feature like in the definition of all dictionary entries. This feature basically states whether an entry is a single word or a multiword. Only entries with *entrytype=multiword* require multiword features.

6.1.2.3 Structure of Multiwords

The following analysis is based on multiword entries from existing dictionaries of several languages. The dictionaries are bilingual, the data are 'real world' data, provided by lexicographers and translators, and collected in such dictionaries. Of course they contain translation and annotation errors, but for the purposes here they are considered as given. The source side entries of the different bilingual dictionaries were used for analysis. Details of the dictionaries are given in Chap. 9.

The challenge is that an analysis rule must be written for each POS pattern of the multiwords, so the interest is to reduce the number of these patterns as much as possible.

1. Length

The first parameter to be considered is the **length** of the multiword pattern, i.e. the number of their elements. Most patterns consist of two, three, and four elements, some five, very few more than five. A compromise between coverage and analysis effort has to be found as the analysis is the more effort-bearing the longer the multiword is. For instance, if there is a seven-part multiword but no rule for it, then this multiword will not be detected in analysis. The pattern length is language-dependent: In Romance languages multiwords tend to be longer than e.g. in German or Turkish. Figure 6.6 shows the relation between multiword length and coverage

Language	ar	de	en	es	fa	fr	it	ps	pt	tr
sw	44,3	90,6	41,5	51	38,9	52,5	69	55,3	67,6	69
+ mw2	86,4	96,7	82,1	65,6	78,9	71,1	83,1	77,4	82	95,6
+ mw3	91,3	99,2	94,6	83,7	90,2	89,2	95,9	89	96,1	99,7
+ mw4	91,6	99,8	98,4	91,4	95	95,3	98,6	94,8	98,7	100
+ mw5	92,1	100	99,5	96	97,2	97,9	99,6	98,1	99,7	100

Fig. 6.6 Multiword pattern coverage: length

	en				de				es			
	total patterns stag	total patterns btag	pattern reduction %	coverage increase %	total patterns stag	total patterns btag	pattern reduction %	coverage increase in %	total patterns stag	total patterns btag	pattern reduction %	coverage increase in %
mw2	667	272	59,2	4,3	263	138	47,5	8	313	179	42,8	1,4
mw3	1053	654	37,9	3,6	430	272	36,7	9	550	388	29,5	2,1
mw4	1114	797	28,5	7,5	292	221	24,3	9,3	632	472	25,3	7,8
mw5	741	605	18,4	8,8	209	123	41,1	6,7	542	418	22,9	8,7
total	3575	2328	34,9	4,4	1194	754	36,9	8,3	2037	1457	28,5	3,1

Fig. 6.7 STAG versus BTAG: pattern reduction and coverage increase

(of dictionary entries). While German covers 99.2% of the dictionary entries if just 3-part multiwords are analysed, for Spanish it would be required to analyse at least 5-part multiwords. In English, after 4-part multiwords no significant increase would be gained. Overall it can be seen that the analysis of multiwords of length 2 and 3 shows good coverage for most languages except for Spanish and French. An extension of multiword length coverage to 4 or 5 is recommendable for those languages. For other languages, esp. Turkish, treatment of multiwords up to length 3 is sufficient. Of course the coverage can be improved if the analysis effort is increased.

2. POS Patterns

The second parameter is the **number of multiword patterns** per length. This is relevant as a rule must be written for each of those POS patterns.

1. Multiword patterns are defined as a **sequence of POS tags**. To identify such patterns, all multiword entries of the dictionaries were lemmatised and POS-assigned using the Lexical Analysis tool described in the previous chapters. Patterns with homograph elements were disambiguated manually. On this basis the frequency of the single patterns was computed.

This task was performed for three languages – English, Spanish (as example of a Romance language) and German (because of its compounding situation) – and four POS tags, namely the content words, i.e. noun, verb, adjective, and adverb.[19]

It can be seen that the number of patterns is huge, both per language and per POS pattern length. It is more than 3000 in English, more than 2000 in Spanish, and more than 1000 in German (cf. Fig. 6.7). This fact has also been observed by (Nissim et al. 2014), criticising the narrow approach in many research papers. It can also be seen that the spread in POS assignments (i.e. the number of patterns) is larger for the longer multiwords, and also for the 'weaker' categories (adjectives, adverbs). Nouns and verbs have less and more systematic patterns.

[19] Multiwords also exist in function word classes, especially in prepositions (en '*up to*', multipart conjunctions ($_{\mathrm{fr}}$ '*parce que*') and numerals ($_{\mathrm{tr}}$ '*on bir defa*'). But content word classes are more relevant for indexing.

Some reasons for this fact are:

- coding errors in the dictionaries used, misassignments of POS categories. Also some codings depend on the context (homographies between prepositions and conjunctions)
- systematic errors in coding principles, resulting from the fact that they are mostly used for translations: Many adjectival entries e.g. in German are translated into prepositional expressions e.g. in English (e.g. $_{de}$ '*vorsichtig*' –> $_{en}$ '*with care*'), so the English side sometimes codes these expressions as adjectives although they never are.
- errors in the creation of lemmata. The dictionaries often include preposition information in the lemma; however this is not part of the lemma but an indication of verb frame subcategorisation ($_{en}$ '*rely on*', $_{de}$ '*sich verlassen auf*').

But even if such phenomena are taken into account, there are many patterns left, too many to be able to code all of them by rules.

2. So the question is how the number of patterns could be reduced. There are two options to do this:

a. **Switch of the tagsets**. Instead of using the STAGs as done in the Lexical Analysis, the basic tagset (BTAG) could be used. The idea is that less tags would lead to less patterns. Multiwords like $_{en}$ '*average industrial standard*' (AdAt AdAt NoC) and '*German industrial standard*' ('AdNa AdAt NoC) could be covered by the same pattern (Ad, Ad, No). Conversion of STAGs into BTAGs are quite easy to perform as the STAG's first bigram is the BTAG.

There would be only few cases where such a reduction of STAGs to BTAGs would lead to incorrect results. An example would be the *DtDd* $_{en}$ '*the red flag*' which is a multiword whereas *DtDi* '*a red flag*' would not be, cf. sentences like $_{en}$ '*to brandish the red card'* ($_{de}$ '*die Rote Karte zücken*') versus $_{en}$ '*to brandish a red flag*' ($_{de}$ '*eine rote Karte zücken*') which would be analysed compositionally. However there are only very few examples of this kind.

A switch of the tagset would be justified by the fact that tags in multiword recognition operate as a filter to validate a multiword entry. They need not provide all possible information for lexical analysis, and for such a filter a BTAG entry could be sufficient.

Figure 6.7 shows the number of multiword patterns when BTAGs are used instead of STAGs. It can be seen that the multiword patterns in the three languages are reduced by about 30% compared to the STAG patterns. The reduction is over-proportional for the short multiwords: Pattern reductions for mw2 and mw3-type multiwords rank above the average reduction, which is important as these multiwords are classes with most members. In particular nouns and verbs profit from the tagset switch. At the same time the coverage of these patterns slightly increases.

b. However, even the remaining multiwords exceed the possibilities of a rule-based description of multiwords, as each pattern would need to be described by a single rule. Therefore an additional option could be examined which would consist of **collapsing all BTAGs of invariant parts** of multiwords into just one tag

	most frequent English Multiword Patterns (BTAG)						
No		1	2	3	4	5	6
mw2	No0 ->	No,No	Ad,No	Vb,No	No,Vb	Ad,Vb	
mw3	No0 ->	No,Ap,No	No,No,No	Ad,No,No	Vb,No,No	No,Vb,No	
mw4	No0 ->	No,Ap,Dt,No	No,Ap,No,No	Ad,No,No,No	No,Ap,Ad,No	No,No,No,No	Ad,No,Ap,No
Vb							
mw2	Vb0 ->	Vb,Pt	Vb,Ad	Vb,Vb	Vb,No	Vb,Av	
mw3	Vb0 ->	Vb,Dt,No	Vb,Ap,No	Vb,Pn,No	Vb,Co,Vb	Vb,Pt,Ap	Vb,Ad,No
mw4	Vb0 ->	Vb,Ap,Dt,No	Vb,Dt,Ad,No	Vb,Ap,Ad,No	Vb,Ap,No,No	Vb,Ap,Ap,No	Vb,Pt,Dt,No
Ad							
mw2	Ad0 ->	Ad,Ad	Ad,Vb	Av,Ad	No,Ad	Ap,No	Ad,No
mw3	Ad0 ->	Ad,Ap,No	Vb,Ap,No	Ap,Dt,No	Ad,Co,Vb	Ap,Ad,No	Ad,Co,Ad
mw4	Ad0 ->	Ad,Ap,Dt,No	Vb,Ap,Dt,No	Ad,Ap,Ad,No	Ap,Dt,Ad,No	Vb,Ap,No,No	
Av							
mw2	Av0 ->	Ap,No	Av,Av	Ad,No	Vb,Pt	Av,Ad	Ap,Av
mw3	Av0 ->	Ap,Dt,No	Ap,Ad,No	Vb,Ap,No	Ap,No,No	Ad,Ap,No	Co,Dt,No
mw4	Av0 ->	Ap,Dt,Ad,No	Ap,No,Ap,No	Ap,Dt,Vb,No	Ap,Dt,No,No	Ad,Ap,Dt,No	Vb,Ap,Dt,No

Fig. 6.8 Top frequent English multiword patterns (BTAG) per pattern length

called '**MWP**' ('multiword part'). MWPs would not change in the various contexts which the multiword could enter. English multiwords like '*letter of intent*' (No,Ap,No) would be described as (No,MwP,MwP). Such an approach would work for post-head prepositional phrases in most languages, and would cover most of the adverbial multiwords like '*per minute*' (ApPr,No), '*roaring drunk*' (VbFG,VbFF), '*single priced*' 'AdAv,VbFP), and their Spanish and German counterparts. They all would enter multiword recognition as 'MwP,MwP'.

However, even if these options are used, the fact should be considered that not all multiwords can be identified as not all of them can be covered with a rule describing their pattern.

c. As a result, a compromise must be found between the **coverage** of the multiword analysis, and the effort to analyse the multiword patterns: The more coverage is intended the more computational effort must be spent.

Such a decision would be based on the frequency of the single patterns. Therefore the top 5 most frequent patterns (in BTAG) were identified for each multiword length and each POS tag; an example of the English patterns is given in Fig. 6.8.[20]

The critical information now is which portion of multiword entries in the dictionaries would be covered by these top POS patterns, and whether or not a restriction to those top patterns (and the rules to be written for them) would analyse an acceptable amount of dictionary entries. This is important as entries not covered by one of the patterns would not be analysed.

[20]The BTag Vb in the patterns for nouns and adjectives refers to participles. To treat them as Ad would be more intuitive; however in indexing there would be a difference.

	lemma no	top5	cov. %	lemma vb	top5	cov. %	lemma ad	top5	cov. %	lemma av	top5	cov. %	lemma all tags	top5	cov. %
en-BTAG															
mw2	83747	80002	95,5	6710	6180	92,1	2423	1347	55,6	2220	1090	49,1	95100	88619	93,2
mw3	19387	14456	74,6	2164	1411	65,2	1007	602	59,8	1898	913	48,1	24456	17382	71,1
mw4	5425	2583	47,6	959	464	48,4	346	170	49,1	815	367	45,0	7545	3584	47,5
mw5	1485	424	28,6	229	80	34,9	120	34	28,3	228	82	36,0	2062	620	30,1
mw>5	762		0,0	105		0,0	35		0,0	105		0,0	1007	0	0,0
total tag	110806	97465	88,0	10167	8135	80,0	3931	2153	54,8	5266	2452	46,6	130170	110205	84,7
de-BTAG															
mw2	6763	5928	87,7	1415	1350	95,4	810	668	82,5	1094	795	72,7	10082	8741	86,7
mw3	2346	1610	68,6	749	642	85,7	266	158	59,4	626	406	64,9	3987	2816	70,6
mw4	690	328	47,5	229	181	79,0	68	40	58,8	117	56	47,9	1104	605	54,8
mw5	157	47	29,9	30	19	63,3	9	9	100,0	29	11	37,9	225	86	38,2
mw>5	67		0,0	2			1		0,0	5		0,0		0	
total tag	10023	7913	78,9	2425	2192	90,4	1154	875	75,8	1871	1268	67,8	15473	12248	79,2
es-BTAG															
mw2	16763	15937	95,1	826	762	92,3	1024	752	73,4	1581	1089	68,9	20194	18540	91,8
mw3	22242	20527	92,3	940	780	83,0	600	399	66,5	1412	922	65,3	25194	22628	89,8
mw4	9670	7344	75,9	307	182	59,3	254	136	53,5	481	233	48,4	10712	7895	73,7
mw5	6139	4604	75,0	104	66	63,5	66	26	39,4	156	62	39,7	6465	4758	73,6
mw>5	5399		0,0	50			30		0,0	66			5545		0,0
total tag	60213	48412	80,4	2227	1790	80,4	1974	1313	66,5	3696	2306	62,4	68110	53821	79,0

Fig. 6.9 Coverage of top 5 rules per POS and multiword length (en, de, es)

The result is shown in Fig. 6.9. It shows what the coverage related to all multiwords of a dictionary would be, if for every main tag rules for only the top 5 patterns would be provided.

The table shows that the implementation of around five POS templates per category and multiword length, i.e. 20–25 rules for each BTag, can show good results. Good coverage of about 90% can be achieved for shorter multiwords, and for nouns and verbs, both representing the bulk of multiword entries. Additional observations would be:

- The recognition rates do not significantly change (by less than 1%) in English and German, if the 5-part multiwords are dropped
- It may be more efficient, instead of implementing adjective and adverb patterns, to spend more rules on nouns and verbs.
- The decision is language specific: While German noun figures reflect the compounding possibility which makes them singlewords and leave less multiwords, Spanish data show that longer multiwords are quite frequent. English figures are a bit better than German and Spanish.

Overall, with the implementation of a manageable number of rules (about 30–40 per language) the coverage of the proposed solution is around 80%, i.e. 80% of the known multiwords can be covered with the strategy presented. Of course the coverage can be improved but only with a significant increase of effort in adding rules for additional patterns. But also the coverage of these additional patterns decreases.

6.1.3 Analysis of Multiwords

The analysis of multiwords compares nodes of the input text, as represented in the Document Model, with nodes coming from the dictionary.

The nodes on the text side are the result of Lexical Analysis. These nodes provide textform, casing (from tokenisation), POS, lemma (from lemmatisation) and possibly additional information as information elements. Dictionaries for multiwords must provide corresponding information as far as POS, lemma, and textforms are concerned. Multiword Analysis must find a match between those two sides.

Unfortunately, none of these information items are able to do the match by themselves:

- The *lemma* of multiwords may contain inflected elements ($_{\text{tr}}$ '*okul servisi*', $_{\text{de}}$ '*regulärer Ausdruck*', $_{\text{es}}$ '*país asociado*') which would not have been produced by the lemmatisation of the textforms.
- The multiword is not just a sequence of *textforms* either, as the head token may be inflected, and some more elements as well.
- The identification of the POS sequence is hampered by the fact that many elements are homographs, providing many alternative interpretations.

This means that the identification of multiwords must be a mix of these criteria. And it depends on the information provided by the dictionary, which forms the basis of the comparison.

The multiword identification will be based on a normalisation of textforms to lemmata, and a comparison of their POS annotations. Lexical analysis will provide the<textform, lemma, POS>information for each input token, and the dictionary will provide the same information for the multiword candidate, in its features *mwlemmalist* and *mwposlist*, providing the lemmata and BTAGs of its members. On this basis, a match can be tried: If the lemmata and POS information of all multiword parts match the input tokens then the multiword can be assumed to be found.

The elements of the multiwords are collected by rules which describe the POS pattern of this multiword. Each multiword pattern would be covered by a rule. As the number of rules is determined by the number of patterns it is important to reduce the number of patterns to a manageable size. This was explained above.

Multiword analysis have been implemented for both LtDeep and LtShallow systems. While the LtDeep implementation is part of a more elaborated analysis component, the LtShallow component is intended to support less resource-intensive tasks, especially query analysis.

6.1.3.1 LtDeep

The integration of a multiword analysis component requires a design of how multiwords would fit into the complete syntactic analysis.

The question therefore is which target structures the multiword analysis should produce. It has been explained above (cf. Fig. 5.25) that the design of the syntax follows an extended X-bar scheme where the X0 nodes are responsible to collect the multiword parts into a single node.

1. Analysis Design: Target Structures

As stated in the beginning a multiword is a semantic unit with internal syntactic structure. Like other semantic units (of singlewords) it would be represented in a **single node**. The result of a multiword analysis should look like the lexical insertion of a single node, to be transparent for the next analysis steps. As far as the multiword's internal structure is concerned this is irrelevant for further analysis: All relevant (morphosyntactic) information must be present at the multiword's head node. For the rest of the analysis the multiword node should look like any other node, present all obligatory features like POS, case, number etc., of course with updated onset and offset features.[21]

Like the other nodes created by lexical insertion, the multiword node would be situated on the **X0 level** in X-bar theory. Like usual, the X1 level would collect modifiers of the multiword, like $_{\text{en}}$ '*new [mobile home]*': The modification holds for the whole multiword, not for parts of it. The X2 level would be reserved for specifiers as usual. As a consequence, if multiword rules should be integrated, the left side of a multiword rule would be a X0 category, e.g. '*N0 –>Ad No*', as opposed to a rule '*N1 –>Ad N0*' which would handle a modification of the whole multiword (Fig. 6.10).

Such a design allows for a seamless integration of a multiword node into the syntactic-semantic analysis. However it assumes that the multiword parts are adjacent, which is mostly the case for non-verb constituents; therefore it needs to cover two deviational phenomena:

- Sometimes modifiers of the multiword occur inside of it, and produce **internal modification**. An example is $_{\text{fr}}$ '*centre de recherche*' versus '*centre Belge de recherche*'. Such cases are not really frequent; they could be covered by allowing slots in the multiword patterns which require certain POS tags as fillers, e.g. N0 ->No [Ad] Ap No.
- The second structure to be investigated are **discontinuous multiwords** as occur in verbs: $_{\text{en}}$ '*he <u>writes</u> it <u>down</u>*', $_{\text{de}}$ '*er <u>schreibt</u> es <u>nieder</u>*'. In both cases the PtVb is in split position to the head element of the verb. Multiword recognition here cannot be done on an adjacency basis but must be part of the complete syntactic analysis. Only once the head node of the verb is completed by the

[21] The internal structure of a multiword is relevant for retrieval in cases where head–modifier relations are exploited. These information items are therefore kept in additional features of the head node. They can also be computed once the position of the head is known. Cf. Sect. 7.3.

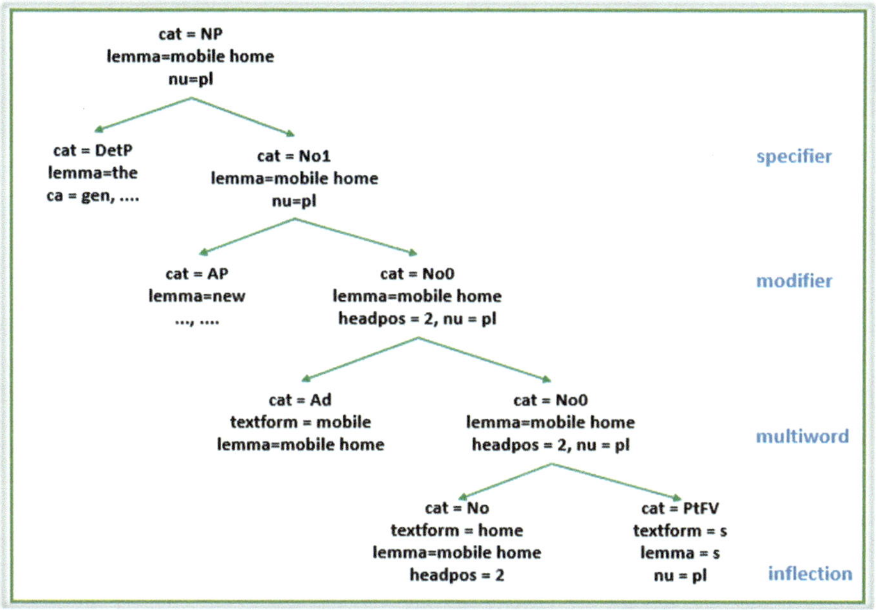

Fig. 6.10 Parse for *‚the new mobile homes'*

particle (PtVb) it would be able to act as multiword, and e.g. provide its own verb frame etc.; the particle however could not be deleted as it plays an important role in syntactic analysis: For instance it helps in analysing the German satzklammer. To mark it as void in the later collection of index terms its lemma feature could be set to zero.

2. Implementation

a. The **instantiation** of the chart first loads all single word entries with their features: There may be readings which compete to the multiword readings. In addition to all single word tokens, an instance of the multiword is created for each part of the multiword: The head (or also simply the leftmost element) instantiates the complete multiword entry, with all annotations of this entry, for the other multiword parts only textform, lemma and POS are required. An example is given in Fig. 6.11, instantiating $_{en}$ '*nuclear power plant*' and all lemmata included.

A verb like $_{en}$ '*bump off*' would instantiate two additional entries: in addition to the textform '*bump*' with lemma '*bump*' a second hypothesis would be inserted into the chart, with textform '*bump*' but lemma '*bump off*'. So, in addition to the preposition and particle '*off*' with lemma '*off*', an entry would be added with textform '*off*' and lemma '*bump off*'.

textform	lemma	(affiliation)
nuclear	nuclear	*1 singleword*
nuclear	nuclear power	*2 multiword-2 (1)*
nuclear	nuclear power plant	*3 multiword-3 (1)*
power	power	*4 singleword*
power	nuclear power	*2 multiword-2 (2)*
power	nuclear power plant	*3 multiword-3 (2)*
power	power plant	*5 multiword-2 (1)*
plant	plant	*6 singleword*
plant	nuclear power plant	*3 multiword-3 (3)*
plant	power plant	*5 multiword-2 (2)*

Fig. 6.11 Multiword instantiation of '*nuclear power plant*'

bump	Vb	lemma = bump
bump	Vb	lemma = bump_off
off	Ap	lemma = off
off	Pt	lemma = off
off	Pt	lemma = bump_off
~~*off*~~	~~*Pt*~~	~~*lemma = switch_off*~~
~~*off*~~	~~*Pt*~~	~~*lemma = turn_off*~~
bump	No	lemma = bump
~~*bump*~~	~~*No*~~	~~*lemma = bump_stop*~~
~~*bump*~~	~~*Vb*~~	~~*lemma = bump_into*~~

Fig. 6.12 Instantiation and lexicon filter

This procedure will create significant overhead as also other multiwords with '*off*' would be instantiated, like '*back/beat/bite/block/bounce/branch … off*'. Therefore at the end of chart instantiation and before the first rules are fired, a component called '*lexicon filter*' is implemented. It eliminates all entries which contain lexical elements that are not found anywhere in the input sentence. Figure 6.12 shows an example of this procedure.

b. The **rules** for the multiword follow the LtDeep rule layout as described in Chap. 2. The phrase structure part of the rule has an X0 category as its left hand side, and one of the multiword patterns on the right.

The test section has four tasks:

- It tests the multiword lemma. The lemma of all right hand side categories must unify. This is the main requirement for the multiword; the POS of all parts is already used in the multiword pattern description.
- it tests the length of the multiword using the *EntryType* feature (with values *mw2, mw3* …). There may be shorter multiwords that could lead to mis-assignments, e.g. '*nuclear power plant*' (*etyp = mw3*) could include '*power plant*' (*etyp = mw2*).

```
ID:    No0-302
PS:    No0 -> Ad  No   N0                    // the multiword pattern
Prob:  0.82
Features:
            equal 1 lemma 2 lemma 3 lemma;   // all parts must have the same lemma
            req 3 etyp mw3;                  // must be a three-part multiword
            req 3 headpos 3;                 // head is the No0 element
            nrq 3 mw true;                   // N0 must not be a multiword itself

            from 3 nu; from 3 ca; from 3 sem; // copy features from the head  node
            set 0 mw true;                   // mark it as multiword node (recursion...)
            prune 1; prune 2; prune 3;       // remove all sub-nodes
Example:  nuclear power plant(s)
Comment: nominal multiword mw3
```

Fig. 6.13 Multiword rule

- It copies all relevant features (syntactic, semantic) from the head of the multiword to the X0 node, using the value of the *headpos* feature.[22]
- Finally, optionally it prunes the subtree below the X0 category and leaves the node as if it was a single word lexical category.

An example is given in Fig. 6.13.

The number of rules depends on the number of multiword patterns to be analysed.

As mentioned the coverage of multiword identification could possibly be extended by using nonterminal categories (like *No0 –>No PP* for '*loss of sight*' instead of *No0 –>No Ap No*); however such an approach would complicate the grammar design considerably. For this reason the patterns here only use terminal categories.[23]

c. The **integration** of the multiword rules into the grammar is determined by the X0 category in relation to modifiers and specifiers. But it must also take care of the inflectional rules as the heads and also some parts of multiwords can be inflected.

Inflection rules also produce X0 nodes (e.g. *N0 –>No PtFN*: *plant+s*). They must fire first, before the multiword is collected. As a consequence the multiword patterns which contain elements that could be inflected must be adapted such that the terminal elements (Ad, No, Vb) are replaced by their X0 counterparts. Instead of a rule '*N0 –>Ad No*' for '*mobile home*' the rule would be '*N0 –>Ad N0*' to cover the inflection: The plural recognition rule fires first and produces an N0 category. Then the multiword rule would take this N0 node to create another N0 node for the multiword.

---

[22] This includes potential morphosyntactic information contributed by the inflection, as described below.

[23] With the exception of inflection.

As both types of rules deal with an X0 node care must be taken that the correct sequence is taken, and that the rule is not applied recursively. This can be achieved by using technical features to prevent recursion.[24]

As for discontinuous elements in verb analysis, the multiword handling would have to be postponed until a more complete sentence analysis is available.

3. Comparison

The multiword analysis as implemented in TINA deviates from other proposals in how a multiword component is integrated into the analysis. Several suggestions exist how to integrate multiwords in an analysis component: Its treatment has been proposed before, after, and during analysis.

a. **Multiword processing after analysis**

Early attempts in MT systems like LMT (McCord 1989) or METAL (Thurmair 1990) could only analyse single words, and used the transfer component for multiword treatment: The transfer lexicon contained entries with tests on the multiword parts, like *'plant → Kernkraftwerk' IF modified by an adjective ,nuclear ' and a noun, power*".

Later proposals use the output of the analysis component, i.e. the analysis tree, try to find a match between subtrees in source and target language, and offer subtree transformation operations in the transfer lexicon for the multiword descriptions, cf. (Deksne et al. 2008). This approach avoids weird transfer lexicon entries like '*plant → Kernkraftwerk*' as in the previous approaches above and provides the multiwords as proper units in the transfer lexicon.

However, the weakness of the attempt to treat multiwords after analysis is that their idiosyncratic properties (like specific verb frames, semantic features) are not taken into account during analysis. This fact increases the probability of incorrect parses (e.g. of wrong PP attachments), which in turn tends to produce failures in the following multiword analysis. In addition, the parse trees themselves must be expressive enough to permit multiword detection.[25]

The consequence is that multiword processing after analysis is too late; the analysis component itself must be aware of it, and use their specific annotations.[26]

---

[24] In previous contributions (Thurmair 2018) it was proposed to enlarge the X-bar scheme and introduce a special level (X1) for multiwords above the lexical X0 level. However the trees for single and multiwords would be different then, and the proposal here is also more in line with the overall X-bar theory.

[25] Indicated in (Bejček et al. 2013) where it is stated that the kind of dependency trees used in their experiments seem to abstract away from some of the information items needed for multiword identification.

[26] Wehrli et al. (2010) They also favour an "*approach in which collocations are identified in a sentence as soon as possible during the analysis of that sentence, rather than at the end of the analysis*" (Wehrli et al. 2010: 28).

b. **Multiword processing before analysis**

An alternative approach is to do multiword processing before the analysis. A special component marks multiword candidates the input string. Analysis then starts with such multiwords as single nodes in the input.

Such approaches have been proposed by (Samaridi and Markantonatou 2014), (Oflazer et al. 2004), and (Hurskainen 2008), where the ‚replace' option of Constraint Grammar rules is used. The result of the multiword treatment is a 'words-with spaces' constituent which serves as input for the parser.

However, there are drawbacks to this approach:

1. The multiword filter in itself needs a kind of formalism to identify multiwords in an input text, for instance in cases of discontinuous constituents; with the result that the input is parsed twice, the first run to identify multiwords, the second one to do the full parsing.
2. In most cases, a found multiword reading is taken as the only input for a given string, and alternative (compositional) readings are removed. However, it is only in the analysis process itself where it can be decided whether a multiword candidate is really an multiword or not, so the removal of alternative readings is premature and can lead to failures.

The conclusion is that multiword treatment must be an integral part of the analysis, and not be treated as a pre- or post-analysis component.[27] This is the way the LtDeep component operates, as described above.

4. Other Languages

Some languages need special consideration in case of multiwords.

a. As the **Chinese** script does not use spaces between words, the difference between single words and multiwords is irrelevant. No multiword component is required, the issue is already covered by the word segmentation component where of course multi-character words are quite common.

b. **Turkish**, although using Latin script, needs special attention. Basically it follows the rules outlined above. Examples for adjectives are $_{tr}$ '*ad düşük değerli*' ($_{en}$ '*inferior*'), for adverbs $_{tr}$ '*gece vakti*' ($_{en}$ '*at night*'), for verbs $_{tr}$ '*borca girmek*' ($_{en}$ '*contract debts*').[28] As many multiwords contain invariant parts, solutions using the MWP shortcut seem recommendable.

As for nominal multiwords, multiwords with adjectival modifiers like $_{tr}$ '*astronomik birim*' ($_{en}$ ‚*astronomic unit*') also follow the standard treatment; however multiwords with nominal modifiers differ in the way that their heads are marked

[27] This conclusion is also taken by (Wehrli 2014).

[28] Turkish verbs may also contain reflexive elements as in other languages, e.g. *çenesini tutmak, maskesini atmak (keep one's mouth shut, pull down one's mask).*

with obligatory possessive suffixes: $_{tr}$ '*gezi*' ($_{en}$ '*trip*') but $_{tr}$ '*vapur gezisi*' ($_{en}$ '*boat trip*'); $_{tr}$ '*üretim*' ($_{en}$ '*production*') but $_{tr}$ '*enerji üretimi*' ($_{en}$ '*energy production*'). This principle is also applied recursively to longer multiwords, e.g. $_{tr}$ '*beyanname* ' ($_{en}$ '*declaration*') versus $_{tr}$ '*vergi beyannamesi*' ($_{en}$ '*tax declaration*') versus $_{tr}$ '*gelir vergisi beyannamesi*' ($_{en}$ '*income tax declaration*').

This phenomenon could be handled by simply adding the possessive particle to the multiword rules, like.

No0	–>	No No PnPoC	// *vapur gezi si*
No0	–>	No No PnPoC No PnPoC	// *gelir vergi si beyanname si*

However if such compounds are put into plural ('*vapur gezisiler*'), the order of rules is a challenge: The inflection must be attached after the possessives. This can be achieved by adding another rule:

No0	–>	No	PtFN	// normal inflection
No0	–>	No0	PtFN	// *multiword noun inflection*

If the grammar prunes all the nodes below the N0 node then the resulting trees would follow the standard tree layout, with No0 as terminal node.

c. Multiword recognition in **Farsi** and **Pashto** need to cope with the peculiarities of the Arabic-Persian script, and would also have to analyse inflections and multiwords in one step: If inflections are split from the stem, their treatment is like the one of multiwords. In Farsi, nominal multiword parts are linked by ezafe, which occurs in writing only in postvocalic position, e.g. $_{fa}$ 'دستگاه های دولتی ($_{fa}$ '*dast.gāh.'hā-.je dow. la.'ti*', $_{en}$ '*government bodies*'). It is more efficient to analyse multiwords and inflected forms in one step, which is easy once LtDeep has been chosen as analysis technology.

Similar to the other languages, the Persian dictionary was analysed for multiword patterns (including split inflections). Human coders had used about 700 multiword patterns in this dictionary.[29] The length of multiwords was restricted to five elements, so multiword rules were created for multiwords consisting of two to five elements. Longer multiwords may exist but have not been found in the corpora. Validation tests showed that with a rule set of about 150 rules, 91.1% of all structures could be covered (cf. Fig. 6.14, split by POS[30]). This set of rules (150) is larger than the ones discussed above (trying to cover the top 5 patterns), and it shows that the coverage can be improved by adding more recognition rules. Pashto shows similar behaviour.

The table also shows that 20% of the multiword patterns detected in the Persian dictionary cover about 90% of all multiword entries. Together with the single words which are covered anyway this seems an acceptable result.

---

[29] This also includes patterns containing split inflections.

[30] FW groups all function words.

POS	no patterns	no entries	patterns covered	entries covered	coverage per POS	coverage of multiwords
Ad	135	1905	19	1538	80,7	8.5
Av	78	355	9	201	56,6	1.1
NoC	204	8587	32	7956	92,7	44.1
NoP	117	1802	25	1561	86,6	8.7
Vb	59	5040	15	4913	97,5	27.2
FW	111	353	48	285	80,7	1.6
total	704	18042	148	16454		91.2

Fig. 6.14 Coverage Persian multiwords

6.1.3.2 LtShallow

In addition to the multiword recognition in LtDeep, also a recogniser for shallow analysis was implemented. There are systems and components for which the launch of a full linguistic analyser would be too much effort, for instance query processing. The LtShallow multiword recogniser would be an alternative option in such situations.

Of course a shallow implementation loses coverage. Multiwords with non-adjacent constituents would not be found. And multiword recognition would be based on a longest match heuristics: It returns always the longest match, without considering overlaps or partial coverage.

The analysis starts from the leftmost constituent of a multiword. If a multiword can be instantiated, this will be done based on a comparison of lemma and POS (here also the BTAG is used). From the first element, the following elements are compared to the input nodes, using the *mwlemmalist* and *mwposlist* features of the multiword lexical entry. If all multiword parts are identified, the multiword is returned.

The identification of a multiword constituent leads to an update of the document model such that onset (of the first element) and offset (of the last element) are collected, the textforms are concatenated into the textform feature, and lemma and POS of the multiword are used as the values of the respective features.

6.2 Named Entity Recognition (LtNamer)

Most sections describing Lexical Analysis for different languages are based on dictionary resources, and they describe how textforms can be linked to dictionary entries. Such an approach by definition has limited success in case of names: Names usually are not part of dictionaries, because they are too numerous; if at all, then they have their own resources, like Wikipedia or DBPEDIA.[31] But as names form a significant part of the terms in indexing it is necessary to investigate them in more detail.

6.2.1 *Design*

6.2.1.1 Definition

The term 'Named Entity' was coined in the nineties in the context of the MUC competitions, and relates to Information Extraction. In linguistic description they are treated as proper nouns. A common definition could be that proper names denote a unique referent in reality, like '*William Shakespeare*' or '*Himalaya*'.

The recognition of Named Entities is important for indexing as beyond their use as an ordinary search term they can be used for all kinds of query expansions, or question answering ('*When was Henry VIII crowned?*'). Links between Named Entities, like '*located_in*', can be established which can improve the search results, for instance in contexts like Semantic Search.

Named Entities are linguistic units like other words. Their properties are described by linguistic annotations: They have a citation form (canonical form), they have a part-of-speech markup. They follow orthographic conventions (they are often capitalised). They need gender annotations ($_{de}$ '*die Schweiz*', '*das (schöne) Frankreich*', '*der Balkan*', '*die Isar*', '*der Lech*') ($_{fr}$ '*la Seine*', '*le Rhône*'). They don't pluralise or singularise respectively ($_{en}$ '*the Hebrides*' *pl.*). In Turkish they need quoted forms if inflected ($_{tr}$ '*Türkiye'den*'). Many of these annotations are idiosyncratic.

Like for other terms, there are homographs between names (like '*Barcelona*' as a city or as a region) or between common and proper nouns: '*BMW*' as a company or a car, '*fisher*' as a person name or a profession. Such cases need to be disambiguated by context analysis.

6.2.1.2 Entity Types

Named Entity recognisers always focus on special types of names. Research papers mostly use a combination of person–location–organisations types, called ENAMEX in MUC7, cf. (Grishman and Sundheim 1996). LtNamer is based on this but extends the list. The following entity types are covered in LtNamer:

[31] DBPEDIA: https://www.dbpedia.org/, Wikipedia.

- PERSONS: Person name expressions are different depending on the language the person name comes from. Sometimes gender information can be given, if indicated by the first name.
- LOCATIONS: The most important and most frequent locations are countries and cities. But there are also regions (‘*Auvergne*’, including US. state names like ‘*Indiana*’) and street names. Further location names include names like continents, rivers, or places like ‘*Kremlin*’, ‘*White House*’ etc.).
- ORGANISATIONS: They are subcategorised following legal distinctions into companies (private law) and institutions (public law). But other organisation exist as well:
  - Companies (‘*Apple*’, ‘*Mitsubishi Heavy Industries Ltd*’).
  - institutions. They comprise governmental or public bodies (‘*National Health Service*’)
  - private: They collect entities like sport clubs (‘*FC Everton*’) societies (‘*American Rifle Association*’) or other institutions (‘*Mafia*’, ‘*Greenpeace*’, ‘*Translators without Borders*’).

- TIMES. Time expressions are dates, partial dates (like ‘*November 1993*’ or ‘*November 16th*’), hour descriptions (’*15:00 CET*’). Also relational information ‘*three days ago*’, ‘*on Wednesday*’) should be accessible, in case someone can make use of it. Time expressions set their respective attributes (‘year’, ‘month’ etc.). These expressions are subcategorised into
  - dates (with unique references) (‘*Christmas 1854*’).
  - partial dates (which are ranges of date expressions) (‘*since May*’).
  - hour expressions.

- UNITS: numeric expressions. They cover
  - percent expressions.
  - prices (currency expressions: ‘*$4,000.00*’).
  - measure expressions, consisting of amounts and units (‘*40kg*’, ‘*5m/sec*’).

- COMMEANS: Communication means. This type identifies URLs, email addresses, phone and fax numbers. Subcategories are
  - phone and fax numbers.
  - email addresses.
  - URLs.

- ENTITIES: There are many other named entities not falling into one of the types above, like ‘*Honda Civic*’, ‘*the Maytag Repairman*’ or the ‘*Clear Water Act*’, ‘*European Monetary System*’ and so on. They are collected here under just ‘*Name*’. This is also the case for products like ‘*Nivea*’, ‘*BMW 317i*’, ‘*Miele 400*’. Products are difficult to distinguish from company names by formal means: ‘*Maytag*’ can be a product / brand name as well as a company. Ambiguous results may be delivered in such cases of homography. It depends on the domain application to which extent such entity types are of interest.

It should be stressed that careful tokenisation is a fundamental requirement for the analysis of TIMES and UNITS, to correctly identify the elements of expressions like '*HK$25,000.-*' or'*25,3m/se* ' Details have been presented above (Sect. 4.2).

Additional entity types have been developed in recent years, extending the coverage and the power of Named Entity recognition significantly. Domains like medical information, cf. (Tepper et al. 2012), bioinformatics as in (Bundschus et al. 2008), or stock price predictions like (Lee et al. 2014) have broadened the scope of Named Entity analysis beyond the news and OSINT domain, and applications like sentiment analysis (Krieger and Declerck 2014) also extend Named Entity techniques.

Temporal analysis and structuring has soon been found to be of special importance for Named Entity analysis. Activities like TempEval, cf. (Verhagen et al. 2009), focused on this aspect. It also was stated that a change in domain (beyond news) has significant effects on the technology needed for temporal analysis, according to (Strötgen and Gertz 2016).

6.2.1.3 Domain Dependency

From many of the above examples it can be seen that NE recognisers are **domain-specific**. Some domains need additional Named Entity types, like bioinformatics, some domains also need extensions to cover the entities in existing Named Entity types: Adaptations in the gazetteers and other resources will significantly improve the coverage of the Named Entity component: For instance a recogniser for Volkswagen should include their most important persons (e.g. the Board of Directors: '*Winterkorn*', '*Blume*'), their production sites (like '*Chattanooga*'), their product names ('*VW Golf*') and other such entities. Without such adaptation an Named Entity recogniser is hardly usable.

However, although if a domain adaptation is required in any case, basic entities should already be in the Named Entity resources to reduce the adaptation effort. Nobody would buy a system with an empty gazetteer.

6.2.1.4 Multilinguality

A component for multilingual indexing must also take care of aspects of multilinguality in Named Entity Recognition. They refer to both the names and the denotated objects themselves.

1. The **structure of names** differs in different cultures. While in Western languages person names usually consist of a first name and a last name, Modern *Chinese* names consist of a mono-symbolic (single-symbol) surname (***míng***; 名), which comes first, followed by a given name (***xìngshì***; 姓氏), which is almost always disyllabic, consisting of two characters. *Arabic* person names consist of

five parts,[32] while place names mostly start with a determiner (Kairo القاهرة, al-Kahira). *Turkish* first names can be taken from anything (*Gök, Gül, Yilmaz, Yildir, Sevgi, Özlem, …*).

So, a Named Entity recogniser is language-dependent and must be able to analyse names in this language, but taking into account that also 'foreign language' names may also occur in the texts and need to be analysed.

2. A second issue is the difference of names denoting the same object. Coreference is an issue also in multilingual Named Entity recognition, cf. (Jacquet et al. 2016). **Multilingual coreference** is easily identifiable for cases where the same name is used in different languages. While '*Santiago di Compostela*' stays mostly the same, there are many cases of divergence: $_{\text{ga}}$ '*Baile Átha Cliath*' versus $_{\text{en}}$ '*Dublin*'; $_{\text{ar}}$ '*al-Quds*' versus $_{\text{en}}$ '*Jerusalem*', $_{\text{pl}}$ '*Wrocław*' versus $_{\text{de}}$ '*Breslau*', $_{\text{tr}}$ '*Mısır*' versus $_{\text{en}}$ '*Egypt*'. Such cases require a translation component for names to be able to have identical referents in the multilingual LtConceptNet. This will be discussed in Sect. 8.2.

Named Entity recognition in translation was first used in Linguatec's Personal Translator, cf. (Thurmair 2005). NE types were used to compute selection restrictions like preposition selection. Later Named Entities were used as anchors to link multilingual documents based on the Named Entities which they contain, cf. (Barker and Gaizauskas 2012).

6.2.1.5 Lexicalisation and Named Entity Identification

Only a fraction of Named Entities can be stored in the dictionary.[33] As a consequence there will be a coordinated strategy in the treatment of names: It needs to be decided which ones should be found in the dictionary (and with which annotations), and which ones stay as 'unknowns' and require special treatment in recognition. The consequence could be two analyser components: one which is (only) dictionary-based, and another one which also cares about the entities which stay unknown after lexical analysis. According to this differentiation there is a lexicalised and a contextual version of the LtNamer, both activated depending on the availability of resources in the respective languages.

The challenge is to find a compromise between a good recognition coverage and a manageable size of the required dictionary. For each type of entity, strategies must be defined which entries should be included in the dictionary and what the resulting coverage would be.

[32] The parts are: first name (ism), patronymic name (nasab), agnomen (laqab), ancestral name (nisbah), epitheton (kunya). cf. https://en.wikipedia.org/wiki/Arabic_name.

[33] Even an ‚empty' country like Afghanistan has more than 70 K location names in the GeoNames database: (https://www.geonames.org/).

The criteria to include names in the Named Entity Dictionary are described in Sect. 9.2.3. For the entries not in the dictionary a special analysis component needs to be implemented.

Non-lexicalised Named Entities are identified using specific analysis heuristics. There are two kinds of knowledge sources: name *parts* and name *indicators*. The difference is that name parts are part of the name (substrings) while the name indicators are not. This is relevant in parsing.

Name **parts** can be endings, or name constituents:

- **person** endings: Strings ending in ‘__*itzky*’, ‘__*dottir*’, ‘__*sić*’ are most likely persons
- **location** name parts: Strings starting with ‘*Bad*’, ‘*San*’, ‘*New*’, ‘*West*’ are mostly locations (except for nationality adjectives like ‘*West German football championship*’)
- location endings: names on ‘*-brück*’, ‘*-ingen*’, ‘*-hill*’, ‘*-town*’, ‘*-abad*’, are most likely place names
- **company** name parts: Strings starting with ‘*Bank of*’ or ending in ‘*Systems*’, ‘*Technology*’ are most likely companies
- company legal forms: Strings ending in ‘*AG*’, ‘*Plc*’, ‘*GmbH*’ are mostly companies (they even turn person names into company names: ‘*Max Schmeling Ltd.*)
- company endings: unknown strings ending in ‘*-tec*’, ‘*-sys*’ are most likely companies.
- **institution** name parts: ‘*Center for*’, ‘*Committee*’, ‘*FC*’ refer to institutions.
- institution legal forms: Strings ending in’*e.V.*‘ are mostly institutions.

Name parts are also units (for amounts) and month names (for dates). For the legal forms, it is unclear whether or not they should be part of the name.[34]

Name **indicators** are not themselves part of the entity, but indicate that the string in their context belongs to a certain entity type.

- **person** indicators: titles (‘*Dr. Skiggs*’), professions (‘*Spokesman Skiggs*’)
- **location** indicators: terms like ‘*region of Skiggs*’, ‘*city of Skiggs*’, ‘*close to Skiggs*’ indicate a location; also ‘*born in Skiggs*’, ‘*located in Skiggs*’ and similar contexts
- **company** indicators: ‘*car maker Skiggs*’, ‘*Internet giant Skiggs*’.

Indicators can refer to any word in the given sentence, even verbs or punctuations.

The analyser of Named Entities must provide rules for both name parts and name indicators, and include or exclude their strings when building the lemma of the Named Entity.

---

[34] The official name would include them, the everyday name would not. In indexing they would probably be left aside.

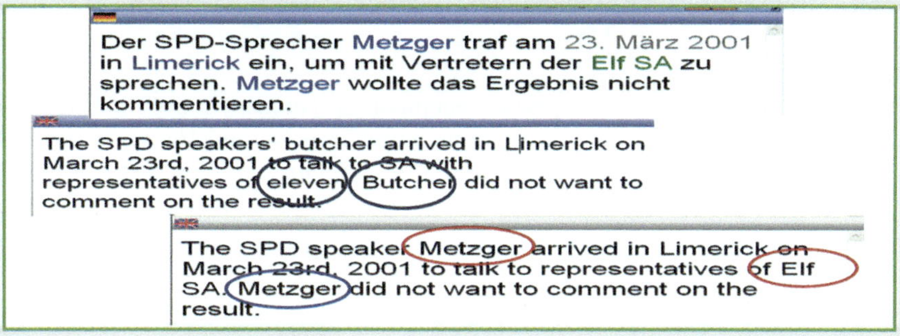

Fig. 6.15 Effect of NE on translation

6.2.1.6 Normalisation and Coreference

The simplest way of recognition consists in marking it up in a text. This is sufficient in translation as it prevents the Named Entities from being translated (cf. Fig. 6.15).

However, this is not sufficient for indexing. Named Entities may occur several times in a document, with slightly different wording. Like lemmatisation for content words, Named Entities textforms need to undergo a **normalisation** step: '*Mitsubishi Heavy Industries, Ltd.*', '*Mitsubishi Heavy*', '*Mitsubishi Heavy Industries (MHI) Group*', '*MHI*' all denote the same entity and should be brought to a common canonical form accordingly. '*Austria*' and '*Austria's*' also refer to the same entity, and so does '*Prof. Albert Skiggs*' and '*A. Skiggs*'. Therefore different occurrences of names in a text need to be **normalised** and brought to a 'canonical' form. This challenge recalls the distinction between textform and lemma in the lexical analysis of normal text.

If such normalisations are not considered, then several index terms will be created for the same object, and the overall accuracy will decrease, except if the index contains pointers from the normalised forms to the textforms, like in the lemma–textform case.

LtNamer normalises persons to pairs of<firstname, lastname>('*Peter M. Skiggs jr.*' –>'*Peter Skiggs*'), companies and institutions to the longest strings except for legal attributes ('*Handelsblatt*' versus '*Handelsblatt GmbH*' versus '*Handelsblatt GmbH & Ko KG*' –>'*Handelsblatt*') '*Siemens AG*' –>'*Siemens*'), and locations to forms without additional annotations ('*Houston, Tx*' –>'*Houston*').

In addition, for each occcurrence of an entity it must be decided if it has already been denoted by a previous one. **Coreference** resolution is essential for multilingual indexing, not just for counting the frequency of the respective entity, like for other terms, but also to collect more information from its different occurrences: A frequently used discourse strategy is to introduce a named entity with more elaborated information, and later just refer to it: '*Prof. M. Skiggs, 43, declared … Skiggs also mentioned …*'. Coreference can be an arbitrarily complex problem in itself, reading from simple repetition ('*Marx*' … '*Marx*') to variations ('*Marx*'–'*K.Marx*'–'*Karl Marx*') and even to paraphrases ('*the author of 'Das Kapital*'').

Coreference is also an important element in Named Entity linking, where the entities are linked to known entities stored in a Named Entity Database, cf. (McNamee et al. 2010); and in slot filling, i.e. finding annotations to entities in texts, and creating full annotated records of entities in an effort to populate knowledge bases. Challenges for slot filling lie in linguistic analysis. cf. (Min and Grishman 2012). Research of this kind is carried out in the context e.g. of the Text Analytics Conference (TAC 2009ff[35]).However, such extended information extraction tasks are not a priori required for Multilingual Indexing.[36]

6.2.1.7 NE Versions

Named Entity recognition is a monolingual task. Each language needs to implement it according to its own conditions (name parts, name indicators etc.). Depending on the available resources, the LtNamer components are different in the different languages:

- **Contextual** recognisers are available for Englisch, German, Spanish, Italian, and French
- **Lexicalised** recognisers of LtNamer identify only Named Entities which are stored in the lexicon (as far as persons, locations, and institutions are involved); such versions were built for Turkish, Farsi and Bulgarian.[37]

The difference lies in the analysis scope: lexicalised versions support only the recognition of names that are stored in the lexicon, while full versions also recognise non-lexicalised names based on contextual indicators.

6.2.2 Analysis

6.2.2.1 Embedding

From the beginning of Research in Named Entity recognition always contained research in processing chains. Approaches range from GATE (one of the earliest frameworks[38]) and UIMA-based systems (cf. (Bank and Schierle 2012)) to flows based on web services as e.g. explored in PANACEA, cf. (Bel et al. 2012). Such

---

[35] https://tac.nist.gov//.

[36] However many interesting aspects of Query Answering could be deteced there, for instance reasoning over place names (cities in regione, regions in countries etc.).

[37] They were developed in a project called MULTISENSOR, cf. (Vrochidis et al., 2015).

[38] (Cunningham 2000; Cunningham et al. 2002).

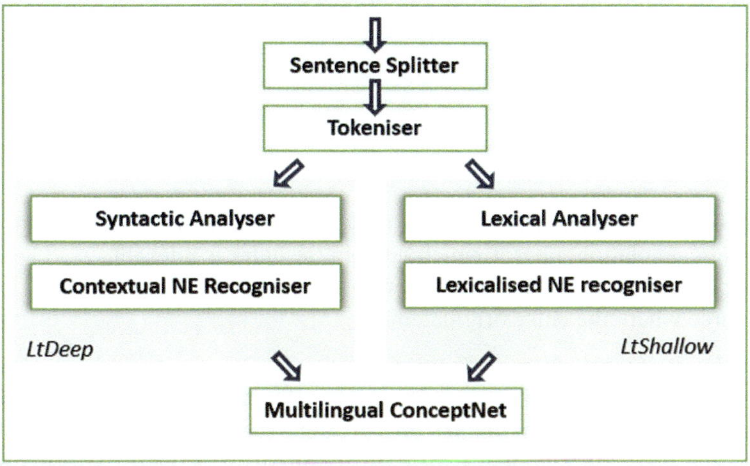

Fig. 6.16 LtNamer versions

approaches require significant integration effort; in addition the quality of some components is a crucial factor for LtNamer, in particular the tokeniser needs a very careful implementation.

The analysis part of the LtNamer has first been implemented on a shallow basis, with a set of finite state automata. Although there are examples of such an approach, e.g. (Neumann et al. 2000), it was abandoned for the LTNamer as the results were not convincing. Therefore it was decided to use the LtDeep framework for a rule-based LtNamer implementation.

The implementation uses the LtSentenceSplitter and LtTokeniser components. In particular the tokeniser is important for recognition of units, currencies, amounts etc. The contextual version of LtNamer uses the LtDeep formalism to write its rules. As it was required to have a stand-alone application of the LtNamer in several projects special output options have been provided.

The lexical version of LtNamer uses the LtShallow processing chain, with a lexicon lookup component in Lexical Analysis (Fig. 6.16).

6.2.2.2 Resources

1. Named Entity Gazetteer (NEG)

Named Entity recognisers always had a lexical component, called gazetteers. This tradition is continued in LtNamer. The lexical resources contain names, name indicators and other material required for Named Entity analysis.

A lexical entry in the Named Entity Gazetteer (NEG) is fully compatible with the Basic Lemma Dictionary (BLD): It consists of the following information elements:

- the textform of the entry (including inflected forms),
- its 'lemma', i.e. its canonical form,
- a part-of-speech tag,
- its semantic description,
- an entry type (the gazetteer contains also multiwords).

In addition an entry contains.

- Named-Entity-related features, esp. '*nesubtype*' and '*nefunction*', explained below,
- domain annotation, i.e. for which domain the entry is relevant,
- other features like gender, casing, or position within a Named Entity,
- a source where the entry originated.

As for the **content** of Named Entity Gazetteer (NEG) it should be noted that the Named Entity Gazetteer is a *lexicon*, containing mostly names, while the Basic Lemma Dictionary is a *dictionary* and contains most content words, or function words. The NEG only contains material required for Named Entity recognition, mainly names but also a few other entries like prepositions, determiners etc., which are needed for Named Entity analysis; other function word forms (like rare prepositions, pronouns, auxiliary verbs etc.) are not part of the NEG. How the Named Entity Gazetteer is collected and composed is described in Sect. 9.2.3.

The Named Entity Gazetteer uses a special **tagset**, designed for the specific requirements of the Named Entity recognition, just like the LtDecomposer has its own tagset designed for decomposition requirements (cf. Sect. 5.2.4). The tags are a mixture of the basic (BTAG) and the standard (STAG) tagset for non-named entities,[39]and the Standard Tagset (STAG) for Named Entities (with tags like *NoPI*, *NoPC* etc.). The Named Entity tags of course are a subset of the STAGs; however many of the STAGs are not needed due to the lack of lexical entries for them. In turn, some STAGs are added needed for indicators of the different NE types. The Named Entity Tags are given in Fig. 6.17. Together with the tags for unknowns (UNK, unk[40]) and for multiword parts (MWP) they form the terminal POS set of the NE grammar.

Beyond the tags, the Named Entity Gazetteer entries contain additional features. Subcategorisation information for the names is collected in the '*nesubtype*' feature, and subcategorisation for the indicators is given in the '*nefunction*' feature (values being '<ne>-part' or '<ne>-indicator'). In addition, features like gender, case, position are available.

2. NE Default list

It has been mentioned that some out-of-vocabulary words could provide information on their Named Entity type by just looking at their string shape. Examples are:

---

[39]The reason for this is that homographs (e.g. between VbFP and VbFF: '*come*') will not be disambiguated by the NE grammar, and would lead to 2 readings therefore. Therefore the basic tagset will do for the non-NE-relevant parts of speech.

[40]The difference between'unk' and 'UNK' is that UNK covers uppercase or capitalised tokens which could be candidates for NE analysis, while unk does not.

STAGs		NoPr-tags	(Proper Nouns)	indicators	
Ad	adjective	NoPF	first name	NoPCI	company indicator
AdNA	adjective-nationality	NoPP	last name	NoPII	institution indicator
ApPr	preposition	NoPs	(full) person name	NoPLI	location indicator
DtDf	definite determiner	NoPL	place name	NoPPI	person indicator
MWP	multiword part	NoPC	company name		
Nm	numeral	NoPI	institution name	Unk	unknown, poss. NEpart
NmCa	cardinal number	NoPB	brand name	unk	unknown no NEpart
NmOr	ordinal number	NoPE	entity name (misc)		
NoCo	common noun	NoPT	time name		
Pc	punctuation	NoPU	unit name		
Vb	verb				

Fig. 6.17 Tagset for the named entity recognition

- words ending in '*_itzky*', '*_dottir*', '*_sić*' '_sson' would most likely be person names.
- words ending in '*_tec*', '*_sys*' would be company names.
- words ending '*_ingen*', '*_hill*', '*_town*', '*_abad*' would be location names.

To exploit such heuristics a special defaulter was written for NE recognition, based on the technology explained in Sect. 5.2.5. An unknown string like '*Skiggs*' is categorised as '*possible person*' OR '*possible place*' OR '*possible company*' when it is sent to the defaulter. Before an UNKnown gets a POS assignment the defaulter is consulted to narrow it down, if possible.[41]

6.2.2.3 Analysis

The Grammar is implemented in LtDeep, described in Sect. 2.2.1.

The parser is a probabilistic Active Chart parser like in (Winograd 1983; Charniak 1993). The grammar has the task to identify Named Entities and normalise their candidates into the same entity. It also includes coreference analysis[42] and maintains a special resource called '*referents*' during analysis to store coreference candidates, to be consulted during parsing.

1. Grammar-Design

The grammar **starts** out from a list of terminal POS tags. Tokens which are not relevant for NE analysis are marked as 'txt'. The other nodes undergo grammatical analysis.

[41] The resulting defaulter, however, has turned out to be too unspecific in many situations (as it defaults *any* ending, not just the clear cases), and a better strategy is to just collect specific endings in a list of 'clear cases'.

[42] Strategies for coreference are described e.g. in (Gaizauskas and Humphreys 2000; Azzam et al. 1998).

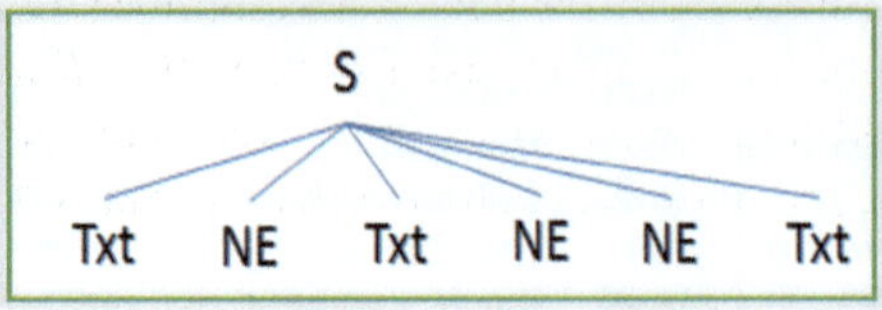

Fig. 6.18 NE target structures

NE	->	NoPF NoPP	// person with first and last name: 'Bob Smith'
NE	->	NoPL	// place name: 'Burundi'
NE	->	NoPL MWP	// multiword place: 'Hong Kong'
NE	->	NoPC NoPCI	// name extensions: 'LR International' + 'Ltd.'

Fig. 6.19 Rules for lexicalised version

The **target** of the grammar is a sentence structure which consists of a tree collecting all nodes, both Named Entity nodes and Txt nodes in a flat structure below an S node, as shown in Fig. 6.18, by recursively using two rules: S –>S NE and S –>S Txt.[43] All intermediate S nodes are deleted transformationally. The NE nodes provide all annotations on types and subtypes required for coreference analysis and output.

Building NE nodes follows two lines:

1. For **lexicalised NEs** there is a special rule set which collects the elements of the NE. Rules are like shown in Fig. 6.19.

Multiwords are analysed as described in Sect. 6.1: The first element contains the NE-POS, the other elements are categorised as MWP and recursively collected, with the value of the lemma feature used for unification.

2. **Contextual NEs** (i.e. non-lexicalised ones) start out from an unknown (Unk) element. Nonterminal categories prefixed by '*POSS_*' are used as seed categories. The grammar then tries to identify the NE type by recursively looking for name parts (*cat=<ne>I; nefunction=<ne>-part;*) and name indicators (*cat=<ne>I; nefunction=<ne>-indic;*). While NE-parts are part of the Canonical form of this NE, the indicators are not. The more such indicators for a given NE are found the more likely is the categorisation as NE with this subtype.

Figure 6.20 gives the nonterminal categories used. The design of an NE node parsing is shown in Fig. 6.21.

Both lexicalised and contextual versions of LtNamer use the same rules for some entity types. This holds for TIMES and UNITS, as well as for communication means. They do not contain unknown elements and are analysed by rules. These rules rely on a proper tokenisation of the input string.

[43] The recursion is initiated by two rules (S–>Txt and S–>NE), both with a left-corner constraint.

PossPersF	possible firstname
PossPersL	possible lastname
PossPers	possible person name
PossLoc	possible location name
PossComp	possible company name
CompLegal	legal form of a company (GmbH & Co OHG)
PossInst	possible institution
PossUnit	possile unit (km/h); forms amounts with NmParts
PossEnt	possible entity
NE	Named Entity; Top category of NEs
Nm	Numeral (from NmCa or NmOr)
NmPart	numeric part of an amount (3, 3 to 4, about 5)
S	terminal symbol of the grammar
TxT	all non-NE nodes

Fig. 6.20 Contextual NEs nonterminal categories

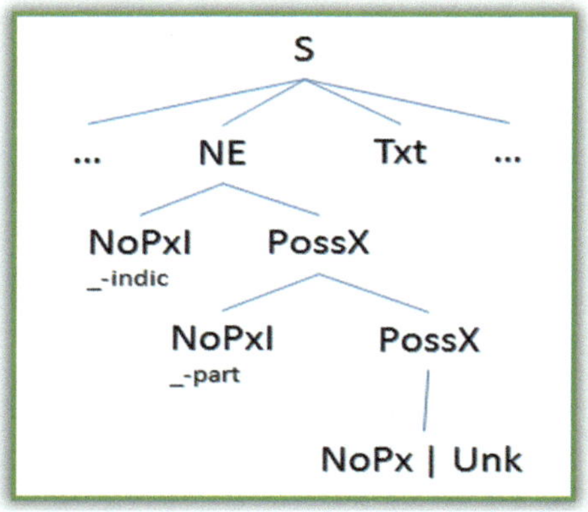

Fig. 6.21 Contextual NE: syntax trees

The grammar rules are organised in groups, one group per entity type. **Person** rules use lexical elements like first names as person-indicators. Special rules cover Chinese person names. **Location** rules use cardinal directions (like '*west-*') as indicators. **Company** rules use legal forms as indicators, and also contain rules for converting e.g. person into company names ('*Mick Skiggs*' –>'*Mick Skiggs Ltd.*'). **Unit** rules first collect the numeric parts ('*between 30 and 50*') and then add units, which are all lexicalised. **Time** rules do not use intermediate node but fire one rule per pattern, both for dates and clock times. However also partial dates must be covered ('*summer 1934*'). **Communication** rules look for indicators like '*http://*' or '*xxxx.co.uk*', characters like '@'; sequences of digits for phone and fax numbers need an indicator like '*phone*' or '*fax*'. Entities of type **Misc** need to be lexicalised to be recognised, they are mainly domain-dependent (like '*MayTag Repairman*').

An example is given in Fig. 6.22. Rule *PossNE-101* turns a non-lowercase Unknown into a possible NE. With *PossComp-315* a company indicator ('*Bank*') in pre-position is identified which is part of a company name; the whole construction is then a candidate for a company name. This is already sufficient for a legal

Rule ID	Phrase Structure	Features	Example	Comment
PossNE-101	PossNE -> Unk	*req 1 casing cp, uc; set 0 netype entity; set 0 src unk;*	We know { Skiggs }	*fallback if defaulter returns no entity type*
PossComp-315	PossComp -> NoPCI ApPr PossNE	*req 1 casing cp, uc; req 1 nesub comp-part; req 1 position pre; set 0 necand comp;*	We know the {Bank of [Skiggs]}	*found a company part: good candidate*
PossComp-320	PossComp -> PossNE NoPCI	*req 2 nesub comp-part; req 2 position post; set 0 necand comp;*	We know { [Skiggs] Energy }. We know { [Skiggs Resin] Energy }	*(NEs could be multiwords)*
PossComp-340	PossComp -> NoPCI ApPr NoPL	*req 1 nesub comp-part; req 1 position pre; set 0 necand comp;*	We know { Bank of Paraguay }	*places as part of companies*
NE-351	NE -> PossComp	*req 1 necand comp; nrq 1 necand pers,loc,inst; set 0 netype organisation; set 0 nesubtype company;*	We know { Matsushita }	*only for 'proven' candidates (necand =comp;)*
NE-352	NE -> NoPCI PossComp	*req 1 nesub comp-indic; req 1 position pre; set 0 netype organisation; set 0 nesubtype company;*	We know { retailer [Skiggs] }	*non-'proven', need further indicators (in pre or post position)*
NE-363	NE -> PossComp NoPCI	*req 2 nesub comp-indic; req 2 position post; set 0 netype organisation; set 0 nesubtype company;*	the { [Bank of Skiggs] board }. the { [Skiggs] management }.	
S-103	S -> S NE	*flatl*		*Recursion. delete intermediate S tree*

Fig. 6.22 NE rules

Named Entity, and built by *NE-351*. Additional evidence can be collected by further company indicators like NE-363, for '*Bank of Skiggs board*'. The resulting Named Entity is added to the S node by rule *S-103*.

2. Scoring

Scoring is used to compare NE candidates with each other and with non-NEs. The scoring mechanism is simple, based on the following heuristics:

- NE candidates should be preferred to textual nodes (Unk versus PossNE).
- lexicalised NEs should be preferred to non-lexicalised ones.
- longer NEs would be preferred to shorter ones.
- NE candidates with more indicators should be preferred to ones with fewer indicators.

In case of several competing S nodes the one with the higher overall score is preferred.

6.2.2.4 NE Collection

1. NE Construction and Normalisation

To generate the output of the LtNamer all NE nodes in a sentence are collected. But the NE grammar has the peculiarity that the **span** covered by the grammar nodes does not match the span of the Named Entities. All NE nodes using NE indicators must decide which elements are part of the NE, and which ones stay outside.

The example in Fig. 6.23 shows that the company-indicating verb (NoPCI) is dominated by the NE node but the verb, although also an indicator, is not part of the NE. Therefore a feature on the NE node must be set which states which constituents belong to the span (cf. Fig. 6.25).

When constructing the basic features of the NE–canonical form, textform, POS, onset, offset–only the elements which really belong to the NE should be addressed. By default NE *indicators* are not included in those feature settings while NE *parts* are.

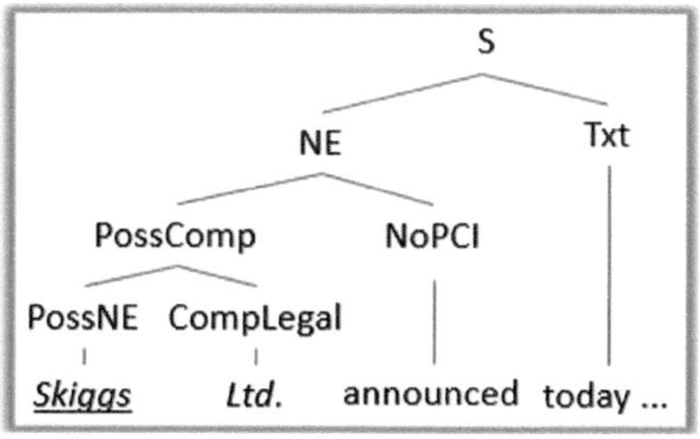

Fig. 6.23 NE construction

<p> ... </p>	Persons	<m%> ... </m%>	Units: percent
<l> ... </l>	Locations	<mc> ... <mc>	Units: currency
<oc> ... </oc>	Organisations: companies	<mu> ... </mu>	Units: amounts
<oi> ... </oi>	Organisations: institutions	<cu> ... </cu>	Communication means: URLs
<op> ... </op>	Organisations: private	<ce> ... </ce>	Communication means: Email-Addresses
<td> ... </td>	Time: Dates (full)	<cf> ... </cf>	Communication means: Phone / Fax numbers
<tp> ... </tp>	Time: Dates (partial)	<e> ... </e>	Entities: other entities (Reports, laws, IBAN)
<th> ... </th>	Time: Hours		

Fig. 6.24 NE markups

```
ID: PossComp-320
PS: PossComp -> PoSSNE  NoPCI
Features:   set 0 span 1_2;
            concatb 0 lemma 1 lemma 2 lemma;
Example: [ Skiggs Energy ]
```

```
ID: NE-357
PS: PossComp -> PoSSComp CompLegal
Features:   set 0 span 1_2;
            from 1 lemma;
Example: [ Skiggs ] Ltd.
```

Fig. 6.25 Span and lemma

This difference is also important in cases where the NEs have to be enclosed by markups: The markups must not contain elements which are needed to identify a NE but are not part of it. To place the markups correctly the NE subtree is traversed recursively down to the left and right element of the span using onset and offset information, and the markups are placed there then.

The markups used are shown in Fig. 6.24. They match the NE types and subtypes.[44]

Of course inserting markups changes the onsets and offsets of the input elements.

The **canonical form** of the NE as defined in the *lemma* feature comprises the NE kernel elements and all<ne>-parts. In such cases the textforms and lemmata of the NE elements are concatenated (using the *concat* and *concatb* operators in LtDeep).

The left example in Fig. 6.25 shows that both rules take both constituents as part of the span (in the phrase structure part), however the left rule builds the concatenated lemma (using the concatb operator) while the right example builds the lemma without the legal annotation (statement '*from 1 lemma*').

2. Coreference

It is not sufficient for an indexing component to just mark the NEs in a sentence. The final task of the NE recogniser is to prepare a list of all names which have been found in the document as index terms. So the results of the sentence-based

[44] The subtypes of LOCATIONS are not further specified here.

analysis must be collected on the document level. For this purpose a resource called **referents list** is used during parsing. It stores all NEs found in the document, together with frequency information.

Each new NE is inspected and compared to the referents list on the basis of their canonical form and annotations. Shorter forms are subsumed under longer ones (‘*Siemens*’ versus ‘*Siemens Gamesa*’).[45] Other annotations are also inspected, to prevent the system from building just one NE for ‘*Paul Skiggs*’ and ‘*Peter Skiggs*’. Also the position in the text of each occurrence of an NE is marked (onset and offset).

This approach follows the well-known discourse strategy already mentioned which observes that new entities are introduced in the discourse by using a more elaborate description (like ‘*the car manufacturer Skiggs Ltd. announced today* ....’), and later mentions just refer to them, without repeating the detailed annotations (‘*Skiggs said* ...’).[46]

The name instances sharing the same canonical form may have different textforms. Therefore for each instance of an entity a list of its occurrences is kept, containing its position and its textform.

3. NAMES in the Document Model

The referents list is used as source for the creation of the LtNamer output. In TINA the output is integrated into the Document Model. As described above (Sect. 2.3.1), the Document Model contains two footer sections, one called ‘NAMES’ for the Named Entities, and one called ‘TERMS’ for the concepts.

LtNamer collects all NEs of a document in the referents list, and presents it as an appendix to the Document Model. Under the heading ‘NAMES’ it adds all found NEs and their instances, grouped by their nametypes. Each NE instance is described by four elements:

- The canonical name of the element (the ‘lemma’),
- the number of occurrences in this document,
- the attributes of this entity type,
- the matches to identify it in the input.

The footer section of Fig. 2.14 above is repeated here (Fig. 6.26) for the NAMES part.

The *atts* section specifies the attributes presented here. The attributes differ depending on the nametype to which they are linked. The main attribute is the ‘nesubtype’ feature which subcategorises the nametype. The features presented

[45] In addition to the canonical forms also the single parts of multiword NEs are kept. For non-relevant name parts like ‘*di*’, ‘*and*’ etc. a special stoplist is created.

[46] Cf. the example in Fig. 6.15 above.

NAMES	::=	(nametype, name*)*
nametype	::=	PERSONS| LOCATIONS | ORGANISATIONS | TIMES |UNITS | COMMEANS | ENTITIES
name	::=	namedef, frequency, atts*, matches+
namedef	::=	lemma, POS, sem
lemma	::=	<string>
POS	::=	< a Btag >
sem	::=	<string>
frequency	::=	<integer>
atts	::=	(feature, value+)+
matches	::=	textform, onset, offset

Fig. 6.26 NAMES in DocModel footer

Nametype	Attributes				Values of nesubtype
PERSONS	firstname=	lastname=	gender=		
LOCATIONS	nesubtype=				country|region|city|street|river|mountain
	located_in=				
ORGANISATIONS	nesubtype=				companies|institutions|private
	legalform=	privtype=			
TIMES	nesubtype=				fulldate|partialdate|hours
	year=	month=	day=	hours=	
UNITS	nesubtype=				percent|currency|amounts
	amount=	currency=	unit=		
COMMEANS	nesubtype=				url|email|phonfax
ENTITIES	entity=				

Fig. 6.27 NE attributes in DocModel footer

in the respective nametypes are shown in Fig. 6.27, including the values of the 'nesubtype' attributes:

- For **persons**, the attributes are '*firstname*', '*lastname*' and '*gender*'.
- For **locations**, the attributes are '*sem*', values being '*country*', '*region*', '*city*', '*street*', '*river*' etc., and '*located_in*', giving a locational hierarchy (*picadilly circus-located_in london-located_in UK* etc.).
- For **institutions** in **organisations**, there is a 'privtype' attribute, values being '*governmental_body*', *football_club*, *religious_institution*, *criminal_institution* ('*Mafia*'), among others.
- **Companies** could specify their *legal_form* attribute.
- **Dates** provide year, month, day, time as attributes, or parts thereof.
- **Units** give the amount ('23.5') and the measurement units ('*km*').

The *matches* section in the NE template allows the system to identify the NE in the input text. Matches contain the following attributes:

- textform: The textform of this entity. It can differ from its (canonical) name: While the NE name could show '*Siemens*' the textform may have been '*Siemens AG*' in one instance, but maybe '*Siemens*' in another instance.

- onset and
- offset: Start and end of the NE.

This way an NE can be uniquely described in a given document.

In applications where only Named Entity is required the NAMES structure can be used in stand-alone mode: One application of the LtNamer is a system called *LtAnonymizer, cf.* (Thurmair 2016), which is used to anonymise documents. It protects private NEs (like names, prices, locations) by replacing them by variable NEs (like '*person_1, person_2*' etc.). So the documents could be sent to external services like Translation Service providers without having to fear information loss. The returned/translated document is later re-anonymised by re-inserting the proper NEs.

In the context of Multilingual Indexing, the LtNamer's task is to fill int the NAMES section of the Document Model.

6.2.3 *Test and Evaluation, Domain Adaptation*

6.2.3.1 Tests

1. The importance of the **distinction between terms and Named Entities** has been shown e.g. in the area of translation where terms need to be translated but names often do not (cf. the example in Fig. 6.15). In particular unresolved homographies between person names and occupational titles (like '*smith*', '*carpenter*', '*fisher*') lead to erroneous translations. Tests in the MT context have shown an increase in translation quality for sentences containing proper names by about 30% on average when NE recognition is activated.[47] The main improvements were:

- no erroneous translations of person names, esp. in coreference positions, where no person indicators are available ('*Pat Fisher was nominated for president. <u>Fisher</u> said …*').
- better contextual adaptations (correct preposition and determiner selection; and correct pronominalisation)
- improvement in parsing, e.g. recognition of dates.

Of course the overall improvement depends on the number of sentences containing NEs.

But also in the area of indexing it is necessary to be able to draw a distinction between the concept '*carpenter* ' and the name '*carpenter*' as quite different search results are expected for both cases.

[47] A total of 1500 sentences was evaluated in three language directions, 15% of which contained proper names, cf. (Thurmair 2005).

Entity-type	total	found	correct	recall	precision
persons	122	148	110	0.90	0.74
locations	310	299	272	0.88	0.91
organisations	221	197	125	0.56	0.63
dates	114	81	81	0.71	1.00
units	88	66	64	0.73	0.97
Total	855	791	652	0.76	0.82

Fig. 6.28 NE tests nametypes

2. The recognition of NEs in text analysis was evaluated in two scenarios in the MULTISENSOR project, described in (Vrochidis et al. 2015).[48]

As for the **test data**, a test corpus containing 60 sentences (starting with English) was composed, covering three use cases (*energy*, *household*, and *dairy*). This corpus was manually annotated, to have a gold standard against which the single components could be evaluated.

Because this is a too small test set, additional test documents were used to broaden the test base for the Named Entity recogniser. From the MULTISENSOR repository, test documents were extracted: For each language, and each use case, about 200 documents were downloaded; of these, 20 documents were randomly selected as test documents. The documents were not further cleaned or processed; they contain formatting, spelling, and other errors. To be able to run evaluations, there documents were manually annotated, in a kind of'gold standard '.

This resulted in 60 documents per language, 20 per use case.

As **test systems** two versions of the LtNamer were made available, one which uses the lexicalised version, and one using the full contextual version. The differences are mainly in the area of person, location, and organisation entities; communication means and units are rather similar.

Different tests were made with the test data mentioned. Tests were always measured with recall and precision.

a. **Entity types**: The recognition quality in relation to the **different entity types** is shown in Fig. 6.28.[49] It is measured from a corpus of 12 English documents from the energy domain, with overall about 1000 NEs, related to a manually produced Gold Standard for these documents.

The table shows that entity types TIMES and UNITS can be recognised rather safely. The 'classical' types (locations, institutions and persons) have lower recognition rates, in particular companies and institutions need more effort to analyse them.

[48] In the TINA context, the MULTISENSOR research way headed by Vera Aleksić.

[49] The COMMEAMS have not been evaluated here.

(en)	lexical NE				contextual NE			
	no-adapt		adapted		no-adapt		adapted	
	recall	precision	recall	precision	recall	precision	recall	precision
dairy	0.29	0.48	0.69	0.73	0.38	0.43	0.71	0.74
energy	0.31	0.58	0.71	0.76	0.37	0.45	0.74	0.79
household	0.18	0.36	0.68	0.69	0.20	0.27	0.70	0.69
Total	0.28	0.51	0.70	0.74	0.33	0.40	0.73	0.75

Fig. 6.29 LtNamer version evaluation

b. **LtNamer versions**: A comparison between **lexical and contextual LtNamer** is shown in Fig. 6.29. It shows that the rule-based contextual version using contextual clues and indicators improves recall but not necessarily precision: It decreases in the non-domain-adapted testset, however slightly increases in the domain adaptation. This shows that NEs detected by contextual indicators are a device to increase recall as they try to identify hitherto unknown NEs. But they always risk to identify NEs which do not belong to the respective entity type.

c. **Domain adaptation**: The data also show that domain-independent NE recognisers cannot hope to be a useful contribution to text analysis. **Domain adaptation** is an absolute necessity for such a component. It intends to optimise the NE recognition results, as it combines the general recognition by domain-specific expertise. This holds in particular for lexical resources.

The lexical resources of LtNamer therefore were adapted to the three domains relevant for MULTISENSOR, which, for short, were called ‚*Energy* ‘ (energy policy, green energy etc.), ‚*Household* ‘ (household appliances, product descriptions, tests etc.), and ‚*Dairy*‘ (activities in the dairy market). Tuning consists of collecting and lexicalising entities from these three domains.

- Lexical resources for ‘*energy*’ were adapted by adding a list of Nuclear power plants and offshore wind parks from Wikipedia, and lists of Energy Actors from Reegle.[50]
- Resources for ‘*household*’ and ‘*dairy*’ could only be adapted by inspecting the MULTISENSOR document base for these domains.

These entries were normalised and prepared for lexicon import by providing required annotations (POS, subtype, entry type, domain, allomorphs etc.). About 4500 entries overall were added to the NE Gazetteer.

Domain adaptation requires a certain amount of bookkeeping as the NE resources may be used in different customisations. In order to keep track of the changes, the lexical feature ‚domain ‘ was used, enriched by the values ‘energy’, ‘household’ and ‘dairy’ for all domain-specific entries. This enables the recogniser to be made sensitive for the respective domain. Runtime domain selection could be automatised by using a global parameter set by e.g. a topic detection component as foreseen in the processing chain (cf. Sect. 3.3).

[50] www.reegle.info

language	domain	Version noadapt		Version adapt	
		recall	precision	recall	precision
en	dairy	0.38	0.43	0.71	0.74
	energy	0.37	0.45	0.74	0.79
	household	0.20	0.27	0.70	0.69
	total	0.33	0.40	0.73	0.75
de	dairy	0.73	0.67	0.82	0.77
	energy	0.64	0.56	0.82	0.72
	household	0.66	0.44	0.78	0.63
	total	0.69	0.57	0.81	0.72
fr	dairy	0.55	0.66	0.51	0.63
	energy	0.58	0.62	0.72	0.74
	household	0.68	0.66	0.80	0.74
	total	0.58	0.65	0.63	0.69
es	dairy	0.47	0.43	0.79	0.84
	energy	0.32	0.35	0.69	0.82
	household	0.45	0.37	0.67	0.63
	total	0.39	0.38	0.71	0.78

Fig. 6.30 Language tests

Figure 6.29 shows that the non-adapted systems, be it lexical or contextual, clearly fail the threshold to usability. Domain adaptation improves the quality of the system in all categories, and ends up in the best category (domain-adapted–contextual) with recall and precision in the mid-seventies.

d. **Language dependency**: The recognition results for four of the LtNamer languages are given in Fig. 6.30. They are all based on domain-adapted versions of the MULTISENSOR corpora.

The most significant improvements can be seen for Spanish and English, while German and French are less significant. In these languages not just recall but also precision increases. The different languages require different recognition components, e.g. recognition for German cannot rely on capitalisation of the entity candidates, and needs a decomposer. Household texts seem to pose more problems to the recogniser than the others; this may have to do with the overall lower quality of these texts (more typos, more structural peculiarities).

6.2.3.2 Comparison

Overall the recognition results are not really satisfactory although they are in line with results published elsewhere. For instance, the CoNLL results show recall between 0.53 and 0.88 and precision between 0.69 and 0.88 for the English data set, and recall between 0.33 and 0.88 for the German data set, cf. (Tjong Kim Sang and De Meulder 2003). Similar results are reported by (Jacquet et al. 2019).

However other authors report much better results, for instance the systems presented in the overview of (Mansouri et al. 2008): They all report recall and precision values of more than 0.85; even higher values are given, e.g. by (Roy 2021). In the contexts of academic research, special setups, resource data and evaluation procedures[51] are used which make comparisons difficult. Several parameters influence the recognition quality, some of them are the following:

a. The selection of the **domain**. Most NER systems use news data. MUC and CONLL use *Reuters* data, CoNLL uses *Frankfurter Rundschau* for German data. This domain has specific properties: Rather standardised structures of the articles, as few redundancy as possible, concise presentation of the content etc.

As could be expected, different domains return different recognition rates.

- Changes in the domain may lead to a drop in recognition performance by 20–40%, as (Nadeau and Sekine 2007) report.
- According to (Balasuriya et al. 2009), analyses of Wikipedia articles reach precision between 0.52 and 0.86 and recall between 0.57 and 0.87, which is quite below their 'home' domain (news).
- (Faruqui and Padó 2010) analysed Europarl texts, cf. (Koehn 2005), and report a drop in recognition quality by about 10% when CoNLL-trained systems are used.
- In turn, biomedical systems based on the GENIA corpus report rather good results, with precision of 0.84 and recall 0.80, cf. (Ju et al. 2011).

For the domains analysed in MULTISENSOR (energy, household, dairy) no reports are known. Household in particular is a difficult domain, consisting of product descriptions of household devices and comparisons of them.

b. The selection of a **test set** is another critical point in evaluation. Nearly all investigations use in-domain data for tests (part of the MUC or CoNLL data). If out-of-domain data are used, the recognition also drops: (Faruqui and Padó 2010) report a drop of recognition by about 10%. In real-world applications, where LtNamer is used in document indexing contexts, data of several domains would have to be analysed, which puts the focus on recognition robustness.

c. A third quality parameter is the **language**. Current NE research is rather English-centric, and attempts to provide NER for other languages produce quite different recognition results, and '*Claims that language and domain independent are based on theoretical considerations rather than on practical tests*': (Tjong Kim Sang and De Meulder 2003) in CoNLL about language-independent NER).

- Klementiev and Roth (2006) analyse Russian texts and use a bilingual corpus (English-Russian) to identify NEs based on temporal alignment and transliteration. They report 63% accuracy.

---

[51] (Nadeau and Sekine 2007) report that different evaluation techniques result in a quality difference of 20–40%.

- Gupta et al. (2012) exploit bilingual corpora for English-Hindi and try to exploit the re-use of text patterns; if full plus partial matches are counted, they report a recall of 0.57 and precision of 0.86.
- Persian NER is also confronted with a lack of training data in Persian. Dashtipour et al. (2017) use rule-based technology and a Persian gazetteer. Results are reported to be comparable to state-of-the-art systems.
- While annotated training data for these languages were not available to the authors, things are different for German. Not only CoNLL provides German data (news texts), also the GermEval initiative[52] collected development and test sets from German News and Wikipedia domains, 31K sentences overall and 590K tokens. Benikova et al. (2014a, b) report recall of 0.63 and precision of 0.76 on average for the participating systems. Note that capitalisation as a feature of NE detection cannot be used in German as all nouns are capitalised.
- For Spanish, (Jacquet et al. 2019) investigate rule-based NER, and Babelnet-derived MW Patterns, and report recall of 0.35 and precision of 0.84.

The above data show comparable results at least for German and Spanish for the LtNamer.

d. The main quality parameter of course is the availability of **training data** in the targeted language and domain. Only few datasets are available, and creating them for supervised learning is very time-consuming and resembles the effort to create knowledge-based NE systems. (Nadeau and Sekine 2007) state that '*the main shortcoming of SL [supervised learning] is the requirement of a large annotated corpus. The unavailability of such resources and the prohibitive cost of creating them*' reduces the general applicability of supervised learning technologies. And it is the quality of the training data which determines the NE recognition quality more than the selection of the learning technique, as investigated by (Roy 2021).

As for the domains analysed in the LtNamer evaluation, only few documents were available per domain (energy, household, dairy), so only knowledge-driven techniques could be applied: linguistic description and generalisation of the few most frequent examples, in the respective languages.

The most important requirement for the success of NE recognisers is the preparation of domain-specific data. Collections like KIND for Italian (Paccosi and Aprosio 2022) ease the pressure, however tools for domain-adaptation are of utmost importance: Commercial retrieval systems must answer the requirements of many customers in many domains. NE developers could offer an adaptation service to such customers. Such an offer will be accepted the faster and more inexpensive such adaptations can be provided. Tools like WEXEA, cf. (Strobl et al. 2022) or TermExtract[53] could serve as a basis for such adaptation work.

e. Some NE recognisers propose to increase the recognition quality by the use of **additional resources**. Several proposals have been forwarded in this respect.

---

[52] Cf. https://germeval.github.io.

[53] Thurmair (2003) and cf. Sect. 9.2.3.

- An obvious solution is to use a **thesaurus** of the domain. For instance, (Steinberger et al. 2002) propose to use Eurovoc for NE improvement; it is the multilingual and multi-domain thesaurus of the European Union.
- Links between NE candidates are provided by resources like **WordNet**; it is used to support unsupervised learning, cf. (Nadeau and Sekine 2007).
- The most used resources however is **Wikipedia**. Several contributions exploit both content and structure of this resource: (Balasuriya et al. 2009) create a Gold Standard for (English) Wikipedia texts, (Bunescu and Paşca 2006) use Wikipedia to disambiguate homograph NEs (i.e. NEs belonging to different types), (Nothman et al. 2013) investigate also the structure of the articles, e.g. their multilingual links.

Access to such resources is prohibitive in multi-customer multi-domain setups, not just for performance but also for licensing reasons. The only alternative option is to provide a local resource in the tradition of the NE Gazetteers as described above. Although (Mikheev et al. 1999) claim that the size of such gazetteers is less important than their coverage for a given domain this observation is weakened in cases where many domains must be targeted; however it supports the claim that domain adaptation is essential for the acceptance of NE recognition.

f. However all this may be, the fact is still that the recognition results given in Fig. 6.29 need improvement to be good enough for practical use. This also holds for comparable results in the literature discussed above. Intuitively, recall and precision of about 0.7 means that three out of four entities are found, and only two of them are correct. This is not a particularly exciting result, and there is room for improvement. Results should be at least in a range above 0.8 in components to be proposed for practical use: Four out of five found, and three of them correct.

Options for improvement could be:

- Providing **more intelligence** for the LtNamer, by adding more context to the NE candidates. It should not only be based on local features like POS and extend the syntactic contexts by semantic ones. The component LtSensExtract (cf. Sect. 7.2) could be used for this purpose. It provides concepts which co-occur frequently with NE types. Such concepts can be provided as an additional resource in the recognition. Similar ideas have been proposed e.g. by (Faruqui and Padó 2010); they call it 'semantic generalization'.
- Improving the **domain adaptation** capabilities to reach an acceptable quality level, esp. for domains where only few training data are available. As such data would have to be inspected manually, tools to improve and speed up such a process would turn out to be quite useful. Also there are collections like JRC-Names, cf. (Steinberger et al. 2011), or DBPedia[54] which support domain adaptation, however the challenge then is to select only relevant entries from them.

The target would be to offer a cost-effective domain-adaptation toolset (or service) to customers that want to use NE components.

[54] https://www.dbpedia.org

6.2.3.3 Assessment

Some conclusions can be drawn from the LtNamer development:

- NE recognition which is only based on 'general-purpose' lexicon lookup is insufficient: Lexicon entries are often missing, especially in new domains, and they are often ambiguous with non-entity concepts. Only if the lexicon can be tuned to a certain domain or even application, and most of the names of this domain can be covered, lexical LtNamer versions can be tried.
- NEs may occur in inflected forms, or even syntagmata, in languages like Arabic ('القاهرة', *Kairo*), Turkish ('*Ankara'daki*'), Russian ('Горбачева', *Gorbatchova*), German ('*Schillers*', '*Lufthansaflugplan*') and others. At least a morphological analyser should be part of the NE component.
- Adding contextual features improves NE recognition significantly. Not only it can find non-lexicalised NEs, it can also better disambiguate the types of NE candidates. In particular recall can be improved this way.
- Contextual analysis should not just cover syntactic environment but should be extended to semantic/conceptual contexts, to make the recognition more stable, and to better disambiguate homograph NE candidates.
- Domain adaptation is a key to the success of NE recognisers. The main source of improvement here is the NE gazetteer. As a consequence, this dictionary would adapt its annotation scheme to store the domain(s) to which a given entity would belong. This way the NE gazetteer would be able to only load and unload the required NEs at runtime, if a given task or domain is required.[55]
- In order to reach an acceptable quality level, tools for easy domain adaptation would play an important role in the whole analysis framework.

The success of NE recognition depends on the recognition quality, not just for known domains but also for new domains and languages.

As for the application of indexing, the consequences are less dramatic: If an entity is not recognised, it still can be indexed, like other out-of-vocabulary terms, and can be searched. Of course aspects of multilingual indexing would be lost, however it could be searched at least in monolingual contexts.

[55] To be able to do this, the internal organisation of the gazetteer needs to be designed carefully, cf. Chap. 9.

References

Acosta, O., Villavicencio, A., & Moreira, V. (2011). Identification and Treatment of Multiword Expressions applied to Information Retrieval. *Proc. Workshop on Multiword Expressions: From Parsing and Generation to the Real World (MWE 2011), Portland, Oregon, USA.*

Amtrup, J., Rad, M., Megerdoomian, K., & Zajac, R. (2000). Persian-English Machine Translation: An Overview of the Shiraz Project. *MCCS-00–319, Univ. New Mexico.*

Appelt, D.E., Hobbs, J.R., Bear, J., Israel D., & Tyson, M. (1993). FASTUS: A Finite-state Processor for Information Extraction from Real-world Text. Proc. Int. Joint Conf. on Artificial Intelligence (IJCAI), Chambéry, France1993.

De Araujo, V., Ramisch, C., & Villavicencio, A. (2011). Fast and Flexible MWE Candidate Generation with the mwetoolkit. *Proc. Workshop on Multiword Expressions (MWE 2011). Portland, Oregon, 2011.*

Attia, M., Toral, A., Tounsi, L., Pecina, P., & van Genabith, J., (2010). Automatic Extraction of Arabic Multiword Expressions. Proc. Workshop of the Multiword Expressions: From Theory to Applications (MWE 2010), Beijing.

Azzam, S., Humphreys, K., and Gaizauskas, R. (1998). Evaluating a Focus-Based Approach to Anaphora Resolution. *Proc. 36th Annual Meeting of the Association for Computational Linguistics (ACL) and 17th Int. Conf. on Computational Linguistics (COLING), Volume 1, pages 74–78, Montreal, Quebec, Canada.*

Balasuriya, D., Ringland, N. Nothman, J., Murphy, T., & Curran, J. (2009). Named Entity Recognition in Wikipedia. *Proc. 2009 Workshop on The People's Web Meets NLP: Collaboratively Constructed Semantic Resources (People's Web), pages 10–18, Suntec, Singapore.*

Bank, M., & Schierle, M. (2012). A Survey of Text Mining Architectures and the UIMA Standard. *Proc. Eighth Int. Conf. on Language Resources and Evaluation (LREC'12), Istanbul, Turkey, pp. 3479–3486.*

Barker, E., & Gaizauskas, R. (2012). Assessing the Comparability of News Texts. *In Proc. of the Eighth Int. Conf. on Language Resources and Evaluation (LREC'12), pages 3996–4003, Istanbul, Turkey.*

Bejček, E., Straňák, P., & Pecina, P. (2013). Syntactic Identification of Occurrences of Multiword Expressions in Text using a Lexicon with Dependency Structures. *Proc. 9th Workshop on Multiword Expressions (MWE 2013), Atlanta, Georgia.*

Bel, N., Papavasiliou, V., Prokopidis, P., Toral, A., & Arranz, V. (2012). Mining and Exploiting Domain-Specific Corpora in the PANACEA Platform. *Proc. 5th Workshop on Building and Using Comparable Corpora at the Eighth Int. Conf. on Language Resources and Evaluation (LREC-2012); 2012 May 23–25; Istanbul, Turkiye.*

Benikova, D., Biemann, Chr., & Reznicek, M. (2014). NoSta-D Named Entity Annotation for German: Guidelines and Dataset. Proc. Int. Conf. on Language Resources and Evaluation (LREC) 2014, Reykjavik, Iceland.

Benikova, D., Biemann, Chr., Kisselew, M., & Padó, S. (2014). GermEval 2014 Named Entity Recognition Shared Task: Companion Paper. *Proc. of the KONVENS GermEval workshop, Hildesheim, Germany, (2014), page 104—112.*

Bond, F., da Costa, L.M., Goodman, M.W., McCrae, J., & Lohk, A. (2020). Some Issues with Building a Multilingual Wordnet. *Proc. 12th Int. Conf. on Language Resources and Evaluation (LREC 2020), Marseille, pp. 3189–3197.*

Bonin, F., Dell'Orletta, F., Venturi, G., & Montemagni, S. (2010). Contrastive Filtering of Domain-Specific Multi-Word Terms from Different Types of Corpora. *Proc. Multiword Expressions: From Theory to Applications (MWE 2010), Beijing, China.*

Bundschus, M., Dejori, M., Stetter, M. *et al.* (2008). Extraction of semantic biomedical relations from text using conditional random fields. *BMC Bioinformatics* ***9****, 207 (2008).*https://doi.org/10.1186/1471-2105-9-207

Bunescu, R., & Paşca, M. (2006). Using Encyclopedic Knowledge for Named Entity Disambiguation *Proc. 11th Conf. of the European Chapter of the Association for Computational Linguistics (EACL 2006), Trento, Italy, pp. 9–16.*

Calzolari, N., Fillmore, Ch., Grishman, R., Ide, N., Lenci, A., MacLeod, C., & Zampolli, A. (2002). Towards best practice for Multiword Expressions in computational lexicons. *Proc. Int. Conf. on Language Resources and Evaluation (LREC) 2002, Las Palmas.*

Charniak, E. (1993). Statistical Language Learning. *MIT Press*

Cunningham, H. (2000). Software Architecture for Language Engineering. *PhD Thesis, University of Sheffield, 2000.*

Cunningham, H., Maynard, D., Bontcheva, K., & Tablan, V. (2002). GATE: A Framework and Graphical Development Environment for Robust NLP Tools and Applications. *Proc. 40th Annual Meeting of the Association for Computational Linguistics (ACL), 2002, Philadelphia, PA, USA.*

Daille, B. (1994). Study and Implementation of Combined Techniques for Automatic Extraction of Terminology. *In: Klavans, J., & Resnik, Ph. (1994) The Balancing Act: Combining Symbolic and Statistical Approaches to Language. Nonantum Inn, Kennebunkport, Maine.*

Dashtipour, K., Gogate, M., Adeel, A., Algarafi, A., & Howard, N. (2017). Persian Named Entity Recognition. *IEEE 16th Int. Conf. on Cognitive Informatics & Cognitive Computing (ICCI*CC).*

Deksne, D., Skadiņš, R., & Skadiņa, I. (2008). Dictionary of Multiword Expressions for Translation into Highly Inflected Languages. *Proc. 6th Int. Conf. on Language Resources and Evaluation (LREC 2008), Marrakech, Marocco.*

Dubremetz, M., Nivre, J. (2014). Extraction of Nominal Multiword Expressions in French. *Proc. 10th Workshop on Multiword Expressions (MWE 2014), Gothenburg, Sweden.*

Escartín, C.P., Losnegaard, G.S., Samdahl, G.I.L., & García, P.P. (2013). Representing Multiword Expressions in Lexical and Terminological Resources: An Analysis for Natural Language Processing Purposes. Proc. Conf. of Electronic Lexicography in the 21st century (eLex) 2013, Tallinn, Estonia.

Family, N. (2006). Explorations of Semantic Space: The Case of Light Verb Constructions in Persian. *Diss. Paris*

Farahmand, M., & Martins, R. (2014). A Supervised Model for Extraction of Multiword Expressions Based on Statistical Context Features. *Proc. 10th Workshop on Multiword Expressions (MWE 2014), Gothenburg, Sweden.*

Faruqui, M., Padó, S. (2010). Training and Evaluating a German Named Entity Recognizer with Semantic Generalization. *Proc. 10th Int. Conf. on Natural Language Processing (KONVENS) 2010, Saarbrücken, Germany.*

Fotopoulou, A., Markantonatou, St., & Giouli, V. (2014). Encoding MWEs in a conceptual lexicon. Proc. 10th Workshop on Multiword Expressions (MWE 2014) Gothenburg, Sweden.

Gaizauskas, R., & Humphreys, K. (2000). Quantitative evaluation of coreference algorithms in an information extraction system, *in: Botley, S., & McEnery, T., eds.: Corpus-based and Computational Approaches to Discourse Anaphora. John Benjamins Publishing Company pp. 145–145.*

Graliński, F., Savary, A., Czerepowicka, M., & Makowiecki, F., (2010). Computational Lexicography of Multi-Word Units: How Efficient Can It Be? Proc. Multiword Expressions: From Theory to Applications (MWE 2010), Beijing.

Grégoire, N. (2009). Untangling Multiword Expressions. A study on the representation and variation of Dutch multiword expressions. *Diss. Utrecht (LOT).*

Grishman R., & Sundheim, B. (1996). Message Understanding Conference - 6: A Brief History. *In Proc. 16th Int. Con. Computational Linguistics (COLING), Copenhagen, Denmark.*

Gupta, P., Singhal, K., & Rosso, P. (2012). Multiword Named Entities Extraction from Cross-Language Text Re-use. *Workshop on Creating Cross-language Resources for Disconnected Languages and Styles (CREDISLAS) Workshop at LREC 2012 (Istanbul, Turkey).*

Hurskainen, A. (2008). Multiword Expressions and Machine Translation. *Univ. Helsinki Technical Reports in Language Technology, Report No 1, 2008.*

Jacquet, G., Ehrmann, M., Piskorski, J., Tanev. H., & Steinberger, R. (2019). Cross-lingual linking of multi-word entities and language-dependent learning of multi-word entity patterns. *In Parmentier, Y. & Waszczuk, J., eds. Representation and parsing of multiword expressions: Current trends, 269–297. Berlin: Language Science Press.*

Jacquet, G., Ehrmann, M., Steinberger, R., & and Väyrynen, J. (2016). Cross-lingual Linking of Multi-word Entities and their corresponding Acronyms. *Proc. 10th Int. Conf. on Language Resources and Evaluation (LREC'16), Portorož, Slovenia, pp. 528–535.*

Jessop, D.M., Adams, S.E., Willighagen, E.L. et al. (2011). OSCAR4: a flexible architecture for chemical text-mining. J Cheminform 3, 41 (2011). https://doi.org/10.1186/1758-2946-3-41

Ju, Z., Wang, J., & Zhu, F. (2011). Named Entity Recognition from Biomedical Text Using SVM. *5th Int. Conf. on Bioinformatics and Biomedical Engineering, Wuhan, China, 2011, pp. 1–4.*

Kahnemuyipour, A. (2014). Revisiting the Persian Ezafe construction: A roll-up movement analysis. *Lingua 150 (2014) 1-24*

Klementiev, A., & Roth, D. (2006). Named entity transliteration and discovery from multilingual comparable corpora. *In Proc. of the Annual Meeting of the North American Association of Computational Linguistics (NAACL), New York.*

Koehn, Ph. (2005). Europarl: A parallel corpus for statistical machine translation. *Proc. Machine Translation Summit X, Phuket, Thailand.*

Krieger, H.U., & Declerck, Th. (2014). TMO — The Federated Ontology of the TrendMiner Project. *Proc. 9th Int. Conf. on Language Resources and Evaluation (LREC'14), Reykjavik, Iceland, pp. 4164–4171.*

Krishnan, V., & Ganapathy, V. (2005). Named Entity Recognition. Stanford Lecture CS229, 2005

Kulkarni, N., & Finlayson, M.A. (2011). jMWE: A Java Toolkit for Detecting Multi-Word Expressions. *Proc. Workshop on Multiword Expressions: from Parsing and Generation to the Real World (MWE 2011), Portland, Oregon, USA.*

Lee, H., Surdeanu, M., MacCartney, B., & and Jurafsky, D. (2014). On the Importance of Text Analysis for Stock Price Prediction. *In Proc. 9th Int. Conf. on Language Resources and Evaluation (LREC'14), Reykjavik, Iceland, pp. 1170–1175.*

Mandl, Th., & Womser-Hacker, Chr. (2005). How Do Named Entities Contribute to Retrieval Effectiveness? in C. Peters et al., eds. : CLEF 2004, LNCS 3491, pp. 833 – 842, 2005 (Springer).

Mansouri, A., Affendey, L.S., & Mamat, A. (2008). Named Entity Recognition Approaches. *in: IJCSNS Int. Journal of Computer Science and Network Security, Vol.8 No.2*

Martens, Sc., & Vandeghinste, V. (2010). An Efficient, Generic Approach to Extracting Multi-Word Expressions from Dependency Trees. *Proc. Multiword Expressions: From Theory to Applications (MWE 2010), Beijing, August 2010.*

Martins, P.H., Marinho, Z., & Martins, A. (2019). Joint Learning of Named Entity Recognition and Entity Linking. Proc. of the 57th Annual Meeting of the Association for Computational Linguistics.

McCord, M. (1989). Design of LMT: a prolog-based machine translation system. *J. Comput. Linguistcs 15, 1, 1989, p.33–53.*

McNamee, P., Hoa, H.T., Simpson, H., Schone, P., & Strassel, St. (2010). An Evaluation of Technologies for Knowledge Base Population. *In Proc. of the Seventh Int. Conf. on Language Resources and Evaluation (LREC'10), Valletta, Malta.*

Mikheev, A., Moens, M., & Grover, C. (1999). Named entity recognition without gazetteers, *Proc. of the 9th Conf. of European Chapter of the Association for Computational Linguistics (EACL), Bergen, Norway, 1999, pp. 1–8.*

Min, B. & Grishman, R. (2012). Compensating for annotation errors in training a relation extractor. *Proc. 13th Conf. of the European Chapter of the Association for Computational Linguistics (EACL), Avignon, France.*

Moreno-Ortiz, A., Pérez-Hernández, Ch., & Del-Olmo, M.A. (2013). Managing Multiword Expressions in a Lexicon-Based Sentiment Analysis System for Spanish. *Proc. 9th Workshop on Multiword Expressions (MWE 2013), Atlanta, Georgia.*

Nadeau, D., Sekine, S. (2007). A Survey of Named Entity Recognition and Classification. *Lingvisticae Investigationes, 30, 3-26*
Neumann, G., Piskorski, J., & Braun, Chr. (2000). An Intelligent Text Extraction and Navigation System. *Proceedings of the 6th Int. Conf. on Computer-Assisted Information Retrieval (RIAO'00).*
Nissim, M., Castagnoli, S., & Masini, Fr. (2014). Extracting MWEs from Italian corpora: A case study for refining the POS-pattern methodology. *Proc. 10th Workshop on Multiword Expressions (MWE 2014), Gothenburg, Sweden.*
Nothman, J., Ringland, N., Radford, W., Murphy, T., & Curran, J. (2013). Learning multilingual named entity recognition from Wikipedia. *Artificial Intelligence 194 (2013) 151–175*
Odijk, J. (2003). Towards a standard for multi-word expressions. ISLE Project Report.
Oflazer, K., Çetinoğlu, Ö., & Say, B. (2004). Integrating Morphology with Multi-word Expression Processing in Turkish. *Proc. 2nd Workshop on Multiword Expressions: Integrating Processing, pp. 64–71, Barcelona, Spain.*
Paccosi, T., & Palmero Aprosio, A. (2022). KIND: an Italian Multi-Domain Dataset for Named-Entity Recognition. *Proc. 13th Int. Conf. on Language Resources and Evaluation (LREC 2022), pp. 501–507 – Marseille.*
Papageorgiou, H., Prokopidis, P., Giuli, V., & Piperidis, St. (2000). A Unified POS Tagging Architecture and its Application to Greek. Proc. 2nd Int. Conf. on Language Resources and Evaluation (LREC 2000), Athens, Greece.
Pazienza, M.T., ed. (1997). Information Extraction. A Multidisciplinary Approach to an Emerging Information Technology. Springer
Quocchi, V., Frontini, F., & Rubino, F., (2012). A MWE Acquisition and Lexicon Builder Web Service. *Proc. Int. Conf. on Computational Linguistics (COLING 2012), Mumbai, India, pp. 2291–2306.*
Riedl, M., & Padó, S. (2018). A Named Entity Recognition Shootout for German. In Proc. of the Annual Meeting of the Association for Computational Linguistics (ACL), Melbourne, Australia.
Roy, A. (2021). Recent Trends in Named Entity Recognition (NER). arXiv:2101.11420
Sag, I., Baldwin, T., Bond, F., Copestake, A., & Flickinger, D. (2002). Multiword Expressions: a Pain in the Neck for NLP. *In: Proc. 3rd Int. Conf. on Intelligent Text Processing and Computational Linguistics. (CICLing), Mexico City, Mexico.*
Sagot, B., & Walther, G. (2010). A morphological lexicon for the Persian language. *In Proc. 7th Int. Conf. on Language Resources and Evaluation (LREC 2010), Valetta, Malta: ELDA.*
Samaridi, N., & Markantonatou, St. (2014). Parsing Modern Greek verb MWEs with LFG/XLE grammars. *Proc. 10th Workshop on Multiword Expressions (MWE 2014), Gothenburg, Sweden.*
Tjong Kim Sang, E.F., & De Meulder, F. (2003). Introduction to the CoNLL-2003 Shared Task: Language-Independent Named Entity Recognition. *Proc. of the 7th Conf. on Natural Language Learning at HLT-NAACL 2003, pages 142–147.*
Savary, A. (2009). Multiflex: A Multilingual Finite-State Tool for Multi-Word Units. Proc. of the 14th Int. Conf. on Implementation and Application of Automata (CIAA), 2009, Sydney, Austrialia.
Schwarz, Chr. (1985). Textrecherche mit Mehrwortbegriffen – *Proc. Jahrestagung der Gesellschaft für Linguistische Datenverarbeitung (GLDV) Sprachverarbeitung in Information und Dokumentation, 1985 – Springer pp. 216–225.*
Steinberger, R., Pouliquen, B., & Hagman, J. (2002). Cross-lingual document similarity calculation using the multilingual thesaurus eurovoc. *Proc. 3rd Int. Conf. on Computational Linguistics and Intelligent Text Processing (CICLing), Mexico, pp. 415–424.*
Steinberger, R., Pouliquen, B., Kabadjov, M., & van der Goot, E. (2011). JRC-Names: A freely available, highly multilingual named entity resource. *Proc. 8th Int. Conf. Recent Advances in Natural Language Processing (RANLP'2011), 104–110. Hissar, Bulgaria.*

Strobl, M., Trabelsi, A., Zaiane, O. (2022). Enhanced Entity Annotations for Multilingual Corpora. *Proc. 13th Int. Conf. on Language Resources and Evaluation (LREC 2022), pp. 3732–3740, Marseille, 20–25 June 2022.*

Strötgen, J., & Gertz, M. (2016). Domain-Sensitive Temporal Tagging. *Springer*

Surdeanu, M., McClosky, D., Smith, M.R., Gusev, A., & Manning, Chr. (2011). Customizing an Information Extraction System to a New Domain. In Proc. of the ACL 2011 Workshop on Relational Models of Semantics (RELMS), Portland, Oregon, USA.

(TAC 2009) Text Analysis Conference (TAC) (nist.gov) https://tac.nist.gov

Tepper, M., Capurro, D., Xia, F., Vanderwende, L., & Yetisgen-Yildiz, M. (2012). Statistical section segmentation in free-text clinical records. *Proc. 8th Int. Conf. on Language Resources and Evaluation (LREC'12), Istanbul, Turkey, pp. 2001–2008.*

Thurmair, Gr. (1990). METAL: Computer Integrated Translation. *Proc. of a Workshop on Machine Translation, July 1990, UMIST.*

Thurmair, Gr. (2003). Making term extraction tools usable. *In EAMT Workshop: Improving MT through other language technology tools: resources and tools for building MT, Budapest, Hungary.*

Thurmair, Gr. (2005). Improving MT Quality: Towards a Hybrid MT Architecture in the linguatec 'Personal Translator'. *In: Proc. of Machine Translation Summit X: Invited paper, Phuket, Thailand.*

Thurmair, Gr. (2016). LtAnonymizer. *Unpublished.*

Thurmair, Gr. (2018). Multiword expressions in multilingual information extraction. *In: Mitkov, R., Monti, J., Corpas Pastor, G. & Seretan, V., eds. Multiword Units in Machine Translation and Translation Technology. John Benjamins.*

Tu, Y., & Roth, D. (2011). Learning English Light Verb Constructions: Contextual or Statistical. *Proc. Workshop on Multiword Expressions: from Parsing and Generation to the Real World (MWE 2011), Portland, Oregon.*

Verhagen, M., Gaizauskas, R., Schilder, F. *et al.* (2009). The TempEval challenge: identifying temporal relations in text. *Language Resources & Evaluation 43, 161–179 (2009).*

Villavicencio, A., Copestake, A., Waldron, B., & Lambeau, F. (2004). Lexical Encoding of MWEs. 2nd ACL Workshop on Multiword Expressions: Integrating Processing, Barcelona.

Vincze, V., Nagy, I., & Berend, G. (2011). Detecting noun compounds and light verb constructions: a contrastive study. *Proc. Workshop on Multiword Expressions: from Parsing and Generation to the Real World (MWE 2011), Portland, Oregon.*

Vossen, P. (2004). EuroWordNet: a multilingual database of autonomous and language-specfic wordnets connected via an Inter-Lingual-Index. *International Journal of Lexicography 17,2*

Vrochidis, St., Kompatsiaris, I., Casamayor, G. et al., (2015). MULTISENSOR: Development of multimedia content integration technologies for journalism, media monitoring and international exporting decision support. *IEEE Int. Conf. on Multimedia & Expo Workshops (ICMEW).*

Wehrli, E., Seretan, V., & Nerima, L. (2010). Sentence Analysis and Collocation Identification. *Proc. Multiword Expressions: From Theory to Applications (MWE 2010), Beijing*

Wehrli, E. (2014). The Relevance of Collocations for Parsing. *Proc. 10th Workshop on Multiword Expressions (MWE 2014), Gothenburg, Sweden.*

Winograd, T. (1983). Language as a Cognitive Process, vol. 1: Syntax. *Addison-Wesley.*

Chapter 7
Disambiguation

The result of Lexical Analysis is a list of terms, defined by<lemma, POS, semantic description>. However in many cases ambiguities will be observed, be it on syntactic or on semantic level. Therefore disambiguation is required, if a proper term list is envisaged. Disambiguation must take the context into account in order to produce terms with one POS and one sense. TINA provides two components for this purpose: LtTagger for syntactic disambiguation, and LtSensExtract for semantic disambiguation.

The purpose of these components is not to try a full syntactic/semantic interpretation of the input text: It is just to disambiguate index term candidates which have been proposed by Lexical Analysis but are still ambiguous.

On the basis of such disambiguated terms, a tool which builds semantic fields by identifying similar terms to a given term can be tried; it computes term associations on the basis of semantic similarity derived from common syntactic head-modifier structures.

All these results must be integrated into the Document Model, as an interface to the IR system backend, i.e. the LtConceptNet.

7.1 Syntactic Disambiguation (LtTagger)

The lexical units produced so far show a specific lemma; however they can differ in their syntactic and semantic description. The syntactic analysis would therefore to have to deal with homographs. Homographs need to be disambiguated by syntactic context analysis. This is the task of a tagger. Therefore a complete text analysis must include a tagging component.

The LtTagger is a component of LtShallow. Deep syntactic parsing would include syntactic disambiguation as part of the parsing mechanism, so in the

G. Thurmair, *Knowledge-Driven Multilingual Text Analysis and Transparent Information Retrieval*, Cognitive Technologies,
https://doi.org/10.1007/978-3-031-91741-7_7

LtDeep processing chain there is no special step like tagging. And of course tagging is language specific. LtTagger is exemplified by implementations for German and English.

There are two purposes of a tagging component.

- from the point of view of **lexical analysis**, tagging is required to disambiguate lexical entries. Some lemmata are homograph and stay ambiguous as long as no context is available (like $_{en}$ *'row'* for *No* or *Vb*); some lemmata are even incomplete as they contain discontinuous elements (like verb particles: $_{en}$ *'stand ... off'*, $_{de}$ *'liest... vor'*) which need to be collected to form the proper lemma–a fact which is often neglected by existing taggers.[1] A tagger would have to identify the right entry both with lemma and POS. As long as this is not done it cannot safely be said of which lemmata ('words') a sentence consists.
- from the point of view of **syntactic analysis**, a tagger is a component of shallow parsing, its purpose being to disambiguate POS candidates and return the proper part-of-speech for a given lemma, with the purpose to feed it into chunkers and/or dependency analysers.

Although both purposes largely overlap there are still differences which affect the design of the tagging component. For instance, a tagger for lexical analysis does not necessarily need to disambiguate function word homographs. Lexical analysis is the main point of view in TINA and takes a prominent role there. But this also implies an extension of the tagging component by procedures of lexical filtering and lexical postprocessing as will be shown.

7.1.1 Design

7.1.1.1 Approaches

While tagsets are language-dependent tagger technologies often are designed to cover other language as well. The main approaches towards tagging[2] have also been applied to German texts.

The earliest taggers were rule-based and did 'homograph analysis', cf. (Haller 1978). One of the latest attempts for a rule-based tagger of German has been reported in (Volk and Schneider 1998). The tagger of EngCG, cf. (Voutilainen 1995) reports excellent results with an accuracy of over 99%; it is a full constraint-based parser, however, and is not a shallow device.

Most taggers developed in recent years are probabilistic, following mainly HMM approaches; the best known examples being TreeTagger, cf. (Schmid 1995),

[1] Numerous examples of Treetagger, (cf. (Schmid 1994; Schmid 1995)) can be found in the WacKy corpus.

[2] Cf. (Jurafsky and Martin 2009:167ff.) for an overview.

and TnT tagger, cf. (Brants 2000). For German they all use the TIGER treebank, cf. (Brants et al. 2002), for training, and the STTS tagset, cf. (Schiller et al. 1995), or a derivation thereof.

German taggers report correct annotations in the high nineties, on the basis of the STTS tagset:

- TreeTagger, HMM-based, reports correct figures of 97.53 (cf. (Schmid 1995)); (Volk and Schneider 1998) give results of 95.27% correctness for TreeTagger.
- TnT (cf. (Brants 2000)) reports accuracy of 96.7%;
- (Giesbrecht and Evert 2009) report 95.82% for TreeTagger, 96.04% for their UIMA-Tagger, 95.71% for TnT, and 97.63% for the Stanford Tagger.

Transformation-based tagging as described in (Brill 1995) combines rule-based and stochastic taggers. The basic approach is to assign probabilities to<lemma, POS>tokens, based on corpus statistics, and apply transformations which change POS tags into others in certain contexts. The tagger cycles over the input and refines POS tags in certain contexts (contexts are described as POS sequences).

7.1.1.2 The LtTagger approach

The LTTagger is part of the toolset developed for lexical analysis. It can be characterised as follows:

- It is a **rule-based** tagger. It uses linguistic knowledge describing patterns of homographs and their disambiguation.
- It is a **shallow** device; It applies rules, specifying certain constraints, to the input. These constraints are used locally, no complete parse of a sentence is attempted/required.
- It was developed for **German** and **English**. Other languages can be added.
- It uses a **standard tagset** (the STAG tagset) with about 90 tags.
- It uses **one pass** through the input, no cyclic processing as e.g. the Brill tagger does.
- It outputs **one solution** per input token for POS and lemma, if possible.
- It is part of the TINA **processing chain**; however stand-alone solutions have also been implemented.

The basic idea for the LtTagger is that tagging consists in **eliminating POS proposals** which do not match a given position in the sentence. Tagging is a task of reduction: All possible tag combinations in the sentence are known at starting time, as the lexical analysis provides them. The job of the tagger is to disambiguate them by eliminating the contextually incorrect ones. Most of them occur in closed-class words anyway, and only few systematic ambiguities are possible in open class words, like finite versus infinite verbs, common versus proper nouns, etc. In German, about 250 such homograph patterns exist, of which about 100 are caused by function words.

The output of the Lexical Analysis component consists in a lattice of entries, with alternatives on some positions. The lattice contains all possible entries and readings of entries which the system is aware of, given its resources (dictionaries and all auxiliary material). At each position of the lattice there is a node list describing the alternatives. The basic idea of the LTTagger is to reduce the ambiguity of readings of an input sentence as provided by the lexical analysis, by removing readings which are implausible or impossible in a given sentential context, such that only one reading per token is left.

The input for the removal decisions is the node list with alternative readings at a given position of the lattice. The context to be used for disambiguation would not just be given by the nodes adjacent to the investigated node, but would also require to look at longer distance dependencies (e.g. German *satzklammer*). All nodes can be accessed by their textform, casing, and POS tag features. The rules of LtTagger are based on the standard tagset (STAG)[3] (Fig. 7.1).

The LtTagger rules consist of condition–action patterns: Conditions describe feature configurations, actions basically remove improper nodes: Nodes of the lattice are deleted under the specified conditions. Each such pattern is described in a tagger rule.

7.1.2 LtTagger Rules

7.1.2.1 Rule Format

All tagger rules have a fixed format. Each rule consists of five parts, identified by keywords:

1. "ID": a rule identifier,
2. "PATTERN": a rule pattern based on the tagset,
3. "CONDITIONS": additional conditions,
4. "ACTIONS": the action to be taken,
5. "EXAMPLE": example and explanations.[4]

IDENTIFIER (ID):
The rule ID consists of two parts, linked by a hyphen: a tag and a number, indicating the rule ordering for this tag. Legal rule IDs are e.g.: NoCo-212, NoCo-341, PtVb-002.

The tag is used to group the rules: All rules for a given tag are fired, if the tag is found in the input. The number is used for sorting. All rules for a given tag are

[3] This restricts the use of morphological information compared to extended tagsets (like the Brill tagset); features like number, case, tense etc. are not available. However it seems sufficient at least for German tagging.

[4] A rule is ended by 10 dashes, for historical reasons.

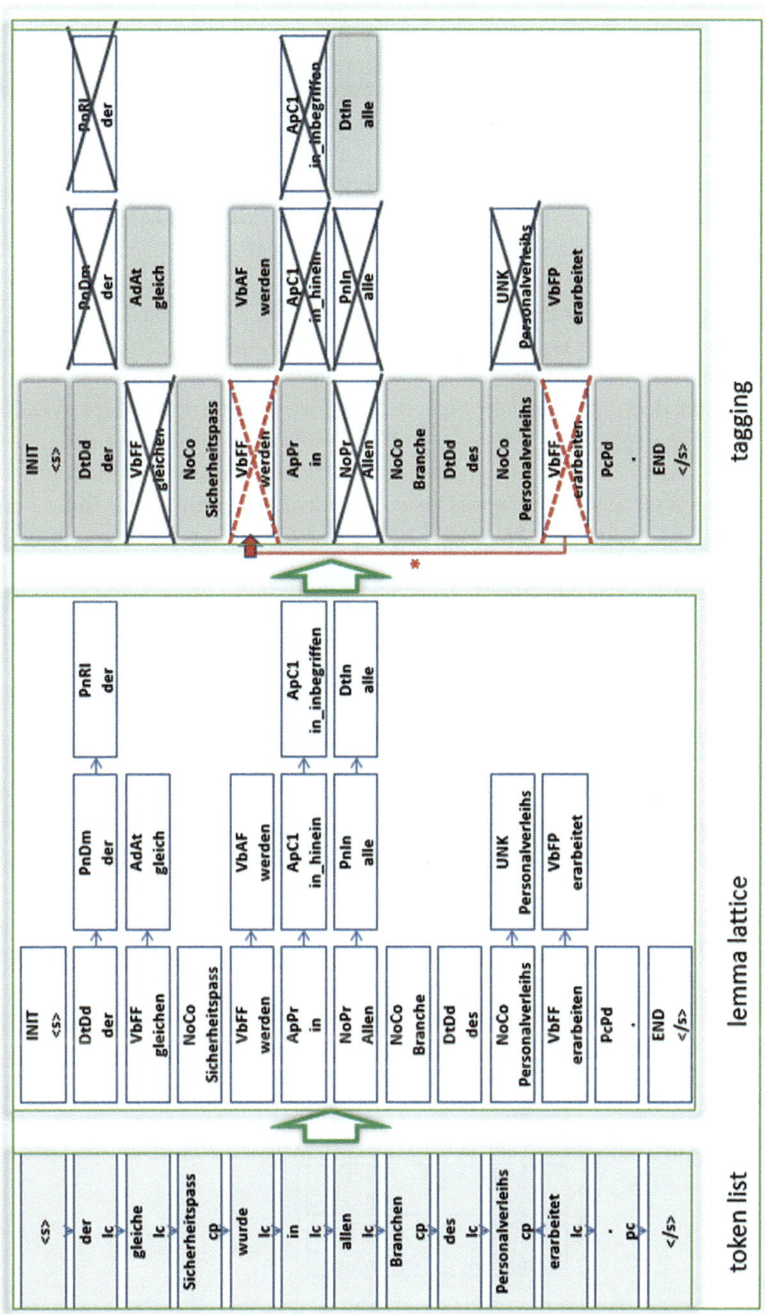

Fig. 7.1 Tagger concept

fired in the order given by the rule numbers (NoCo-212 fires before NoCo-341). The ordering of the rules is in the responsibility of the rule author.

PATTERN:
The pattern gives a sequence of lattice positions. Lattice positions are separated by semicolon (*A; B; C*). Example: '*PATTERN: DtDi; AdAt; NoCo*' describes a sequence of three tags. As a lattice position can contain several nodes (homographs) nodes on the same position are put into brackets and separated by comma (*A; (B,D); C*). Figure 7.2 shows a configuration of *PnPs; (AdAt, VbFI); (VbFF, VbAF)*: e.g. *'(weil er) es stillen wird'*. There is no restriction in the length of the pattern.

The POS descriptions may contain wildcards ('*'). A wildcard in a tag allows for any extension of the tag: Example: "*PATTERN: ApPr; Dt*; (VbFQ, VbFF); No**" asks for *any* determiner and *any* noun in the given pattern, and the pattern description does not need to be repeated for all determiner tags. Wildcards describe the subcategories of a basic tag (BTAG): *Dt* or *No* in the example.

Wildcards may also stand for whole tags. There are cases where longer distance phenomena need to be solved. In this case, a wildcard matching an infinite number of tags is available. Example: "*(VbAF, VbFF); *; VbFP*" asks for a structure where a homograph must be disambiguated in dependence of an occurrence of a participle ('*... hat ... [es]... gesehen*').

CONDITIONS:
The condition part is optional and can be empty. It gives additional tests on the nodes in the pattern. As the underlying data structure consists of a lattice of (textform) nodes provided by the tokeniser, all features produced by the tokeniser (i.e. textform and casing) and by lexical analysis (i.e. lemma and tag) available in the node can be used for testing. While textform and casing are identical for all nodes of a given lattice position, lemma, tag, and possibly other features are different per node.

Several conditions, separated by semicolon, can be tested in one rule; example: '*CONDITIONS*: *2 casing lc; 3 lemma zu*' which tests a lowercase form in the

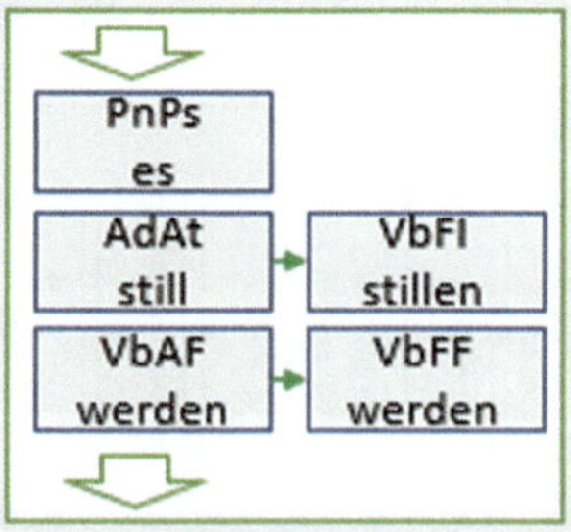

Fig. 7.2 Tagger configuration

second pattern element and a specific lemma on the third. Textforms and lemma tests test for strings as values of these features: '*CONDITIONS: 2 textform dog*'. They can also give ***lists*** of strings, in brackets and separated by comma, like '*CONDITIONS: 2 lemma (tag, stunde, woche)*'. If one of them is found, the test succeeds. They can also contain wildcards in the string tests: '*CONDITIONS: 2 textform *tens*' succeeds, if the textform ends in '*tens*', like '*viertens*', '*achtens*' etc., and '*CONDITIONS: 2 lemma kauf**' succeeds, if the lemma starts with '*kauf*'[5] ('*kauft*', '*kaufen*', '*kaufmann*').

Casing is a feature provided by the LtTokeniser, and can have the values 'lc' (lowercase), 'cp' (capitalised), 'uc' (uppercase), 'mx' (mixed), 'pc' (punctuation), and 'dg' (digit). Example: '*CONDITIONS: 2 textform dog; 2 casing uc*' distinguishes the common noun 'dog' and the proper name '*DOG*'.

No feature annotations can be given for wildcard elements in the pattern.

ACTIONS:
There are three types of actions: KEEP, DELETE and SET.

- KEEP keeps the tag mentioned in the statement, and **deletes all other tags** of this vector. Example: '*ACTIONS: 3 PtVb KEEP*' deletes all readings at the third vector except the PtVb reading. KEEP not only deletes the tags mentioned in the patterns but *all* tags, also tags *not* explicitly mentioned in the pattern. It is used if linguists are sure that the given tag is the correct one.
- DELETE **deletes the tag mentioned** in the rule but leaves all other tags in the vector in place. So, '*ACTIONS: 3 PtVb DELETE*' leaves all readings on the third node unchanged except for the PtVb which is deleted. (If the tag to be deleted is the only one on this position, then the deletion is not performed).
- SET is used to **assign a tag** to a node. This action contradicts to the general statement that tagging consists only in the elimination of nodes. But SET is only used to assign a tag to an 'Unknown' input node. It does not create a new node but just overwrites the tag in an existing one: '*ACTIONS: 1 Unk SET NoPr*' sets the tag NoPr into the node defined by the previous 'Unk' tag, i.e. leaves all other features (textform, casing, lemma) untouched.

Several actions, separated by semicolon, can be given in one ACTIONS statement, like '*ACTIONS: 1 VbAF DELETE; 3 PtVb KEEP*'.

EXAMPLE:
This field is for grammar writers and gives examples for which the rule is intended to fire, or hints for rule writing.
Examples of rules are given in Fig. 7.3.

[5] Normalforms and lemmata have been lowercased before by the Normaliser.

```
ID: AdAt-023
PATTERN: (AdAt, VbFI); (VbFF, VbAF)
CONDITIONS: 1 casing lc; 2 casing lc                    (1)
ACTIONS: 1 VbFI KEEP; 2 VbAF KEEP
EXAMPLE: (weil er es) stillen wird
----------
ID: AdAt-100
PATTERN: (AdAt, NoCo)
CONDITIONS: 1 casing lc                                 (2)
ACTIONS: 1 NoCo DELETE
EXAMPLE: reiche - delete noun reading if lc
----------
ID: AdAv-087
PATTERN: AdAv; (AvVb, AvMF)
CONDITIONS:  2 lemma genug                              (3)
ACTIONS: 2 AvMF KEEP
EXAMPLE: groß genug
----------
ID: DtIn-311
PATTERN: Dt*; (AdAt, AdAv); NoCo                        (4)
ACTIONS: 2 AdAv DELETE
EXAMPLE: ADJ-disambig: ein braver Hund
----------
ID: VbAF-103
PATTERN: (VbFF, VbAF); *; (VbAI, VbAF); VbMI           (5)
CONDITIONS: 1 lemma werden; 3 casing lc; 4 casing lc
ACTIONS: 1 VbAF KEEP; 3 VbAF DELETE; 4 VbMI KEEP
EXAMPLE: Satzkl.: (er) wird (es) sein koennen
----------
```

Fig. 7.3 Tagger rule examples

- Rule 1 prefers an infinitive and an auxiliary over adjective and two finite forms, keeps the infinitive-auxiliary combination.
- Rule 2 excludes a noun reading as the node is lowercase.
- Rule 3 disambiguates a post-constituent adverbial (*lemma 'genug'*) as opposed to an adverb position.
- Rule 4 disambiguates adjective-adverb homographies between determiner and noun.
- Rule 5 covers aspects of German satzklammer in sentences like *[er] wird [im nächsten Jahr vielleicht Lehrer] werden können*. All verbal parts are ambiguous between full and auxiliary verbs, and between finite and nonfinite forms.

There is a risk in the scope of the wildcard. It must be farsighted enough for long-distance nodes, however it usually is non-greedy and takes the next-possible matching node. Attempts to restrict the scope to the next (subclause-indicating) comma had to be dropped because of enumerations (*[er] kauft [Äpfel, Birnen und Kekse] ein*). So only a MAXDISTANCE parameter was introduced to mark the longest possible distance between nodes.

Special symbols are used for sentence-initial and sentence-final nodes.

The LtTagger is implemented for German and English. The whole grammar comprises about 400 rules for German and about 300 rules for English.[6] Parts of it were developed in the context of the PANACEA project, cf. (Bel 2010). The development time for such tagger grammars is weeks rather than months. The software supports it by providing tracing facilities.

7.1.2.2 Rule Application

The rules of LtTagger are constructed such that they expand to the right from a given lattice position, starting from the sentence-initial symbol (*$A*). This design enables the tagger to analyse a sentence in one pass, without having to cycle over it. A consequence is that the current position of rule application has already been disambiguated by previous rules.

The tagger rules are compiled into an internal processing format at loading time, depending on the language to be analysed. The software itself is language-independent.

The main loop calls a method *applyRule*, moving from the first to the last position. This method performs three main tasks:

1. For each lattice position it collects the rules to be fired here. All rules for a given tag will be collected and **sorted** according to their rule ID. In addition, as there may be several nodes at a given lattice position, several rule sets must be prepared, and it needs to be decided which set fires first. Basically the weighting is *No>Ad>Vb>function words*. First the tags at a given position are ranked, and then the rules for the ranked tags are fired in their ranking order.
2. For each rule in the current ruleset a **pattern match** is tried.
   a. A rule pattern matches, if all nodes specified in the rule pattern are found in the sentence lattice position. All tags given in the rule must be identified,[7] which means that the lattice can contain *more* tags at this position but not less. If the pattern contains a wildcard, then the first configuration following the wildcard is searched, the closest match is taken: a non-greedy strategy. If none can be found, the pattern fails.
   b. For each matching node, the **conditions** are evaluated. While textform and casing are identical for all nodes on a given lattice position, the lemma is node specific.[8]
   c. If the whole pattern matches, and all conditions are satisfied, the **actions** are executed. They consist in deleting readings on the specified node: For KEEP, all other nodes at this position are deleted, for DELETE, the current

[6] This is much less than the 3.700 rules reported for EngCG (Voutilainen 1995): (Jurafsky and Martin 2009:172).

[7] So the tags in the rule pattern are AND-ed, not OR-ed.

[8] Tests for *textform* are executed on the *normalised textform* (case-independent).

node is deleted, and for SET, the tag of the current node is reset. (A statement like '*2 NoCo DELETE*' would delete a node with POS *NoCo* at the lattice position 2 (relative to the current position).

3. *All* rules prepared on a given position are tried. If a rule succeeds, then the following rules fire on a *modified* lattice.

When all rules have fired the parser moves to the next position in the lattice and calls *applyRule* from there. Note that this *applyRule* again operates over a *modified* lattice. This procedure is repeated for all positions in the sentence lattice, until it reaches the sentence end.

This is a rather unsophisticated procedure. It gives full responsibility for rule firings and side-effects to the rule designer. For a shallow device this may be adequate, and more complicated devices would not improve accuracy significantly, and would call for a technology switch to LtDeep.

7.1.3 LtTagger Component

Although the rule application is the core component of the tagger other tasks have to be performed before and after rule application.

a. **Before** the core tagger starts the following tasks are performed:

- Add **syntactic defaults**. It is known that verbs (infinitives and some participles) and adjectives can be nominalised. This fact is presupposed in dictionaries and would not be made explicit there: Infinitives are marked as verbs but not as homographs between nouns and verbs. Before tagging however the nominal readings for such nodes must be added to the lattice to complete the selection options (in German with *casing=capitalised*).
- Apply a **multiword filter**. Multiwords are instantiated with the allomorph of just one of their lemmata. The filter checks if the input contains the other parts of the multiword as well. For instance, a lemma '*doomsday*' would instantiate the single word '*doomsday*' but also the multiword '*doomsday scenario*'. The lexicon filter would remove the multiword reading, if the corresponding '*scenario*' could not be found in the input sentence. Details are given in Sect. 6.4. In the example Fig. 7.1 above the filter would remove the circumpositions (*in_hinein* and *in_inbegriffen*) at position '*in*' as their second parts are not in the input.
- Apply a **prefix filter** for homographs of prepositions and verb particles. These lexemes are quite numerous, like $_{\text{de}}$ *an, auf, um, ein, mit,* $_{\text{en}}$ *on, off, about* etc. To remove ballast from the tagger a filter is applied which removes the verb particle reading, if the lemma of the main verb does not ask for it. For instance, in $_{\text{de}}$ '*Er erreicht um 5 Uhr Berlin*' a verb particle '*um*' instantiating a verb '*umerreichen*' is not lexicalised and can be filtered away, leaving only the preposition reading.

- Apply a **defaulter** to set default POS for the unknowns. If the input contains an UNK node, then the defaulter is called to provide a rudimentary POS annotation (very often common or proper nouns). The defaulter is described in Sect. 5.2.5.

b. **After** the rule application the lattice is inspected for positions which still have **alternative nodes**. To disambiguate the remaining positions a simple POS-based weighting scheme is applied (like *No>Ad>Vb>function words*). With this step the task of the tagger is completed: to remove ambiguities produced by lexical analysis.

When all nodes have been disambiguated a **cleanup** component takes care of proper lemmatisation, by creating proper lemmata for split constructions: Lemma of '*kommt ... an*' is '*ankommen*', not '*kommen*'.[9] It also removes lemmata from empty nodes (multiword parts, verb particles).

Finally the resulting list is handed to the Document Model, with all features of the lexical analysis: Lemma, POS, and additional features as built during analysis. Of course other forms of output are also possible like stand-alone tagging applications.

7.1.4 Tests

LtTagger was built in versions for German and English texts.

7.1.4.1 Test Design

A test is based on the comparison between the results of the tagger and a reference analysis, nowadays called Gold Standard, for a set of test texts. The percentage of matches between tagger results and Gold Standard, both for POS tags and for lemmata,[10] determines the tagger quality. Error rates are measured by token, evaluating whether a given token has a correct<POS, lemma>assignment or not.

7.1.4.2 Test Data

A tagger test requires two sets of data: A training set containing the Gold Standard sentences for component development, and a test set for the final test. Both sets should cover different domains.

Providing texts for the **Gold Standard** is a challenge. Most tagger tests use the 'usual suspects', i.e. the PENN treebank (English) or the TIGER treebank

---

[9] Many taggers for German return the wrong lemma ‚*kommen*' due to missing particle treatment.

[10] The produced lemmata are usually not subject to evaluation in tagger tests, e.g. incorrectly prefixed verbs resulting from satzklammer are often not flagged.

(German). This option, however, cannot be used for the LtTagger tests as both collections use different tagsets: an extended tagset (Penn treebank tagset[11]) for English, and a different standard tagset (STTS) for German.[12] The option to convert these tagsets into the LtTagset was abandoned as it would introduce new uncertainties, in particular as some tags cannot be easily mapped. Instead a bootstrapping procedure was used and the test corpus was built up sequentially: New texts were analysed by a pre-version of the tagger, then manually corrected, and added to the test set as 'Gold Standard' sentences for regression tests. For a final test a new text corpus was selected and evaluated by human experts. The drawback that the amount of evaluation data is limited in such an approach is compensated by the possibility to collect test data from different domains.

7.1.4.3 Test Results

For final tests, texts from Health&Safety, Sports, and Food were selected for German, texts from news, software manuals, Europarl speeches, and environment were chosen for English. The test texts were not cleaned and contain errors (typos, punctuation etc.). They were tagged and manually evaluated by two experts. The results are reported in Fig. 7.4.

7.1.4.4 Evaluation

The results show that the LtTagger is competitive in terms of accuracy. The results for English are slightly lower than other taggers; this may be due to the fact that

	sentences	tokens	errors	accuracy %
English				
News (reuters)	76	1349	52	96.15
Software (ibm)	124	1259	74	94.12
Environment	65	1341	101	92.47
Speeches (europarl)	123	3261	115	96.47
total	388	7210	342	95.26
German				
Health&Safety	130	3993	77	98.07
Sports (sz)	98	1427	39	97.27
Food (sz) (Milk)	100	1385	27	98.05
total	328	6805	143	97.90

Fig. 7.4 Tagger test results

[11] Marcus et al. (1993), discussed in Sect. 2.3.2 above.

[12] Schiller et al. (1995), also discussed in Sect. 2.3.2 above.

other English taggers mainly use an extended tagset (Penn Treebank tags) while LtTagger uses a standard tagset (without morphological features). The results also show slightly different results in different domains for both languages.
The factors which influence the tagger accuracy seem to be the following:
1. Selection of the **tagset**. Extended tagsets provide more information for disambiguation; and while the accuracy for the *extended* tagset is lower (as more features have to be disambiguated), the *standard* tagset profits; an observation also made by (Tufiş 2000) and (Toutanova et al. 2003). Schmid and Laws (2008) report 97.9% on a testset close to the training set but a drop by about 10% to 86–88% with an extended tagset. But the *design* of the tagset is important as well, e.g. the use of tags for multi-part elements (like circumpositions or two-part conjunctions), or subcategorisation of adverbs into NP and/or VP modifiers.
2. The closeness of the text to the training **domain**. Most taggers are trained and tested on the same domain, usually news text. All taggers mentioned above have at least 3% lower accuracy, if the domain is changed. (Giesbrecht and Evert 2009) show that TreeTagger drops to 93.7%, UIMA to 91.7% Stanford Tagger to 92.6% and TnT to 92.7%, if the domain is changed. Also the Brill tagger reaches only 91.6% when other text types (web texts) are tried. The same has been stated for a rule-based German tagger (TBL, cf. (Volk and Schneider 1998)), based on the Brill tagger: It also drops by 3% for out-of-domain text. Note that it is difficult to train such taggers for other domains, as usually, the required training data are not available and would have to be created manually, which requires significant effort for their development, possibly more than writing a rule-based tagger would need.
The LtTagger is rule-based; so it profits from the fact that the *rules* of a language are domain-independent; the results show that it is more robust towards different types of text and domains.
3. The amount of **out-of-vocabulary words**. Unknown words are always a source of error. In LT-Tagger, this holds for proper nouns in particular, which are sometimes analysed as common nouns. Domain adaptation therefore is an important quality factor to reduce unknown words. LtTagger runs after the Named Entity recogniser (LtNamer, cf. Sect. 6.2 above) in the processing chain so it can take care of unknown names. In addition there is a special component for the recognition of out-of-vocabulary words: The LtDefaulter described in Sect. 5.2.5 above tries to tackle this problem in a more systematic way.

If the results of taggers are compared with human performance, the results are similar: Both show a 3–4% error rate, according to (Jurafsky and Martin 2009:189). Compared to other approaches the results of the LtTagger seem to be competitive. For a developer of a tagger this may be satisfactory.

However the comparison could also look at another direction: Fig. 7.4 above shows that the average sentence length in the test corpus was 18.5 words for English, 20.7 for German. So, given the above-cited accuracy rates, about four out of five sentences contain a tagging error. As tagging errors are seldom corrected by later components in shallow parsing, major disturbances can be expected in chunkers or parsers. This fact gave rise to considerations to include tagging in a deep parsing component with full syntactic analysis (like LtDeep). For an indexing application this problem is less severe as the purpose of tagging is just lexical disambiguation: No complete sentence coverage is required for this task, and no function words need to be disambiguated (they show the highest ambiguity and error rates).

7.2 Semantic Disambiguation (LtSensExtract)

The previous component (LtTagger) had the task to disambiguate syntactic (POS) annotations. This component (LtSensExtract) must disambiguate semantic annotations. Only if terms can be semantically identified, they can be used in multilingual indexing, as they must correspond to their equivalents in other languages. In cases where a term is semantically ambiguous it is necessary to identify the correct sense of the term in a given context, in order to translate it and link it to the proper index term.

The identification of the correct sense (or reading) is the task of a Word Sense Disambiguation (WSD) component. Word Sense Disambiguation is an 'intermediate task', according to (Ide and Véronis 1995), i.e. required for many linguistic applications, like Information Retrieval, Content Analysis, Machine Translation, Speech Understanding,[13] and others, cf. (Correa et al. 2018). It has the objective to assign the correct sense to a term in a given context, cf. (AlMousa et al. 2021), (Kwon et al. 2021).

However, as already noticed by (Specia and Motta 2006), the focus in WSD research abstracts away from the specific application and the requirements which a given application poses for the sense disambiguation. This can be seen by the two main research lines taken: The 'all words' line tries to find the best sense for each word in a sentence while the 'lexical sample' line identifies a predefined couple of 50 to 100 terms based on POS (nouns, verbs, adjectives), frequency (in the BNC corpus), and polysemy.[14] Both lines of research do not fulfil the requirements of the sense disambiguation of text analysis: Such applications need to disambiguate all terms with more than one sense, neither *all* terms (in the 'all words' case), nor only some specially *selected* terms (in the 'lexical sample' case): They need to find proper word senses for translations or correspondences in a multilingual setup.

7.2.1 Sense Definition

Talking about Word Sense Disambiguation requires first talking about word senses. Although there is common agreement that e.g. '*menu*' in gastronomy and in computer science are two different senses the definition of what a sense is

[13] e.g. to determine stress: $_{\text{en}}$ *désert* versus *desért*, $_{\text{de}}$ *dúrchsetzen* versus *durchsétzen*, etc.
[14] Cf. the descriptions of SensEval in (Kilgarriff and Rosenzweig 2000), (Correa et al. 2018).

would be the task of philosophers, not computational linguists. So far there is no agreement what a sense is, cf. (Ide et al. 2002), and if human experts are asked to assign senses to terms, there is not really a common opinion on this, cf. (Kilgarriff and Rosenzweig 2000). Also, when looking up different (human-edited) dictionaries different sense descriptions can be found. As a result, if the assumption of senses is questionable, then also their disambiguation is unstable. On the other hand, different senses should differ in their linguistic behaviour and show different linguistic properties. Claiming that the results of some clustering activity would produce different senses[15] is not sufficient as long as such senses cannot be described, i.e. what the members of such clusters have in common. What is discussed as ‘labelling’ is in fact the human achievement of abstraction.[16]

An often observed consequence is to take an existing resource and work with the sense inventories offered there. The resource most often used is WordNet, cf. (Fellbaum 1998). It contains terms, senses with synsets for each sense, links like hyponyms/hyperonyms, and other useful information. The concept of WordNet was transferred to many other languages, like Italian, cf. (Florian and Wicentowski 2002), Dutch, cf. (Postma et al. 2016), Swedish, cf. (Merkel and Andersson 2001), Persian, (Rouhizadeh et al. 2021), and many others.

However, WordNet senses are based on psycholinguistic research, not on computational requirements. As a consequence, some sense differentiations are either irrelevant or not distinguishable by computational means. Also the WordNet synsets differ in explicitness (according to (Wiedemann et al. 2019)) and give sometimes only very small contexts, cf. (Banerjee and Pedersen 2002), which is a drawback, if synsets are to be used as disambiguation material, cf. (Siemiński 2011).

Other attempts to sense disambiguation use lexical dictionaries, cf. (AlMousa et al. 2021), BabelNet, cf. (Kwon et al. 2021), ConceptNet, cf. (Havasi et al. 2007) or (Speer et al. 2017), or an Associative Concept Dictionary containing senses found by association research.[17]

All these decisions are of limited use in the context of LtSensExtract.[18] From a computational point of view, senses must have properties which must be computationally tractable such that they can be identified by computational procedures, e.g. different verb frames, different passivation, different morphology or other linguistic clues. For multilingual indexing, this requirement can be defined in a rather

[15] (Wiedemann et al. 2019) assume no predefined senses and propose to use a system parameter for the number of senses to be used. It is unclear what the use of such clusters could be.

[16] If it was just labelling, then a simple enumeration of integer values could do.

[17] (Tsutsumida et al. 2010), cf. also (Kwong 2012).

[18] The component was called LtXfr in previous contexts (Thurmair 2014a), related only to Machine Translation but explaining the same idea.

straightforward way: The system must disambiguate senses which lead to different translations, and therefore also to different index terms.[19] So, roughly speaking, a term has as many senses as it has translations. The task then is to find a disambiguation procedure for these terms. An approach towards combining sense indicators and transfer entries in lexicography has been presented by (Atkins and Bouillon 2006), where domain information is used as one of the indicators for sense discrimination. An example is given in Fig. 7.5. Corpus-based techniques can help to improve these results, according to (Kwong 2012).

The complexity of the task can be estimated by a comparison between the number of source language terms and the number of their target equivalents. Figure 7.6 shows the average number of translations for a given source term in the TINA resources, i.e. the disambiguation effort to be undertaken. It is between 1.3 and 2.4 senses per word on average.

If different senses are assumed for a term, then the **naming** of the sense may be an issue. Dictionaries often give numbers (like ‘*cell*^{1}’, ‘*cell*^{2}’, ‘*cell*^{3}’) to differentiate them. This strategy is adopted in some LtLemmatiser dictionaries which use reading numbers. In multilingual contexts this approach is a bit clumsy as different translations need to be assigned to different senses. But then the senses could be named after its translation right away. This strategy is adopted in LtSensExtract. However the naming is less important as long as the names refer to the right meaningful multilingual clusters.

Winde, f
1. TECH winch, windlast, hoist
2 BOT bindweed

Fig. 7.5 Domain as sense disambiguation means

lang direction	no SL terms	no TL terms	senses/word	lang direction	no SL terms	no TL terms	senses/word
ar->de	38,7	50,1	**1,29**	es->en	118,8	221,1	1,86
de->en	150,0	222,2	1,48	fa->de	59,1	82,3	1,39
de->fr	59,5	84,7	1,42	fr->de	51,0	84,7	1,66
en->de	138,6	222,1	1,60	fr->en	82,1	165,6	2,02
en->es	128,5	221,0	1,72	it->en	76,1	169,8	2,23
en->fr	85,4	166,0	1,94	pt->en	63,8	152,3	**2,39**
en->it	78,2	169,6	2,17	ps->de	42,5	60,8	1,43
en->pt	67,5	152,0	2,25	tr->de	62,0	89,2	1,44

Fig. 7.6 Number of senses in different bilingual dictionaries

19 Marsi et al. (2011) take the opposite view: They assume a dictionary which is already word-sense annotated, and try to identify transfers for the (pre-existing) senses. They shift the sense definition issue to their input resource.

7.2.2 *Features for WSD Analysis*

Word Sense Disambiguation (WSD) assigns sense tags to terms depending on the contexts in which such terms occur. The contexts must provide the information which allows to disambiguate the senses.

7.2.2.1 Reference Data

It depends on the WSD approach which data can be used.

- In case of **supervised** approaches, significant amounts of hand-labelled training data are required. The quantity of such data is one of the factors which most affect the system quality, according to (Banerjee and Pedersen 2002). In turn, the lack of sufficient labelled training data is the central problem for the supervised learning approach, according to (Yuan et al. 2016). Therefore most contributions focus on few publicly available training data, mostly SensEval-1 and Senseval-2, cf. (Kilgarriff and Rosenzweig 2000), (Kilgarriff 2001). Other available data are the contexts of senses in WordNet as presented in the synsets, e.g. (Siemiński 2011) although the contexts are rather small, cf. (Banerjee and Pedersen 2002) and their distribution over the senses is somewhat unbalanced, cf. (Wiedemann et al. 2019).
- Approaches of **unsupervised** learning are less restricted in the selection of their training data, and for instance use Wikipedia, as in (Devlin et al. 2019).
- **Knowledge-based** approaches also do not require pre-annotated training data and are not limited in the number of words they are supposed to analyse, according to (Kwon et al. 2021).

LtSensExtract follows an approach proposed by (Ide et al. 2002) and uses crosslingual data. This fits to the way senses are defined: If different senses lead to different translations, then bilingual data would be the best source for training data.

The corpus used by LtSensExtract consisted of parallel sentences collected from different domains: Europarl, JRC, news, health&safety, automotive, and others. In total, 3.8Mio parallel sentences German English were used. Many of them are aligned and delivered in TMX format,[20] and therefore the context window for the detection of a given term-sense candidate are sentences (instead of a fixed length context window). The restriction to sentence contexts is a small drawback as paragraphs would be a more natural unit for sense analysis[21]; however paragraph information is not available in TMX.

[20] TMX V1.4b is defined in https://www.gala-global.org/tmx-14b.

[21] An approach based on context of paragraph size has been used in (Thurmair 2005).

7.2.2.2 Features for WSD Analysis

It depends on the WSD approach which features are selected for disambiguation. Unsupervised learning only uses word embeddings, i.e. only looks at tokens and their contexts. Supervised learning uses two kinds of features:

- **Local** features look at the term and its immediate context, and according to e.g. (Correa et al. 2018), can be divided further:
  - **word related** features: In addition to unigram/bigram information, using words as features requires lemmatisation/base form production and decomposition (in German).[22] This is a very important operation for languages with inflection, like German or Italian: (Florian and Wicentowski 2002) report that a ten times larger training corpus would be needed for Italian, if no lemmatisation was used. Base forms instead of tokens improve the analysis results significantly, according to cf. (Lindén and Piitulainen 2004).
  - **syntactic features**, esp. syntactic relations. They seem to increase precision but lower the recall.[23] Much more analysis rules would be required to cover all the syntactic variants which can be found in texts. Also, as syntactic analysers have a certain amount of erroneous analysis results syntactic features often lack robustness.[24]
  - **semantic features**, e.g. Named Entities, cf. (Dang and Palmer 2002), additional relations like offered in WordNet, and other sources.

There is significant literature which features could be used, which features are suited best for which parts of speech, and what is the minimum of features required for acceptable analysis results, cf. (Mohammad and Pedersen 2004).

- **Topical** features address the larger context of a word and collect words which frequently co-occur with a given sense of a word in a contextual window of a certain size, like (Ide and Véronis 1998). For instance, (Leacock and Chodorow 1998) investigated a word sense assignment procedure based on a window of 6 content words. The results are usually collected in a 'bag of words' structure[25] and contain the most probable context terms for a given sense-term.

  While the bag-of-words technique does not provide information of the contextual sequence of such senses,[26] the focus shifted from there to approaches using word embeddings (Iacobacci et al. 2016) in language models, and proposing systems like BERT, cf. (Devlin et al. 2019) and unsupervised training.

---

[22] A non-decomposed list like given in (Marsi et al. 2011) contains entries like 'bank*feiertag*', '*blut*bank', '*daten*bank' which have nothing to do with banks and point to quite different senses.

[23] As steted e.g. by (Martínez et al. 2002; Lindén and Piitulainen 2004).

[24] As a consequence, required syntactic contexts are not found, and just default assumptions can be given, cf. (Thurmair 2014a).

[25] Examples are (Kwong 2012; Correa et al. 2018; Dang and Palmer 2002).

[26] As observed by e.g. (AlMousa et al. 2021; Yuan et al. 2016).

Topical features are found to improve the recognition significantly as compared to local features, according to (Correa et al. 2018), and have more impact on the analysis result than local features. This also holds for the use of word embeddings, cf. (Iacobacci et al. 2016). However the setups of such comparisons specialize more and more as the focus of attention shifts to advanced learning methods, using word sense disambiguation just as an example application.

An important issue not considered by most such approaches is the treatment of multiwords.

Papers on word sense disambiguation and multiwords often focus on the identification of multiwords using WSD techniques: (Tanner and Hoffmann 2023) identify verbal multiwords, (Hey et al. 2021) search for nominal multiwords (names and compounds) in software requirements texts, (Arranz et al. 2005) also focus on multiword identification in running texts. However this is not word sense disambiguation for multiwords, but in turn identifying multiwords using word sense disambiguation.

Intuitively, word sense disambiguation for multiwords should be identical to the one of single words: Both are concepts with specific semantics, their difference being that the lemma of multiwords contains white spaces. So no extra effort should be required for multiwords once they are identified as such (and do not try to do WSD for their single word elements).

However most of the proposed WSD techniques rely on single word analysis, in particular if they analyse token contexts for sense identification. In cases like $_{en}$ ‘*high street bank*’ obviously ‘*high street*’ is not a context of ‘*bank*’ but part of the concept itself, and so is ‘*blood bank*’: No sense for ‘*bank*’ would match this term. ‘*Street price*’ is a stock exchange term and has nothing to do with ‘*street*’, ‘*street name*’ is a dummy company[27] and neither contributes to a sense of ‘*street*’ nor to a sense of ‘*name*’. There are thousands of such examples, especially in special domains: ‘*exposure value*’ in photography, ‘*limit bridge*’ in sensor technology, ‘*stability moment*’ in shipbuilding all are examples for mistakes, if systems try to assign senses to just parts of multiwords. It is obvious that techniques relying only on the proximity of tokens are misguided in cases where several tokens belong to a multiword and drop the performance, even if large corpora are available. It has been shown above that about a third of all terms are multiwords.

LtSensExtract uses terms consisting of triples of<textform, POS, lemma>as local features for word sense disambiguation, as produced by the Lexical Analysis components. They represent single word and multiword concepts the same way.

[27] Also ‚*bogus firm* ‘, $_{de}$ ‚*Scheinfirma* ‘ (cf. (Muret & Sanders, 2003 (Langenscheidt)).

7.2.3 *LtSensExtract*

7.2.3.1 Design

The basic idea is based on the following considerations:

In a multilingual text analysis setup where in a final step monolingual concepts are linked to a multilingual index, it is plausible to exploit the potential of multilinguality:

1. The sense inventory of an entry in the source language is defined along the different transfers of this entry into the target language: A sense of an entry corresponds to one of its transfers.
2. In order to find training data to detect the topical features for these senses, two sorting operations are performed:

- First the corpus is sorted on the *source* side. This operation creates groups of terms sharing the same source term.
- Next, each of these groups is sorted according to the *target* side. This operation creates sense groups, i.e. groups of source terms sharing the same target term.

3. The *source* side sentences of the resulting sense groups are collected into subcorpora. These subcorpora represent contexts of the word senses and contain the data to extract topical features for the sense identification.
4. The source side is analysed to find the topical features, the target side is only needed to define the training corpus. The resulting features form a resource to identify the sense of the term; the collection of features for a sense is labelled by the target term.
5. The result is a resource which for every sense of a term provides source language context terms, labelled by a translation. This resource is used at runtime for sense disambiguation.

So, the bilingual corpora are used to identify training data for the different senses of the source term. It should be noted that the target terms are only used as a means to define the training data to extract sense data on the source side, and to subdivide the source side sentences along the target terms/senses.

This approach is not a multilingual sense disambiguation: The target words used as source sense descriptions may themselves have additional meanings and other contexts. This is not in the focus. Target data are only relevant inasmuch they can support the sense distinction on the source side. (Bonansinga and Bond 2016) detect senses in bilingual sentences by using the sense intersections of WordNets in several languages. In contrast, LtSensEval disambiguates word senses in one language by using topical features found with the aid of another language.

7.2.3.2 Implementation

LtSensExtract was implemented for German, with English as 'sense-defining' language, in the context of a transfer selection task in the PANACEA project, cf. (Bel 2010).

part of speech	no. entries	no. senses	no. senses / entry
adjectives	6900	18200	2.83
nouns	15600	35400	2.27
verbs	4500	17800	3.26
total	27000	71400	2.63

Fig. 7.7 Word sense repository data

1. Data

a. **Dictionary data**: A bilingual dictionary was taken for the LtSensEval implementation as used for human lookup: LinguaDict as a publicly available dictionary[28] was selected in order to have a realistic size of a dictionary. It containes 145 K German lemmata and 213 K English translations, i.e. about 1.5 translations per term.

All terms with just one transfer were removed, as it is implied that that there is only one sense for this term. This is the bulk of the terms, mainly special domain vocabulary, but also many general vocabulary terms. Only terms with more than one transfer, indicating more than one sense, were kept. Of those, all transfers with non-identical POS tags on source and target were also removed.

After this operation, 27.000 terms (de) with 71.400 sense candidates (en) remained. The English terms are used as a sense inventory for the German terms. Figure 7.7 gives the details on the dictionary used for the following analysis.

b. **Corpus data**: As the idea is to use the target language as sense inventory for the source, the relevant data would be collections of aligned bilingual texts.

The corpus used for the experiments consisted of parallel sentences collected from different domains: Europarl, JRC, news, health&safety, automotive, and others, cf. Figure 7.8. In total, 3.8 m parallel sentences German-English were collected.

Domain	No sentences
automotive	47k
administration (dgt)	530k
europarl	1,739k
health&safety	57k
legal (jrc)	1,239k
literature (kindle)	82k
news (statmt_dev)	15k
news (statmt_news)	136k
total	**3,848k**

Fig. 7.8 WSD corpus data

[28] The Linguadict dictionary, under http://www.linguatecapps.com/linguadict. Meanwhile no longer available.

1. Processing of the Training Corpus

Preprocessing

The corpus data were processed in the following way:
1. **Format conversion** into a uniform processable format: The LtText format as described in Sect. 3.1 was used for this purpose.
2. **Lemmatisation and tagging**, using the Lexical Analysis tools described in the above chapters.
3. **Monolingual indexing**: all terms (i.e. <lemma, POS> combinations) were indexed with the IDs of the sentences in which they occurred, both for the source (de) and the target (en) side.
4. **Bilingual indexing**: The two monolingual index files were merged such that subcorpora of all sentence pairs which contained the term on the source side and one of its senses on the target side were built, using the two-step sorting just explained, and the common sentence IDs were stored.

There were many cases where no bilingual sentences were found, obviously because dictionary and corpora were two independent sources. Terms which had no correspondences in the corpus had to be removed.

Feature Selection

The bilingual index file was processed in the following way:

Step 1: **Subcorpus selection**: Based on the bilingual indexing, subcorpora were extracted such that for each source term all sentences containing the same target term were collected. This operation left 8.12Mio[29] aligned sentences with relevant <term,sense> pairs.

Step 2: **Word alignment**: It could happen that a sense candidate of an term is found in the target sentence, however it is not a translation of this term but just happens to occur there. In order to avoid accidental co-occurrence of a<term,sense>pairs, a filtering step was inserted: The subcorpora were word aligned, using GIZA++[30]. This operation removed 40% of the context sentences, leaving 4.9 Mio bilingual sentences.[31] Figure 7.9 shows the remaining data sets.

---

[29] One sentence may contain several source terms, therefore this number exceeds the total number of sentences.

[30] GIZA++ (Och and Ney 2004). A problem was that Giza++ aligns only single words.

[31] The amount of rejections here is a bit surprising, and may need further inspection of the Giza++ results.

Part of speech	original pairs <entry,sense>	after bilingual indexing	after word alignment
adjectives	6,900	4,670	1,240
nouns	15,600	11,360	3,690
verbs	4,500	3,930	1,680
total	27,000	19,960	6,610

Fig. 7.9 Data set size during preparation

As a result, 6,600<term,sense>pairs out of the original 27,000 could be used for word sense disambiguation. So, parallel data for sense disambiguation could be provided for only 25% of the terms even in a large parallel corpus.
After this operation, the target side information is no longer needed for word sense disambiguation.

Step 3: **Classification**: The resulting subcorpora were used on the *source* side as training data to find topical features. To cope with data sparsity, each source sense-subcorpus was further subdivided into packages according to the data which were available for feature selection: Class 1: all sense candidates of an term have at least five sentence contexts; Class 2: each sense candidate of an term has at least one sentence context; Class 3: an term contains also sense candidates with no parallel sentence contexts.

Figure 7.10 shows the result. Only one third of the remaining subcorpora showed more than five sentences per term (class 1). Nouns are slightly better represented than verbs and adjectives. Most terms have just one example sentence per sense candidate.

For each<term,sense>subcorpus of class 1, the most frequent context words of the respective term were extracted from the source side. To select the best features a simple frequency count was used. More elaborate measures could be used but the main problem was not the measure but the sparsity of the training data.

Part of Speech	total <term, sense> pairs	class 1 (> 5 contexts)	class 2 (>= 1 context)	class 3 (no context)
adjective	1244	301	667	276
		24.2%	53.6%	22.2%
noun	3693	1370	1789	534
		37.1%	48.4%	14.5%
verb	1677	447	859	371
		26.6%	51.2%	22.2%
total	6614	2118	3315	1181
		32.0%	50.1%	17.9%

Fig. 7.10 Data for feature extraction

Therefore experiments to further qualify the resulting features, e.g. by computing probabilities or tf·idf values, were abandoned, given the sparsity of the data.

The result is a collection of contextual (topical) terms which indicate a sense for this term.

From the subcorpora, about 1000 parallel sentences of class 1 were subtracted to be used as a test set; the rest was used for building the WS repository.

The LtSensExtract Repository

The results of the word sense disambiguation of LtSensEval were stored and maintained in a special resource: the Word Sense Repository (WSR). It is language-specific of course. It contains the topical features for each of the senses of an term. Each entry has the following structure:

- the term, in the form of lemma and POS (BTAG),
- the sense, its lemma being a foreign language transfer (and its BTAG),[32]
- a comma-separated list of topical features which frequently cooccur with the term, sorted by frequency.

An example is given in Fig. 7.11.

Fallback Strategy

The analysis of the package coverage showed that sufficiently many contexts for feature collection would be available only for about one third of the terms. So there are several reasons why a fallback strategy is recommendable:

- Not all senses of an term can be covered with training data, as can be seen in Fig. 7.11.
- For many terms, no training data can be collected at all.
- Even if WSR entries can be constructed, it may be that a given context at runtime does not contain any of the topical features.

In all these cases of data sparseness a fallback strategy is applied consisting of a simple frequency check. Determining senses by simple frequency is a good means of disambiguation, cf. (Marsi et al. 2011).

In LtSensExtract frequency information is collected from two sources:

- All terms with >1 senses are assigned probabilities related to the number of training sentences. Senses occurring in more sentences score better.
- If no sentence can be assigned, then the corpus frequency of the different target terms (i.e. the senses) is taken. This heuristics is rather weak, however, because such transfers can also be translations of other terms. However there are cases

[32] The tag is only needed in translation applications. It is only marginally relevant for sense disambiguation.

stärke	No	strength	europäisch, union, liegen, europa, land, politisch, schwäche, neu, groß, wirtschaftlich
stärke	No	forte	angehen, drogenhandel, abscheulich, machen, fern, loyalistisch, terrorist, organisieren, verbrechen
		force	groß, politik, krieg, instrument, teil, bringen, pakistan, unterseeboot, kriegsschiff, echt, wirtschaftlich, verleihen, protest, einsetzen, international, militärisch, fall, gewiss, grenze, schließung
		starch	erzeugnis, gehalt, erstattung, menge, enthalten, stärke, grunderzeugnis, charakteristisch, einzeln, gewährend, staffeln, rohfaser, asche, spelze, fett, tatsächlich, protein, verarbeitungserzeugnis, herstellung

dichtung	No	seal	integriert, einreichen, absaugung, wirksamkeit, erforderlich, einzelheit, unversehrtheit, einzeln, drücken, einfuellstutzen, dichtung, fest, schließstellung, verschluss, gehalt, position, sicher, ausliefern, einbaufertig, fetten
dichtung	No	gasket	sem, semicarbazid, verschlossen, dicht, sterilisation, treibmittel, kunststoffdichtung, metalldeckel, schließe, glasgefäß, verwenden, neu, erkenntnis, zerfallen, azodicarbonamid, hitzeeinwirkung, herstellung, aufgeschäumt
dichtung	No	packing	
dichtung	No	poetry	einführung, werk, bringen, gang, zivilgesellschaft, russisch, spielen, katalysatorenrolle, internet, massenmedien, buchdruck, wichtig, kapitalismus, finanzinstrument, belletristik, osmanisch, erbe, klassisch, islamisch

Fig. 7.11 Examples of word sense repository

where transfers of dictionaries have no occurrence at all in the corpus, so the target probability has a certain amount of justification.

The two probabilities are stored in a separate resource.

3. Runtime Analysis

At runtime, each term of the text (lemma) is looked up in the LtSensExtract WSR repository. If it has an entry there, it is compared to the topical features by means of a modified Lesk algorithm.[33] The best sense is returned.

In case no result can be found the fallback resource is consulted and the sense with the highest probability is returned.

7.2.3.3 Tests

It is a challenge to test a component like LtSensExtract as no comparison data are available, a special sense inventory was used, etc.

The test is based on the above-mentioned consideration that the word sense assigned to a term would trigger a specific translation in a multilingual setting, differing from a compositional translation. Therefore a correctly assigned word sense would return a correct translation. This indirect method could be used for testing: The word sense disambiguation component is supposed to return the proper transfer in a translation environment. While a correct translation may be the result of other reasons the opposite obviously holds: A different translation indicates a different word sense.

So the test would take a bilingual dictionary (LinguaDict) and try to find out which word sense was selected in a given sentence context by inspecting the chosen translation.

1. Test Corpus

The test corpus was taken from the subcorpora used for the training. Before the cooccurrence was computed, one test sentence each was extracted from all packages of class 1 (i.e. translation has more than 5 example sentences), for each POS: from nouns (694 sentences), verbs (205 sentences), and adjectives (694 sentences); overall the test corpus consists of 1043 sentences. The sentences were manually annotated with the correct sense/transfer. The test sentences themselves were not cleaned; they contain different kinds of errors (sentence segmentation, tagging etc.); they are just as found in the corpus. Each test sentence is a triple of<the

[33] (Lesk 1986), cf. also (Kilgarriff and Rosenzweig 2000).

source lemma to be tested, the target lemma (i.e. sense) expected, the source sentence as context>. The idea is to analyse the test sentence using the LtSensExtract component, and compare its proposal with the reference target lemma.

2. Ranking

The word sense selection component is tested by translating a test word in a given sentence context, and comparing its translation with the reference translation. In the best case, all translations proposed by the LtSensExtract component are identical with the transfers selected in the reference translations.

However, as the LinguaDict dictionary contains many near translations, which can hardly be distinguished, a special evaluation procedure was adopted, consisting of three ranks:

- **Rank 1**: the translation proposed by the system is identical to the one in the test reference sentence
- **Rank 2**: the proposed translation is close/synonym to the one in the test reference sentence. This is the case if
  - the proposed translation belongs to the same WordNet synset as the reference
  - the proposed translation is orthographically similar to the reference (like: '*electric*' versus '*electrical*', '*agglutinating*' versus ,*agglutinative*', '*dialogue*' (UK) versus '*dialog*' (US) etc.
- **Rank 3**: the two translations are (still) different.

The evaluation would count rank1 and rank2, and reject rank3 results.
Two additional resources were produced for the ranking (esp. rank2: similarity):

- an indexed version of WordNet V3, whereby for a given input lemma a list of possible synonyms was retrieved (i.e. the synset lemmata).
  It should be noted that WordNet entries differ considerably from the LinguaDict entries; WordNet has 155,000 entries while LinguaDict has 136,000 English lemmata (from 210,000 transfers); however the two resources have only 45,200 entries in common.
- a resource for orthographic similarity. For all parts of speech, a resource was used which unifies US and UK spelling (This list contains about 4,700 entries). Additional patterns were collected for adjectives, conflating terms with the same meaning but different suffixation, cf. Fig. 7.12.

The test frame applies pattern matching for the strings, and simple lookup for the differences in locale.

adj + -ed	*'abstract' vs. 'abstracted'*
adj-ic + al	*'acoustic' vs. 'acoustical'*
adj+able + ive	*'adaptable' vs. 'adaptive'*
adj+ating + ative	*'agglutinating' vs. 'agglutivative'*
adj+ated + ative	*'agitated' vs. 'agitative'*
adj + ous	*'amorph' vs. 'amorphous'*
adj+ant + ing	*'ascendant' vs. 'ascending'*
adj+ist + istic	*'capitalist' vs. 'capitalistic'*

Fig. 7.12 Adjective endings treated as equivalent

3. Test Frame

A special test system was written which has an term (source lemma) and its sentence context as input, and returns the 'best matching' sense (target lemma). This return lemma is compared to the gold standard sense assignment, and ranked: In case they are not identical (rank 1), the system checks if they are both in the same WordNet synset, or are orthographically similar (rank 2).

For a comparative evaluation all publicly available translation systems were used: four rule-based[34] and one statistical[35] system. All test sentences were translated, and the transfers for the terms to be tested were collected and compared to the reference translation. The total (rank1) and partial (rank2) identity were computed, as well as the overall scores.

4. Test Results

Figure 7.13 shows the evaluation result. It shows that the results of LtSensExtract are significantly better than the baseline (col. 1) which would be built by a random selection of senses. Already the simple frequency count strategy (col. 3) is better than the baseline, and the full word sense disambiguation (col. 2) is again significantly better than the frequency fallback.

The comparative evaluation in Fig. 7.13 also shows that the results of LtSensExtract outperform the available translation systems, both the best rule-based (RMT) and statistical (SMT). Of course all systems have different dictionaries with different transfers/sense inventories. But the evaluation result is an indication that the proposed method for word sense disambiguation is a valid option for quality improvement.

5. Discussion

It has turned out that **data sparseness** is the main challenge for LtSensExtract. Starting from a corpus of 8 Mio parallel sentences, only a minority of transfers

[34] Linguatec 'Personal Translator', Langenscheidt 'T1', Systran, ProMT.

[35] Google Translate in a version of 2014.

	Baseline	Xfr- full	Xfr- frq	SMT	Best RMT
Noun	41.6				
r.1		61.2	49.3	55.3	44.2
r.1+2		75.2	66.7	69.3	61.4
Adj.	43.1				
r.1		58.6	49.7	53.1	40.0
r1+2		71.7	66.2	64.1	54.5
Verb	39.0				
r.1		61.3	50.5	47.5	44.6
r1+2		79.4	68.1	66.2	65.7
Total					
r.1		60.9	49.6	53.5	43.7
r1+2		**75.6**	**66.9**	68.0	61.3

Fig. 7.13 LtSensExtract acuracy (in %)

described in the dictionary was found in the corpus. Aside from the fact that this sheds a light on the scope of corpus-based approaches it shows the need for an option for human experts to adapt the Word Sense Repository (WSR) by adding/deleting entries manually to increase precision. The requirement to have an option to manually edit the repository has an impact on the sophistication of the statistical annotations: More and better WSR terms seem to be more important than the sophistication of the statistical computation.

Even if the results of LtSensExtract show the best performance of all compared systems, there are many parameters which may influence the outcome: Language direction, corpus selection and size, dictionary baselines and size, test set selection, and others. So there is significant room for improvement of the approach presented. However it seems to be a promising strategy for sense selection, in particular as it shows the following features:

- It is **robust**, inasmuch it does not react of parsing failures or problems of syntactic analysis. It does not use syntactic environments but just conceptual ones.
- It considers only **inner-sentence** contexts, i.e. sentences or at most paragraphs. This fact allows for easy integration into existing MT architectures, which are also sentence based.
- The WSR repository can also be used to disambiguate **Named Entities** (in cases like '*White House*', as building (place) or as government (institution)). Moreover it can represent both single word and multiword terms.
- It is **modular**. It can be added to many different system architectures as an add-on component. Adding new entries does not require re-computation of all transfer conditions. The dictionary can be loaded and unloaded as a special module depending on the analysis flow: If an application (e.g. tagging) does not need word sense analysis, the dictionary simply is not loaded and does not complicate the workflow

- It is **expandable**. It supports corpus-based lexicon extension, e.g. in cases where and extended data source leads to updates in the Word Sense Repository, and a new version must be built: Only the data resource of the system needs to be exchanged. This also holds for analysing results of post-editing by which new contexts may be created. It also allows for human-edited adaptation in cases of corpus gaps, which will always be found in special-domain environments.

However, the idea to use transfer entries as sense inventories has the disadvantage that it depends on the bilingual dictionary used, and on the target language: If senses are conflated in the target language (like $_{en}$ '*cell*' for biology, religion/monastery, prison, IT-Excel), then the corresponding $_{de}$ '*Zelle*' will not be disambiguated in German-English whereas in Italian $_{it}$ '*cellula*' and '*cella*' would have to be distinguished and would have created two senses in the source. The component would have to be re-run in such cases. But as long as there is no human interaction required it would be a feasible easy option.

7.3 Syntactic Analysis and Knowledge-Based Term Association

So far, the focus was on the creation and identification of index terms as such terms are the elementary units in indexing. Beyond term creation there are additional options how knowledge based analysis could contribute to the quality of indexing.

The tradition of computational linguistics would turn now from term analysis to syntactic and semantic components, probably with the focus on dependency analysis (like in the CONLL competitions, e.g. (CoNLL 2006)). Before doing so, it would be worthwhile to consider what a syntax analysis component could contribute to the indexing task. Syntactic analysis as such is in most cases a device for precision improvement, and recall often is reduced.[36] The reason is that syntactic analysis would have to cope with the fact that the same semantic relations can be expressed in a variety of morphosyntactic structures like

car import
car imports
import of cars
importing cars
imports of US cars

But syntax could also be considered as a means for determining semantic similarity. This is the approach taken in the TINA Term Association component.

[36] An example is given in the word sense section above (Sect.7.2).

Semantic analysis is important in indexing as it reflects the well-known fact that users do not always use the most appropriate terms for search, cf. (Salton and McGill 1983). Providing semantically similar terms would be a means to support the indexing process, usually in a query expansion situation. In consequence, the task consists in the identification of terms which are semantically similar to a given index term.

7.3.1 Comparisons

7.3.1.1 Lexical Features

One option to define sematic similarity is to consult dictionaries or resources like WordNet. However there are two phenomena to be taken into account:

- The feature set used to describe semantic relations differs from resource to resource: Features like '*synonym*', '*antonym*' contrast to features like '*is_a*', '*part_of*', '*instrument_for*' or semantic markers like '*buys*', '*sells*'. A concise set of semantic relations is not yet in industrial use.
- The terms described in such resources do not match the terms used in different domains, in particular as such domains often coin terms which are based on general-purpose meanings (like '*window*', '*handle*', '*thread*' and so on in IT). It has already been stated (in Sect. 7.2) that the 155 k entries of WordNet contain only 45 k lemmata of the 210 k entries of the LinguaDict dictionary, which is not even domain-specific. So a large amount of non-overlap has to be taken into account when using WordNet-type resources.

A domain-independent general description of semantic relations would be required.

7.3.1.2 Statistical Cooccurrence

Term Association has been proposed since the beginning of search system production. Most approaches are statistically based and look on co-occurrence phenomena. Co-occurrence, however, neglects two important observations:

- The **size of the context window** in which cooccurrence is investigated is not fixed: Cases like

  milk storage
  storage of milk
  storage of white US imported farm milk

  show that between two and seven terms would have to be considered to properly associate '*storage*' and '*milk*'. If the context is too small, important cooccurrences are not found; if it is too big, then too much noise is collected.

- The **order of terms** is an important aspect: Although terms like *'pump oil'* and *'oil pump'* are adjacent their meaning is quite different, the first one talking about *oil*, the second one about *pumps*. Similar cases are *'frequency control'* versus *'control frequency'*, *'specification language'* versus *'language specification'* or *'array transducer'* versus *'transducer array'*.

Such cases, if not considered, add noise to a purely statistical analysis and contribute to the sometimes unpredictable results of such components. Statistical results could be improved by taking such observations into account.

7.3.2 TINA Term Association

7.3.2.1 Design

The Term Association component of TINA was developed by Christoph Schwarz and Gerda Ruge; the description in this section is based on their work.[37] It is the result of the following considerations:
1. The focus of attention is on **noun phrases**. This is because most queries in retrieval are noun phrases. Verbs come into play in constructions where gerunds and participles are involved (like *'US imported cars'*).
2. Noun phrases have **variable length**. The boundaries of the phrases, and so their length, can be determined on the basis of linguistic clues (punctuations, conjunctions, casing etc.). The phrases form the units for the association detection. In contrast, most statistical techniques use fixed length contexts.
3. The most general semantic relation between terms is the **dependency** relation, expressed by head-modifier relations. Therefore the TINA Term Association was based on this relation. Each index term is identified in its context of head and modifier terms.
4. The head-modifier relation can be expressed in **many syntactic variations**. It is the task of a syntactic analysis to extract the underlying head-modifier relation from the syntactic surface structures. As large amounts of text must be processed a shallow syntactic analysis (LtShallow) will be most appropriate. The German analysis must also analyse compounds ($_{\text{de}}$ *'Import von Autos'* versus *'Autoimporte'*).
5. All heads and modifiers of a given term in a text base can be collected; they can serve as a detailed semantic description of this term in the given domain. All terms with a similar head-modifier description are considered to be semantically similar, and form a **semantic field**. Terms of the same semantic field can be presented as strongly related for query expansion.

[37] It is documentend among others in (Ruge and Schwarz 1991; Ruge 1995). This research was not continued, however it is worth remembering as it can be a means for better statistical text analysis and word embedding.

7.3.2.2 Implementation

The component was developed using 200,000 abstracts from the US Patent and Trademark Office (PTO), overall 130 Mbyte of text, and additional material from the Food Science Technology Abstracts (FSTA[38]), provided for research purposes.

The first analysis step consisted in the segmentation of text into sentences and phrases, similar to the LtSSplit component described in Sect. 4.1.

The goal of the syntactic analysis was to identify head-modifier structures within phrases, above all noun phrases. The main heuristics are based on the well-known fact that in NPs all (content) words modify from left to right, with the head being the rightmost element, while after a preposition the modification direction is from right to left. Examples are given in Fig. 7.14.

This basic principle of course undergoes many modifications and exceptions, esp. for participles and gerunds, which leads to complications in the syntactic descriptions.

The syntactic analysis is driven by syntactic patterns, based on of part-of-speech tags similar to the STAG set described in Sect. 2.3.2 but with less complexity in case of verbal elements. Rules describe adjective-noun combinations '(*high temperature*': *temperature* ← *high*), participle-nouns ('*chlorinated polyethylene*': *polyethylene* ← *chlorinated*; '*interconnected by lines*': *interconnected*[39] ← *lines*), NPs with gerunds ('*means for controlling the position*': *means* ← *control* ← *position*) etc. More examples are given in (Ruge 1995: 127). Some syntactic phenomena need more detailed descriptions, like the treatment of gerunds which requires about 10 patterns. Overall about 50 such rules were developed for the English syntactic analysis. They were applied in a one-pass left-to-right manner in a sequence defined by the grammar writer.

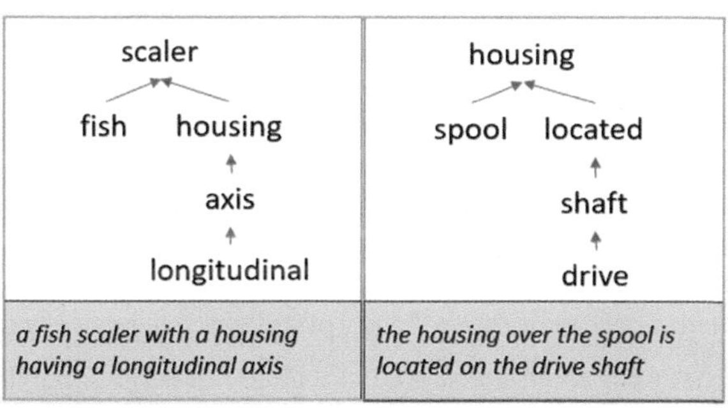

Fig. 7.14 Head-modifier relations

[38] https://www.ifis.org/fsta

[39] It was decided to use the participle form as base form, not the verb (‚*clorinate* '), for better readability.

storage			
modifers	**frq**	**heads**	**frq**
data	258	tank	512
heat	215	device	499
energy	202	container	252
charge	117	position	239
water	106	chamber	224
information	104	unit	206
main	86	battery	189
memory	60	system	184
liquid	58	compartment	173
image	54	area	160
material	39	capacitor	158
signal	39	cell	146

Fig. 7.15 Heads and modifiers for ‚storage' (Ruge 1995: 154)

After syntactic analysis all heads and modifiers for a given term were collected. About 8250 terms were found to provide sufficient links, i.e. terms with at least 10 heads or at least 10 modifiers.[40] The average number of heads and modifiers for a term were 94.4 and 92.6 respectively. Examples are given in Fig. 7.15.

To evaluate the recognition quality the standard measures recall and precision were used, however not for the terms themselves but for their links: how many links were found, and how many of them were correct. This way, link recall and link precision could be computed. About 1000 links collected from a subcorpus were given to human experts for annotation, and the links produced by the human experts were compared to the links produced by the system. A recall of 0.85 and a precision of 0.84 was found for the TINA syntactic analysis.[41]

The final step was to try to determine semantic similarities between terms. The hypothesis used was the following: '*The higher the correlation between the heads and modifiers of two terms, the higher the probability that the two terms belong to a common semantic field*', cf. (Ruge and Schwarz 1991:23), similar (Lavelli et al. 2004)).

As the basis of similarity calculation for the terms only their most frequent heads and modifiers were used, with more than 30 types on heads and/or modifiers; some terms had up to 3000 links. These terms were used to compute their semantic similarity; the selected similarity measure is discussed in (Ruge 1995: 158ff.). Some examples of the resulting semantic fields (the top 10 terms) are shown in Fig. 7.16.

[40] For details see (Ruge 1995: 127ff.).

[41] (Ruge and Schwarz 1991). The main phenomena for irritations were homographs with different sets of head-modifier links but mostly problems on term level (typing errors, lemmatisation issues etc.).

cable		acceleration		efficient		container	
terms	*similarity*	*terms*	*similarity*	*terms*	*similarity*	*terms*	*similarity*
cable	1.000	acceleration	1.000	efficient	1.000	container	1.000
conductor	0.333	deceleration	0.416	economical	0.466	enclosure	0.466
connector	0.263	speed	0.283	simple	0.466	bottle	0.466
wire	0.283	velocity	0.250	effective	0.433	receptacle	0.433
rope	0.266	inclination	0.200	easy	0.433	cavity	0.433
rod	0.250	movement	0.166	compact	0.433	vessel	0.433
line	0.233	correction	0.150	simultaneous	0.416	tank	0.416
pipe	0.216	rotation	0.150	direct	0.400	pouch	0.400
unit	0.216	engine	0.083	low	0.383	housing	0.383
chain	0.200	exhaust	0.050	utilizable	0.366	compartment	0.366

Fig. 7.16 Associations for *cable, acceleration, efficient, container*: Ruge & Schwarz (1991):25

It can be observed that these results show less noise than comparable association tables. The reason is that the *quality* of the cooccurrence is taken into account, in the form of head-modifier relations, and not just the pure cooccurrence itself.

7.3.2.3 Discussion

The purpose of the syntactic analysis for Term Association is to find additional search terms, usable in a query expansion phase: In cases where the original term selection is not the best option then semantically similar terms of a given domain could be presented. As semantic similarity is not clearly definable, and its use depends on the intuition of the searching person, the approach was to create supportive tools from which search terms could be picked. Such a supporting system, called REALIST (‘*RE*trieval *A*ids with *LI*nguistics and *St*atistics’, cf. (Thurmair et al. 1986)) was developed in an experimental setup where the list of index terms was enriched with proposals for query expansion based on term morphology, linguistic term associations, and statistical term cooccurrence.

Not only semantically similar terms could be detected with the Term Association but the head-modifier relations could also be used to some extent as support for the navigation of broader/narrower terms: General terms are often not very specific and therefore not the best choice for a search term, as shown above, however in connection with their modifiers they can return more precise search results.

7.4 Output of the Document Analysis: LtDoc FOOTER

The results of the text analysis are communicated to the downstream components in the footer section of the Document Model. While its body section contains the analysis results of the single sentences the footer section collects the results on document level.

The footer is divided into two groups, one comprises information on ‘TERMS’, the other one on ‘NAMES’. This distinction reflects the two main analysis methods for concepts, Named Entity analysis for names/proper nouns, and conceptual analysis for terms/common words. From an indexing point of view, a single source of reference could be envisaged, as both can be given to the indexing component as concepts; for this purpose they must share information like lemma, frequency, and position in the text; however entities can be used for additional applications like question answering or applications like the LtAnonymiser, moreover entities have additional semantic attributes which could be communicated. Therefore TERMS and NAMES are given in separate sections in the footer. Their structure has been given in Fig. 2.14, repeated here as Fig. 7.17.

Footer	**::=**	**TERMS?, NAMES?**
TERMS	**::=**	**term***
term	**::=**	**termdef, frequency, atts*, matches+**
termdef	**::=**	**lemma, POS, sem**
NAMES	**::=**	**(nametype, name*)***
nametype	**::=**	**PERSONS| LOCATIONS | ORGANISATIONS | TIMES |UNITS | COMMEANS | ENTITIES**
name	**::=**	**namedef, frequency, atts*, matches+**
namedef	**::=**	**lemma, POS, sem**
lemma	**::=**	**<string>**
POS	**::=**	**< a Btag >**
sem	**::=**	**<string>**
frequency	**::=**	**<integer>**
atts	**::=**	**(feature, value+)+**
matches	**::=**	**textform, onset, offset**

Fig. 7.17 = 2–14: LtDoc footer section

7.4.1 Terms

Terms contain the concepts of the document. Their **definition** (*termdef*) consists in a lemma, a part-of-speech tag, and a semantic description. In case a sense is determined it would be stored here in the *sem* attribute. Multiword terms are described by a multiword lemma. Different senses of a word result in different terms here (i.e. different *termdef* records).

The **frequency** gives the number of matches. It counts lemmata, not textforms. Members of multiwords are not counted separately.

The **attributes** section is used to carry additional attributes to the concept. For the time being, the following additional information items are considered:

- a 'domain' feature, in case such a marker is lexicalised, the value being one of the topics defined in Sect. 3.3.
- a 'stem' feature to carry the stem in case a concept could be stemmed, the value being the stem string.
- a 'members' feature, for multiword concepts or (German) compounds, containing the mwlemmalist and the mwposlist of the entry (cf. Sect. 6.1).
- a 'heads' feature giving all head lemmata and their frequency[42]
- a 'modifiers' feature giving all modifier lemmata and their frequency.

[42] This also holds for German compounds: '*Haustor*' would have '*haus*' as modifier and '*tor*' as head.

In order to stay flexible the Document Model uses the generic features '*feature*' and '*description*', with '*description*' giving the value and its frequency. They can take the respective feature names and values; multiple values are given in comma-separated form.

All occurrences of the terms are noted in the '**matches**' section. It needs to be mentioned that the occurrences must give the different onsets and offsets of the *text forms* of the term (not the lemmata or the normalised forms), in case the matches need to be highlighted by the retrieval system.

It is an open discussion whether or not **stop words** should be eliminated from the list of concepts before indexing. They import significant noise, and in the first IR systems they were removed for reasons of storage size and processing overhead. However, they do carry meaning: There is a difference between '*bicycles with electric motors*' and '*bicycles without electric motors*', and precision decreases if this difference is neglected. Stopwords also play an important role in multiwords, like '*American foundation for the Blind*' or '*European Code against Cancer*'. In systems with indices consisting only of single words such multiwords could not be safely identified, if the prepositions had been stopped away.

A solution could be to restrict the stopword list on purely grammatical entries like auxiliaries, inflections, entries which do not have an easy reference, like pronouns,[43] and other such parts of speech. Other function words like prepositions, adverbs, and determiners, would be kept.

In case stopwords should be used there are many stopword lists available for many languages; also TINA offers its stop word lists for the languages which it supports.

Beyond such global stopword lists, sometimes domain adaptation could be helpful to tag close-to-non-meaningful terms, like '*engine*' in a collection of automotive texts. Such domain-specific stop lists would be kept separate from the global stopword list, and added at runtime in cases where the respective domain text collection is to be processed.

The treatment of stopwords in collections for machine learning and word embedding is discussed e.g. in (Hvitfeld and Silge 2022).

7.4.2 Names

The structure of the NAMES section differs from the TERMS by an additional subcategorisation level subsuming all entities of a certain entity-type, e.g. "*persons*".

The NAMES definition is like in the TERMS section. The entities are grouped along the *nametype*. feature. Additional information to the *nametype* is given in the *nesubtype* feature in the 'atts' section.

[43] Pronoun resolution could however increase the retrieval accuracy for the pronominalised terms.

The features and values of the '*atts*' section depend on the entity type. They give more detailed information about a name, first of all a '*nesubtype*' (like '*unit|currency|percent*' for units), but also more detailed information in some nametypes, like '*located_in*' for place names, '*gender*' for person names etc. Details for the 'atts' feature are given in Sect. 6.2 above.

In addition to the TERMS and NAMES sections, the footer could contain additional information types, like templates of Named Entity recognition, triplets of semantic relations, and others. It can be communicated to other software components e.g. by JSON schemata.

References

Agirre, E., & Edmonds, Ph., eds. (2007). Word Sense Disambiguation, Algorithms and Applications. Springer.

Agirre, E., Lopez de Lacalle, O., Magnini, B., Otegi, A., Rigau, G., & Vossen, P. (2007). SemEval-2007 Task 01: Evaluating WSD on Cross-Language Information Retrieval. In: Peters, C., et al, (2007). Advances in Multilingual and Multimodal Information Retrieval. CLEF 2007. Lecture Notes in Computer Science, vol. 5152. Springer, Berlin, Heidelberg.

AlMousa, M., Benlamri, R., & Khoury, R. (2021). A Novel Word Sense Disambiguation Approach Using WordNet Knowledge Graph. arXiv:2101.02875

Arranz, V., Atserias, J., & Castillo, M. (2005). Multiwords and Word Sense Disambiguation. *Lecture Notes in Computer Science. 3406. 250-262.* https://doi.org/10.1007/978-3-540-30586-6_28.

Atkins, B. T. S., & Bouillon, P. (2006). Relevance in dictionary making: Sense indicators in the bilingual entry. *In L. Bowker, ed. (2006). Lexicography, terminology, and translation: Text-based studies in honour of Ingrid Meyer (pp. 25–43). Ottawa: University of Ottawa Press.*

Banerjee, S., Pedersen, T. (2002). An Adapted Lesk Algorithm for Word Sense Disambiguation Using WordNet. *CICLing '02: Proc. of the Third Int. Conf. on Computational Linguistics and Intelligent Text Processing, Pages 136–145.*

Bel, N. (2010). Platform for Automatic, Normalized Annotation and Cost-Effective Acquisition of Language Resources for Human Language Technologies: PANACEA. *Proc. Intern. Conf. of the Spanish Society For Natural Language Processing (SEPLN), Valencia*

Bevilacqua, M., Pasini, T., Raganato, A., & Navigli, R. (2021). Recent Trends in Word Sense Disambiguation: A Survey. Proc. Thirtieth Int. Joint Conf. on Artificial Intelligence, Survey Track. pp. 4330–4338.

Bonansinga, G., & Bond, F. (2016). Multilingual Sense Intersection in a Parallel Corpus with Diverse Language Families. *In Proc. 8th Global WordNet Conference (GWC), Bucharest, Romania, pp. 44–49.*

Brants, S., Dipper, St., Hansen, S., Lezius , W., & Smith, G. (2002). The TIGER Treebank. *In: Hinrichs, E., & Simov. K., eds. Proc. on First Workshop on Treebanks and Linguistic Theories (TlT) 2002, Sozopol, Bulgaria.*

Brants, Th. (2000). TnT-A Statistical Part-of-Speech Tagger. *Proc. of the sixth Applied natural language processing Conf., ANLP-2000, Seattle, USA, pp. 224–231.*

Brill, E. (1992). A simple rule-based part of speech tagger. Proc. third Conf. on Applied Natural Language Processing (ANLC '92). Association for Computational Linguistics, Stroudsburg, PA, USA, pp. 152–155. https://doi.org/10.3115/974499.974526

Brill, E. (1995). Unsupervised learning of disambiguation rules for part of speech tagging. *Proc. 3rd ACL Workshop on Very Large Corpora.*

CoNLL-X (2006). CoNLL-X Shared Task: Multilingual Dependency Parsing. *Proc. 10th Conf. on Computational Natural Language Learning - New York City, 2006.*

Correa, E.A., de Andrade Lopes, A., Amancio, D.R. (2018). Word sense disambiguation via bipartite representation of complex networks. Information Sciences 442-443, pp. 103-113

Dang H.T., & Palmer, M. (2002). Combining Contextual Features for Word Sense Disambiguation. *Proc. of the ACL-02 Workshop on Word Sense Disambiguation: Recent Successes and Future Directions, pages 88–94.*

Devlin, J., Chang, M.W., Lee, K., & Toutanova, K. (2019). BERT: Pre-training of Deep Bidirectional Transformers for Language Understanding. *Proc. of the 2019 Conf. of the North American Chapter of the Association for Computational Linguistics: Human Language Technologies, (NAACL) Volume 1 (Long and Short Papers), Minneapolis, Minnesota, pp. 4171–4186.*

Fagan, J.L. (1987). Automatic phrase indexing for document retrieval: An examination of syntactic and non-syntactic methods. Proc. 10th Annual Int. ACM SIGIR Conf. on Research and Development in Information Retrieval, New York, pp. 91–101.

Fellbaum, Chr., ed., (1998). WordNet, An Electronic Lexical Database. *MIT Press*

Florian, R., & Wicentowski, R. (2002). Unsupervised Italian Word Sense Disambiguation using WordNets and Unlabeled Corpora. *Proc. of the ACL-02 Workshop on Word Sense Disambiguation: Recent Successes and Future Directions, pp. 67–73.*

Giesbrecht, E., & Evert, St. (2009). Is Part-of-Speech Tagging a Solved task? An Evaluation of POS Taggers for the German Web as Corpus. *Proc 5th Workshop Web as a Corpus,(WAC5), San Sebastian, Spain.*

Haller, J. (1978). Ein Algorithmus zur Reduktion syntaktischer Homographien in einem Informationssystem. *Diss. Regensburg*

Havasi, C., Speer, R., & Alonso, J. (2007). ConceptNet 3: A flexible, multilingual semantic network for common sense knowledge. *MIT Media Lab* https://www.media.mit.edu/publications/conceptnet-3-a-flexible-multilingual-semantic-network-for-common-sense-knowledge/

Hey, T., Keim, J., **&** Tichy, W. (2021). Knowledge-based Sense Disambiguation of Multiword Expressions in Requirements Documents. https://doi.org/10.5445/IR/1000139591/post; https://doi.org/10.1109/REW53955.2021.00017

Hvitfeldt, E., **&** Silge, J. (2022). Supervised Machine Learning for Text Analysis *in: R. Chapman & Hall/Crc Data Science.*

Iacobacci, I., Pilevar, M.T., & Navigli, R. (2016). Embeddings for Word Sense Disambiguation: An Evaluation Study. *Proc. 54th Annual Meeting of the Association for Computational Linguistics (ACL), Berlin, Germany (Volume 1: Long Papers), pp. 897–907.*

Ide, N., Erjavec, T., & Tufiş, D. (2002). Sense Discrimination with Parallel Corpora. *Proc. SIGLEX/SENSEVAL Workshop on Word Sense Disambiguation: Recent Successes and Future Directions, Philadelphia, pp. 54–60.*

Ide, N., & Véronis, J., eds., (1995). The Text Encoding Initiative: Background and Context. *Dordrecht: Kluwer Academic Publishers, 181-190*

Ide, N. & Véronis, J. (1998). Word sense disambiguation: The state of the art. *Computational Linguistics. 24. 1-41.*

Jurafsky, D. & Martin, J.H. (2009). Speech and Language Processing. *Pearson Int. Edition (2nd ed.).*

Kilgarriff, A., & Rosenzweig, J. (2000). English Senseval: Report and Results. *Proc. 2nd Int. Conf. on Language Resources and Evaluation (LREC), Athens, Greece.*

Kilgarriff, A. (2001). English lexical sample task description. *P.roceedings of the 2nd Int. Workshop on Evaluating Word Sense Disambiguation Systems (SENSEVAL '01), pp. 17–20.*

Kwon, S., Oh, D., Ko, Y. (2021). Word sense disambiguation based on context selection using knowledge-based word similarity. *Information Processing & Management Volume 58, Issue 4, July 2021.*

Kwong, O.Y. (2012). Psycholinguistics, Lexicography, and Word Sense Disambiguation. *Proc. 26th Pacific Asia Conf. on Language, Information, and Computation, Bali, Indonesia, pp. 408–417.*

Lavelli, A., Sebastiani, F., & Zanoli, R. (2004). Distributional term representations: an experimental comparison. *Proc. 13th ACM Int. Conf. on Information and Knowledge Management. pp. 615–624.*

Leacock, Cl., & Chodorow, M. (1998). Combining local context and WordNet similarity for Word Sense Identification. *in: Fellbaum, Chr., ed.: WordNet, An Electronic Lexical Database. MIT Press.*

Lesk, M. (1986). Automatic sense disambiguation using machine readable dictionaries: how to tell a pine cone from an ice cream cone. *Proc. 5th Annual Int. Conf. on Systems Documentation (SIGDOC '86), pp. 24–26*

Lindén, K., & Piitulainen, J.O. (2004). Discovering synonyms and other related words. *Proc. 3rd Int. Workshop on Computational Terminology (CompuTerm 2004), 63–70.*

Marcus, M.P., Santorini, B., & Marcinkiewicz, M.A. (1993). Building a Large Annotated Corpus of English: The Penn Treebank, *in Computational Linguistics, Vol. 19, No. 2 (June 1993), pp. 313--330 (Special Issue on Using Large Corpora).*

Marsi, E., Lynum, A., Bungum, L. & Gambäck, B. (2011). Word Translation Disambiguation without Parallel Texts. *Int. Workshop on Using Linguistic Information for Hybrid Machine Translation (LIHMT-2011).*

Martínez, D., Agirre, E., & Màrquez, L. (2002). Syntactic features for high precision word sense disambiguation. *Proc. 19th Int. Conf. on Computational linguistics (COLING '02) Vol.1, pp. 1–7.*

Merkel, M., & Andersson, M. (2001). Combination of Contextual Features for Word Sense Disambiguation: LIU-WSD. *In Proc. of SENSEVAL-2 Second Int. Workshop on Evaluating Word Sense Disambiguation Systems, pp. 123–126, Toulouse, France.*

Mohammad, S., & Pedersen, T. (2004). Combining Lexical and Syntactic Features for Supervised Word Sense Disambiguation. *Proc. 8th Conf. on Computational Natural Language Learning (CoNLL-2004) at HLT-NAACL 2004, pp. 25–32, Boston, Massachusetts, USA.*

Muret, E., & Sanders, D. (2003). Langenscheidts Grossworterbuch Deutsch-Englisch *Langenscheidt 2003.*

Och, F., Ney, H. (2004). The Alignment Template Approach to Statistical Machine Translation. *Computational Linguistics 30,4*

Pasini, T., Raganato, A., & Navigli, R. (2021). XL-WSD: An Extra-Large and Cross-Lingual Evaluation Framework for Word Sense Disambiguation. The Thirty-Fifth AAAI Conf. on Artificial Intelligence (AAAI-21).

Postma, M., van Miltenburg, E., Segers, R., Schoen, A., & Vossen, P. (2016). Open Dutch WordNet. *In Proc. of the 8th Global WordNet Conf. (GWC), pages 302–310, Bucharest, Romania.*

Rouhizadeh, H., Shamsfard, M., Tajalli, V., Rouhziadeh, M. (2021). Persian-WSD-Corpus: A Sense Annotated Corpus for Persian All-words Word Sense Disambiguation. arXiv:2107.01540

Ruge, G., & Schwarz, Chr. (1991). Term Associations and Computational Linguistics. *in: Int Classif. 18(1991) No.1, pp.19–25.*

Ruge, G. (1992). Experiments on linguistically-based term associations. Information Processing & Management Volume 28, Issue 3, 1992, Pages 317–332.

Ruge, G. (1995). Wortbedeutung *und* Termassoziation. Methoden zur automatischen semantischen Klassifikation. *Hildesheim 1995 (Olms).*

Salton, G., & McGill, M.J. (1983). Introduction to Modern Information Retrieval. *New York, NY (McGraw-Hill)*

Samuelson, Chr., & Voutlinainen, A. (1997). Comparing a Linguistic and a Stochastic Tagger. Proc 8th Conf. of the European Chapter of the Association for Computational Linguistics (EACL), Madrid, Spain.

Schiller, A., Teufel, S., Thielen, C., (1995). Guidelines für das Tagging deutscher Textcorpora mit STTS. *Technical Report, IMS-CL, University Stuttgart, 1995.*

Schmid, H., & Laws, Fl. (2008). Estimation of Conditional Probabilities with Decision Trees and an Application to Fine-Grained POS Tagging. *Proc 22nd Int. Conf. on Computational Linguistics(COLING), Manchester, UK*

Schmid, H. (1994). Probabilistic Part-of-Speech Tagging Using Decision Trees. *Proc. of Int. Conf. on New Methods in Language Processing, Manchester, UK.*

Schmid, H. (1995). Improvements in Part-of-Speech Tagging with an Application to German. *Proc. of the ACL SIGDAT-Workshop. Dublin, Ireland.*

Schwarz, Chr. (1990a). Content Based Text Handling. In: Inform. Process. & Management 26(1990) No.2,219–226.

Shamsfard, M., & Fadaei, H. (2008). A hybrid morphology-based POS tagger for Persian. In Proc. of 8th Int. Conf. on Language Resources and Evaluation (LREC), Marrakech, Morocco.

Siemiński, A. (2011). WordNet Based Word Sense Disambiguation. *In: Jędrzejowicz, P., Nguyen, N.T., Hoang, K., eds. (2011). Computational Collective Intelligence. Technologies and Applications. ICCCI 2011.*

Specia, L., & Motta, E. (2006). A hybrid approach for extracting semantic relations from texts. *Proc. of the 2nd Workshop on Ontology Learning and Population: Bridging the Gap between Text and Knowledge. 2006.*

Speer, R., Chin, J., Havasi, C. (2017). ConceptNet 5.5: An Open Multilingual Graph of General Knowledge. *AAAI 31 (2017) 4444–4451.*

Tanner, J., and Hoffman, J. (2023). MWE as WSD: Solving Multiword Expression Identification with Word Sense Disambiguation. *Proc. Findings of the Association for Computational Linguistics: EMNLP 2023, pages 181–193, Singapore.*

Thurmair, Gr., Niedermair, G., Schwarz, Chr., & Wessel, A. (1986). Retrieval Aids by Linguistics and Statistics (REALIST): Systemkonzeption. *in: Schwarz, Chr., Thurmair, Gr. eds. (1986). Informationslinguistische Texterschließung. Hildesheim (Olms), pp. 251–265.*

Thurmair, Gr. (2005). Improving MT Quality: Towards a Hybrid MT Architecture in the linguatec 'Personal Translator'. *In: Proc. of Machine Translation Summit X: Invited paper, Phuket, Thailand.*

Thurmair, Gr. (2014). Conceptual Transfer: Using local classifiers for transfer selection. Proc. 9th Int. Conf. on Language Resources and Evaluation (LREC 2014), Reykjavik, Iceland

Toutanova, Kr., Klein, D., Manning, Chr., Singer, Y. (2003). Feature-Rich Part-of-Speech Tagging with a Cyclic Dependency Network. *Proc. 2003 Human Language Technology Conference of the North American Chapter of the Association for Computational Linguistics NAACL), pp. 252–259.*

Tsutsumida, K., Okamoto, J., Ishizaki, Sh., Nakatsuji, M., Tanaka, A., & Uchiyama, T. (2010). Study of Word Sense Disambiguation System that uses Contextual Features - Approach of Combining Associative Concept Dictionary and Corpus -. *Proc. of the 7th Int. Conf. on Language Resources and Evaluation (LREC'10), Valletta, Malta.*

Tufiş, D. (2000). Using a Large Set of EAGLES-compliant Morpho-Syntactic Descriptors as a Tagset for Probabilistic Tagging. *Proc. 2nd Int. Conf. on Language Resources and Evaluation (LREC 2000), Athens, Greece.*

Volk, M., & Schneider, G. (1998). Comparing a statistical and a rule-based tagger for German. *Proc. KONVENS-98, Bonn.*

Voutilainen, A. (1995). A syntax-based part-of-speech analyser. *Proc Conf. of the European Chapter of the Association for Computational Linguistics (EACL), Dublin, Ireland.*

Wiedemann, Gr., Remus, St., Chawla, A., & Biemann, Chr. (2019). Does BERT Make Any Sense? Interpretable Word Sense Disambiguation with Contextualized Embeddings. arXiv:1909.10430

Yuan, D., Richardson, J., Doherty, R., Evans, C., and Altendorf, E. (2016). Semi-supervised Word Sense Disambiguation with Neural Models. *Proc. 26th Int. Conf. on Computational Linguistics (COLING 2016): Technical Papers, pp. 1374–1385, Osaka, Japan.*

Chapter 8
Transparent Information Retrieval (TIR) and the LtConceptNet

One of the applications of Text Analysis is Multilingual Search. The TINA system and its components have been developed in this context, in a series of projects for multilingual text analysis and retrieval.[1] The concept of Transparent Information Retrieval was developed in such projects.

8.1 Transparent Information Retrieval (TIR)

8.1.1 Definition

Transparent Information Retrieval (TIR) is an extension of Cross-Language Information Retrieval (CLIR). The idea is that users may search a text base in their native language, and retrieve the answer (documents) in their native language. It is transparent to the users in which language the documents are written. All translation activities don't demand their attention.

The idea is illustrated in Fig. 8.1.

While CLIR may return a set of possibly multilanguage documents the requirement for Transparent Information Retrieval is to return the search results in the query language. Recent projects also put a focus on this requirement; an example is the MATERIAL project which aims at TIR: "*This system will allow analysts to submit queries in English and receive short English summaries of relevant foreign language items*".[2]

[1] Aventinus (Thurmair 1997a), Sensus (Thurmair 2004b), Panacea (Bel 2010), Multisensor (Vrochidis et al. 2015), Eastin (Andrich 2011).

[2] https://www.iarpa.gov/research-programs/material.

G. Thurmair, *Knowledge-Driven Multilingual Text Analysis and Transparent Information Retrieval*, Cognitive Technologies,
https://doi.org/10.1007/978-3-031-91741-7_8

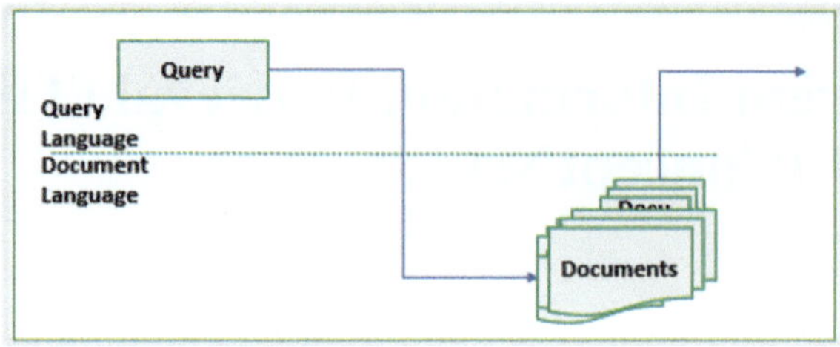

Fig. 8.1 Transparent information retrieval

Compared to CLIR, TIR poses additional constraints to the retrieval process as will be shown.

8.1.2 System Design

Classical IR systems are monolingual. They store the documents, and provide access via an index[3]; the index terms point to the documents in which they can be found. The query is mapped to the index. If many documents are returned, a ranking process decides in which sequence they are displayed.

The main problem in retrieval has been described as a conceptual mismatch between the documents ('authors') and the searchers: The terms and the conceptual context which the users choose do not easily match the terms and conceptual contexts which are represented in the documents.

Solutions that have been proposed follow basically two lines:

- Normalisation: query and document terms are normalised to become comparable. Normalisation steps are spelling, casing, lemmatisation, stemming etc.
- Expansion: Query terms are expanded by adding additional potentially relevant terms to the search. Expansion techniques can be linguistic (synonyms, narrower terms etc.) or statistical (term association), i.e. proposing terms which are frequently found in the context of the search term.

Among many other discussion points, one relevant issue here is which textual objects should be used for search, and stored in the index. Options are to use inflected forms, lemmata, stems, or ngrams as possible search objects; overviews are given by Galuščaková et al. (2020), Galuščaková et al. (2021); other contributions also propose byte pair encodings i.e. subword units, cf. (Senrich et al. 2016).

[3] E.g.: "*When a document is stored, it is indexed and fully searchable in near real-time–within 1 s. Elasticsearch uses a data structure called an inverted index that supports very fast full-text searches. An inverted index lists every unique word that appears in any document and identifies all of the documents each word occurs in.*" (https://www.elastic.co), cf. (Gormley and Tong 2015; Bugara 2020).

The whole picture changes significantly once multilinguality is introduced into the system. Atreya et al. (2012) state that most of the frameworks for developing search engines are not suitable for issues of multilinguality.

8.1.2.1 Multilingual Information Retrieval (MLIR)

This kind of retrieval deals with a multilingual text base and considers several languages in parallel. The immediate architecture decision is whether the terms of all languages should be collected in one index, or if each language should stay in a separate index. This decision influences the complete system setup. While systems like *Autonomy*[4] do not maintain several language on one server,[5] systems like *ElasticSearch* offer two options: Either one index, and parallel document fields for the different languages, or one index per language.[6] For correct assignment the system offers Language Identification tools similar to the ones provided by TINA (cf. Sect. 3.2).

A common index for several languages introduces noise in the search process. There are sufficiently many terms occurring in more than one language with quite different meaning, e.g.:

- en-de: *abort, also, bald, gift, fast, handy, kind, lack, mist, note, spend*
- en-fr: *but, caution, bride, grave, menu, pair, parent, patron, nurse*

The more languages are involved the more noise is added to the system. As a result, a common index for all languages is only an option, if the number of foreign-language documents is insignificant.

Multilingual Information Retrieval treats the respective languages in parallel: It is a 'several-times monolingual' retrieval. Queries and documents are in the same language, no translation is intended. The searchers decide in which language they want to search.

Such an approach can be taken in cases where multilingual parallel documents are available, like in some government Institutions: Multilingual parallel documents are required by legal regulations there, and search is possible in either language.

It should be noted that unlike in CLIR, out-of-vocabulary items are not an issue in IR, nor in MLIR: Unknown terms are simply added to the index, and can be searched like other terms. Once the index is complete enough most unknown terms are Named Entities.

---

[4] Autonomy: https://docs.oracle.com/cd/E13155_01/wlp/docs103/search/.

[5] "*The default configuration for an Autonomy server is one language and one encoding across all repositories using that server. When you use this configuration for multiple languages, you need separate Autonomy servers for each language.*" https://docs.oracle.com/cd/E13155_01/wlp/docs103/search/multilang.html.

[6] "*There are two basic indexing strategies that we'll be using: language per-field and language per-index*" elasticsearch. A discussion of the pros and cons follows there.

8.1.2.2 Cross-Language Information Retrieval (CLIR)

The characteristics of Cross-Language Information Retrieval is that query and document base are written in different languages. This fact requires a translation step somewhere in the processing. The immediate question is where the translation should be done, and how it should be done. The challenge is the selection of adequate translations for the search terms; Nie (2010) considers this to be the main problem in CLIR. Several options exist for translation:

a. The most profound translation expertise has been collected in traditional bilingual dictionaries based on lexicographic knowledge. **Human readable dictionaries** collect all possible translations for a source language term, and add selection restrictions, usage instructions, domain and register information and many other useful knowledge items. Modern lexicography uses corpus-based technology to update them, and collects dictionary entries from corpus data, cf. (Atkins et al. 1997), or recently (Afentouolidou and Christofidou 2021) on neologisms. The aspiration of this lexicography is to cover all possible translations of a source term and guide users to select the best one in a given context. So, bilingual dictionaries built by lexicographers define the 'upper bound' of potential translations of a given term.[7]

The bilingual dictionaries of this type do not contain frequency information. One reason is that frequency is always related to a given corpus, and to a given domain.[8] For instance, the translation of $_{de}$' *Bereitschaft*' would be translated as $_{en}$ '*readiness*' in general texts with a higher frequency than the more technical $_{en}$ '*stand by*'; in technical texts the opposite relation would be found. Therefore it would be difficult to assign frequencies to the different translation alternatives in such dictionaries. Other disambiguation means, like topic identification, would have to be applied before the translation step.

b. **Machine Translation dictionaries** (in the rule-based setups) are mostly derived from the human dictionaries; as such they are just a subset of such ditionaries. They try to formalise the selection of translations for a given term by performing contextual tests.

Transfer tests can be global settings (like the selection of a domain or of a lexicon module for a text), or local or contextual tests. Local tests would refer to the attributes of the term in question (like: its gender[9]), contextual tests would inspect the context (e.g. an instantiated verb frame), e.g. $_{en}$ '*eat*': {if *subject=human*} ->$_{de}$ '*essen*'; {if *subject is animated*) ->$_{de}$ '*fressen*'}. Some systems also provide cooccurrence tests which look for terms occurring in the same sentence.

[7] It should be mentioned that there is a decline in this type of lexicography. Their dictionary production is challenged by commercial points of view, the publishing companies losing ground to dictionaries in the internet which come for free but without quality validation.

[8] Balanced corpora like the British National Corpus (http://corpora.lancs.ac.uk/bnc2014/)are too small to cover the variance of translations found in the large dictionaries.

[9] E.g.: $_{de}$ '*See*' if *gender=masc* ->$_{en}$ '*lake*'; but $_{de}$ '*See*' if *gender=fem* ->$_{en}$ '*sea*'.

An overview is given in Thurmair (2014a); examples for contextual tests can be found in Sect. 9.2.5.

At runtime, such tests are executed such that the most specific tests are tried first, less specific ones thereafter. If no test matches, a default translation is presented. However, such tests require sufficient context for validation, but such context is not always available e.g. in query translation. In addition they presuppose valid structures as basis, i.e. successful parses of the input sentences.

Comparisons between the use of Machine Translation and the use of dictionaries seem to favour the dictionary approach (Nie 2010), possibly because these dictionaries offer more coverage than the MT dictionaries. Also, MT dictionaries often do not provide the granularity of translations as given in the Human Readable Dictionaries.

c. **Data driven MT** systems try to identify translation correspondences by identifying cooccurrence contexts from bilingual training data. Such contexts, organised in phrase tables, are used at runtime to determine the best matching target-language phrase.

This procedure is corpus and domain dependent. It is difficult to adapt a given system to a specific (customer) terminology, or to a new domain; examples are given in Pecina et al. (2012). Also, bilingual aligned corpora required for learning are not available for many language directions and many domains. From the point of usage, there are two main challenges:

- The translations are not transparent to users. It cannot be said why a given output had been produced, and why the same sentence was translated quite differently after a new training run. This leaves a hard task to posteditors in the translation industry.
- The translations introduce noise: Not all statistical correspondences are correct. This fact leads to translation errors, which are rather difficult to spot due to the intransparency of the technology. As a result, it is not clear whether a given translation is correct or not. Several methods to identify translation errors have been proposed, cf. (Popović and Ney 2011), however a validation of the terms selected by the MT is only possible vis-à-vis a conventional bilingual dictionary.

d. Compared to phrase-based MT, neural approaches recently have gained attention. Translations using **crosslingual word embedding** have been researched in the last years. Overviews can be found in Galuščaková et al. (2020), Yu and Allan (2020), Ruder et al. (2019).

The rationale is that statistical patterns reflect semantic information of terms, according to e.g. (Vinokourov et al. 2002); in particular the context of a word can influence its semantic content, cf. (Lavelli et al. 2004); so by selecting the contexts and computing their similarity the similarity of terms can be found out. The contexts are represented by vectors, their similarities are computed based e.g. on cosine similarity.

Crosslingual Word Embedding were explored by creating a joint embedding space from the monolingual embedding models, usually using word-aligned data, and computing the nearest vectors. A survey on crosslingual word embedding is given in Ruder et al. (2019). Attempts with a common embedding space, aside from statistic computing, have been explored by using a hierarchical structure as represented in the links e.g. of BabelNet, and populating its nodes by bilingual terms, cf. (Franco-Salvador et al. 2014).

It should be noted that research in this field has moved away from the cross-lingual retrieval into the Machine Learning domain, with particular constraints, like fixed sets of queries and documents, special evaluation procedures, huge available computational power,[10] huge available data resources.[11] The support for CLIR applications has lost priority, and many problems, like multiwords, domain adaptation, sense distinction etc. are still to be solved, according to Ruder et al. (2019). Airio et al. (2003) also noted that language with rich morphology score worse than English-based models, and Ruder et al. (2019) commented that the embedding space in two languages should be approximately isomorphic to show useful results, which is a strong hypothesis.

It should also be noted that the assumption that a term's semantics is determined by its context is a circular reasoning: The terms surrounding a declaredly semantically empty term are also defined contextually, i.e. themselves semantically empty as well, and so are *their* contexts. As a result, *all* terms are empty, and it can never be decided whether two terms sharing the same context are synonyms or noise. This is the way how noise enters the system, up to a point where it cannot be determined, if a translation based on statistical techniques is correct.[12] But as humans are well able to distinguish semantically similar terms from noise although both have been produced by the same analysis, there must be semantic content in the terms, independent of the contexts. Contexts would be able to *modify* it, but not *create* it.

8.1.2.3 Transparent Information Retrieval (TIR)

Transparent Information Retrieval adds an additional constraint to Cross-Language Information Retrieval: Users need not be aware of the fact that the documents they are looking for are in foreign language(s), i.e. the translation dimension should be transparent for them. This constraint of course requires that after the translation of the query into the document language, the found documents are re-translated into the query language.

[10] "*The training took one week using a Telsa V100 GPU*" (Ruder et al. 2019).

[11] Yu and Allan (2020) state that more data are needed on the query side than usually are available.

[12] This fact turned out to be a problem in Language Industry: Translation post-editors have increased efforts in comparing source text to nice-looking translations.

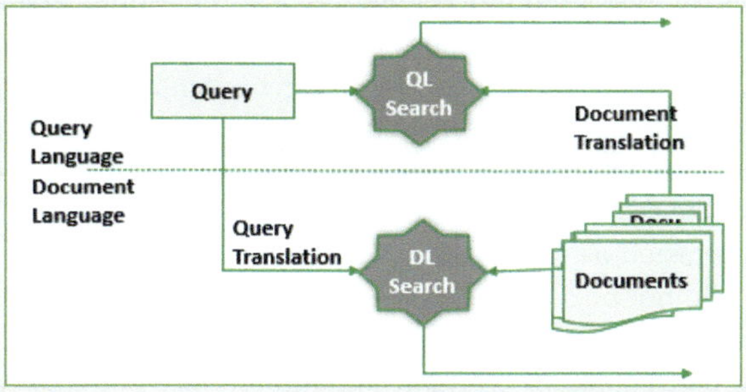

Fig. 8.2 Translation options in CLIR

However, as opposed to CLIR approaches based on document translation and query language search[13] here not the whole document collection needs to be translated, but only the documents which match a given search request.

Requirements of this kind have been forwarded e.g. in the MATERIAL project presented in Rubino (2020), where end-to-end CLIR systems, among other things[14] should accept queries in English and provide the found evidence also in English.

8.1.3 Translation Integration in CLIR

The classical architectures as given in Peters et al. (2012) or Nie (2010) discuss two options to integrate translation components, cf. Fig. 8.2:

- query translation and search in the document language, result presentation in document language,
- document translation and search in the query language, result presentation in the query language.

A third option is to translate on both sides into a third (pivot) language like English, or into a crosslingual word embedding space.

8.1.3.1 MT for Query Translation

Using MT for Query Translation faces several challenges,[15] all of which result in the recommendation to refrain from using it for this purpose:

---

[13] Cf. the overview in e.g. (Peters et al. 2012).

[14] Like Speech Retrieval or Information Extraction.

[15] Even if several MT systems are applied in cascaded form, as proposed by Savoy (2003).

- The analysis of the structure of queries e.g. by Spink et al. (2001) or Roy (2013) shows that they are usually short, in a keyword type manner. There is not much **context**. Therefore the MT systems tend to return only the default translation for a given input term, due to the lack of contexts required for disambiguation. But once the default translation is given it is not clear whether this is due to the fact that the context tests failed, or to the fact that there was no context to be tested as the input is too short. As a result, (rule-based) MT dictionaries would not be a good help in CLIR contexts. In addition, many systems do not handle multiwords entries properly.
- For data driven translation the same situation holds: The translation model may fail to take advantage of the contextual information due to the lack of input material (Nie 2010).
- There may be **polysemy** in the input terms. Again there is not enough context to do proper disambiguation. The only help is to use global parameter settings like a domain tag. The alternative to simply try *all* translation options which the dictionary offers, as proposed by Pirkola (1998), can be sure to add noise as the translations represent different senses; and taking just one may pick the wrong one.
- MT is unable to do **query expansion**. It returns just one hit. So in cases of unclear source term selection or in cases of dictionary gaps there is no translation. This lack of robustness cannot be easily overcome. Adding expansion terms to the input would confuse the MT system even more. Also dictionary coverage is a critical issue, cf. (Xu et al. 2001).
- One of the biggest drawbacks is that dictionaries, be it machine-readable or machine-translation based, do not return **probabilities**, as also noted by Zhang et al. (2009). Nearly all retrieval models use probabilities in computing similarities to search and rank documents. Attempts to provide such information are based on parallel corpora like in Zhang et al. (2009), or apply methods like probabilistic structured query (PSQ, cf. (Montazeralghaem et al. 2019)), combining document frequency and translation frequency, however requiring corpora of queries which usually are not available in practical settings.[16]

The result is that Query Translation requires special treatment and cannot be done by MT or by simple Dictionary Lookup. What is needed is a resource which combines translation information with information on query term handling like query expansion. Query Translation therefore requires a special component, accessing such a special resource. Such a resource, the LtConceptNet, is described below (Sect. 8.2).

8.1.3.2 Index Term Translation

An analysis of the CLIR query processing shows that what really should be translated are concepts, not simply words. So all operations described above, like tokenisation, normalisation, spelling correction, lexical analysis and lemmatisation, multiword

[16] They have been produced in projects like MATERIAL for research purposes.

recognition etc. are required to provide good index terms. Such index terms would form a good basis for multilingual operations: Crosslingual retrieval would exploit translation links between such terms. This is the idea of index translation.

Index translation can provide specific answers to the CLIR challenges described above:

1. Non-matching Terms

The main challenge in retrieval, **non-matching terms**, of course holds for the index translation as well. However, the two main strategies to reduce such gaps, normalisation and expansion, are applied as follows:

- **Normalisation** is applied to the queries in a query analysis component. This component performs text analysis like described in the previous chapters:
  - Normalisation and spelling correction
  - Lemmatisation and base form production
  - Multiword recognition

  So a query analysis component is required before index translation can be called, to bring the query terms into the state of the document terms. As such a query analysis must be carried out at runtime it is time-critical. An option would be, given a given document collection, to produce a static resource for this collection, with lemmatisers, decomposers, multiword analysers all compiled into a static form, and to replace the runtime analysis by a hash-table access.
  Most current system use only stemming as normalisation technique, as a surrogate technology for proper word form analysis. Such a component adds noise and produces results which are much worse, in particular in inflectional languages.[17] Stemming has been discussed in Sect. 5.5 above.
  Normalisation could also be considered for proper names. Early systems offered functions like SOUNDEX[18] which normalised homophones into normalised sound patterns. Name variants like '*Henderson*' '*Hendersson*' '*Henderssohn*' etc. could be normalised by SOUNDEX. As this procedure is language-specific variants for other languages were developed.[19] Such an option could be considered in cases where the focus of name search lies on recall.[20]
  In CLIR more sophisticated procedures were proposed to transliterate source language names into their target variants: Knight and Graehl (1998) presented a proposal for transliteration between English and Japanese; Merhav and Ash (2018) use several bilingual named entity transliteration datasets in their design study.

[17] "*stemmed indexes give much worse results than morphologically normalized indexes*" (Airio et al. 2003).

[18] https://en.wikipedia.org/wiki/soundex.

[19] Modules for French, German, and Italian, are found in https://github.com/corentin-ryr/py-phonem. For German cf. (Hassel 2018).

[20] E.g. in the SENSUS system, cf. (Thurmair 2004b).

- **Query Expansion** is the main method to broaden the conceptual base of a query. Whether the expansion should be done manually, interactively, or automatic, cf. (Chandra and Dwivedi 2019) is less relevant compared to the question which resources could be used. There are different kinds of resources at hand:
  - There are synonym dictionaries like English '*Oxford Dictionary of Synonyms and Antonyms*' (Oxford 2014), German '*Duden Das Wörterbuch der Synonyme*' (Duden 2024) etc. They can be used for query term lookup.
  - Repositories like WordNet or BabelNet provide not just synonym links but also hierarchical relations like broader and narrower terms. They can be used for Query Expansion. Using narrower terms for expansion seems to be less noisy than using broader terms. Of course such resources are only useful if the terms are not ambiguous.
  - An early version of TINA, called REALIST (Retrieval Aids for Statistics and Linguistics, cf. (Thurmair et al. 1986), offered expansion tables for a given term. They proposed expansion based on statistical co-occurrence, semantic co-occurrence based on head-modifier analysis as described in Sect. 7.3 above, and morphological expansion based on derivational and compositional morphology as in Thurmair (1984). The idea was that users could select the best terms for their search in an interactive retrieval setup.
  - The document base itself can be used for Query Expansion, in approaches of relevance feedback. Azad and Deepak (2019) give an overview of the techniques to weight and rank query expansion terms. They construct statistical models for the query terms and chose the expansion terms with the highest probability. The top scoring terms can be added to the query.

  An often discussed question is whether Query Expansion should be done on the source or on the target side. Source term query expansion could result in many terms being translated into the same target term, this way contradicting the intention. Target query expansion may introduce noise, if not properly translated terms are expanded. Expansion on both sides seems to deliver the best results, according to Azad and Deepak (2019).

  Query Expansion tends to add noise to the query. There is a range from rather 'safe' additional terms (like inflected forms) via 'rather safe' terms like synonyms to 'less safe' terms, like broader/narrower terms or relevance feedback terms. Which ones of those should be accepted depends on the amount of noise that is considered to be acceptable in a given CLIR system.

As a result, index term translation requires a query analysis component which normalises the query and returns the proper index terms, and enables the system to provide support for (source side) query expansion.

2. Canonical Forms of Terms

Index translation deals with terms presented in their canonical/base form. The previous chapters have shown how the text analysis can produce such forms. Query analysis will have produced them with a possibly simplified technique. Terms can be single words, compounds, or multiwords. Index translation just take the canonical forms for translation, and produces the corresponding term, be it single or multiword.

Not using canonical forms influences also word embedding systems; Yu and Allan (2020) report 'vocabulary mismatches ' for words like '*telefónicos teléfono telefónica telefónia telefóno*', which in fact are just textform variants, and could have been avoided by using lemmatisation.

Not analysing compound or multiword terms decreases the retrieval quality: Searching for $_{en}$ '*nuclear power plants*' ($_{de}$ '*Atomkraftwerk*') fails if '*nuclear*' (->'*Kern-*'), '*power*' (->'*Macht*') and '*plant*' (->'*Pflanze*') are translated in isolation. Also, cases like $_{en}$ '*money laundering*' ->$_{es}$ '*lavado del dinero*' ->$_{de}$ '*Geldwäsche*' correlate quite different canonical forms.[21]

To overcome the multiword problem, Pirkola (1998) proposes distance operators. Such a technique dates back to the first retrieval systems (IBM STAIRS) where operators like NEAR or ADJACENT could be used. Pirkola also proposes proximity operators (like '*uw3*' for three-words distance). However it is not just the proximity but also the specification direction which is relevant,[22] otherwise the terms '*oil pump*' (a kind of pump) and '*pump oil*' (a kind of oil) would return the same document set, which of course they should not.

Another surrogate technology to handle multiword terms is the use of ngrams, like (Mikolov et al. 2013b). Its use does not only depend on the participating languages (the more distant they are the less useful is the approach, cf. (Franco-Salvador et al. 2014); it also introduces a significant amount of noise to the system as many ngrams do not represent valid terms.

3. Polysemy

Using all possible translations of a polysemous term nearly inevitably introduces noise as only one sense is intended in the retrieval query. Missing sense distinctions also deteriorate word embeddings as noted by Camacho-Collados and Pilehvar (2018).

Sense-based information is stored in the SEM fields of the concepts resulting from text analysis, be it a domain marker or a sense marker (cf. Sect. 7.2 above). If one of the neighbouring query terms happens to be a member of the Word Sense Dictionary, then the senses can be disambiguated. Also a global variable setting like 'domain' could help in disambiguation, cf. (Xiong et al. 2016).

[21] In addition, the English and Spanish example show a contrarian head-modifier direction.

[22] Cf. Chapter 7.3 above or Schwarz (1990b). They also give examples that their distance is variable in a noun phrase, and is hard to determine by context operators.

An option is to have users decide which sense, i.e. which set of translations, they want to use. This would require an interactive retrieval setup. Such an approach was taken e.g. in the SENSUS project, cf. (Thurmair 2004b).

4. Unknown Words

Unknown terms cannot be translated. However they are added to the document language index, and can be searched in monolingual search. This is useful because after a certain saturation of the dictionary resources most unknown terms are Named Entities (names of persons, places etc.). In this respect they are not 'out-of-vocabulary' as they would be accessible for the search. But they are not translatable, if not covered by a Named Entity Gazetteers.

As lexical coverage indeed is a critical factor for CLIR, cf. (Xu et al. 2001), a practical solution consists in periodic maintenance runs to update the lexical resources of the system (cf. Sect. 9.3). The concern that the dictionary size could reach an upper limit of usability for searching (as mentioned by Nie (2010)) does not hold if additional domains are involved; for a given domain indeed the vocabulary plateaus, except for Named Entities.

It has been proposed to use subword units to mitigate the out-of-vocabulary or rare words problem. Galuščaková et al. (2020) report on Byte Pair Encoding or Wordpiece. As just mentioned, subword vocabularies would only be helpful in closely related languages; in German—Turkish pairs they might not show good results; otherwise they just introduce noise.

5. Probabilities

The Document Model provides frequency information for the index terms of the document language. However, index translation is not able to provide probabilities for CLIR. As explained above, the selection of translations depends on several factors, like the text domain, the availability of translation alternatives, the ambiguity/polysemy of the target language term and so on. For instance, the probability to translate $_{en}$ '*monitor*' by $_{de}$ '*Ausgabegerät*' is rather high in IT texts while its translation as $_{de}$ '*Abhörgerät*' is high in texts on wiretapping. There is no point in assigning a general probability to such terms. Computing translation probabilities for each domain would be a significant effort, and would require training data which usually are not available.

8.1.3.3 Document Translation

Besides query translation and index translation there is the option to base CLIR systems on document translation (cf. Fig. 8.2 above). Search is done on the query language side, and the document base would be translated into the query language for this purpose.

Chen and Gey (2004) and Nie (2010) report that tests have shown that document translation, or even both translation steps, produce better results than query translation alone. This may be due to the special requirements of query translation and expansion as explained above. An exception may be CLIR in specialised domains, like medical, cf. Galuščaková et al. (2020), Pecina (2016).

However, in reality document translation is nearly never an option: The whole document collection would have to be translated into all languages involved in a CLIR application, stored and maintained in parallel. This is a significant logistic effort, as the document base constantly undergoes changes. Multi-language documents are only available if legal or administrative regulations require it: English-French in Quebec, Dutch-French in Belgium, German-French-Italian in Switzerland. The European Commission translates into all official languages (however sometimes only on demand), institutions like the European Patent Office or the United Nations require multilingual documents. But the normal situation is that documents are stored once, in the language they come in, and must be searched in the language in which they are provided.

If documents are to be translated, then Machine Translation is the main option, due to the amount of data. This fact influences also the search as the MT systems apply special strategies to select transfers for a given term: If the query translation components picks a different translation, then rather poor search results will be observed. So consistency between query and document translation is an important requirement.

8.1.4 Transparent Information Retrieval (TIR)

Transparent Information Retrieval uses both document translation and query translation. Query language terms are translated into the document language, search is done in the document language, and the search results are translated back into the query language. For document translation MT would be the means of choice, as just stated.

8.1.4.1 Translation

The fact that Transparent Information Retrieval includes two translation steps poses an additional requirement to a CLIR scenario: The terminology used in the query must match the terminology returned in the translated documents, otherwise the conceptual space of the users is not consistent (Fig. 8.3).

This fact will influence the whole architecture of the translation components: The MT system will have to react on translation requests in different domains, and must provide domain-specific terminology. As users may want to search for different domains the MT system would have to be modular and load/unload dictionary modules for the domains in question. Such modules would also be of use for the

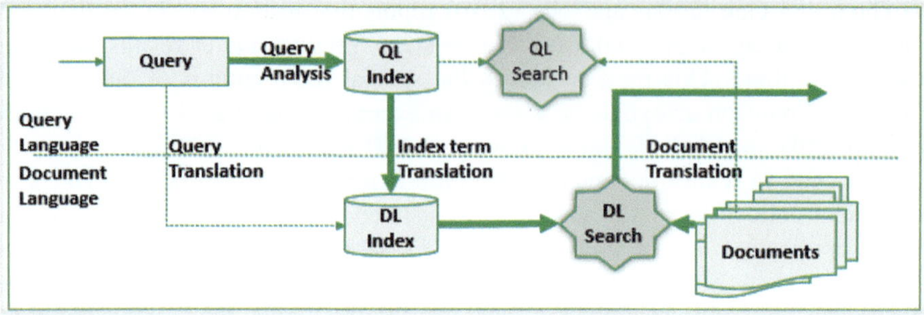

Fig. 8.3 Transparent information retrieval: details

query translation component, this way enabling consistent translation support for TIR. Domain adaptation[23] is a challenge in MT.[24]

8.1.4.2 Ranking

Transparent Information Retrieval has also an effect on the ranking of potentially relevant documents. While fusion in CLIR faces the problem to compare different-language documents, and come forward with solutions like round robin, raw score, Dataset Size based (like Airio et al. 2003) or z-score, cf. (Savoy 2003), none of which is really satisfactory. Recently, ranking using word embedding with BERT has been proposed, cf. Jiang et al. (2020) also for crosslingual contexts. In Transparent Information Retrieval the known standard ranking techniques can be used as all document language results are returnd in the query language, and can be compared therein.

Of course the MT translation influences the ranking, in particular if term consistency is not observed.

8.1.5 The Retrieval Resource: LtConceptNet

The result of these considerations is that Query Translation and Expansion (QTE) must be carried out by a special component, using a dedicated resource. The requirements for such a resource can be derived from the architectural descriptions as follows:

1. It should provide all **index terms** available in a given document collection, just like in all IR systems. Terms which cannot be found in the index cannot be

[23] "*Learning from translations of tractor manuals might not be wise when the goal is to search poetry*" (Galuščaková et al. 2020).

[24] Cf. (Pecina et al. 2012), evaluation in Thurmair (2015).

searched. The index terms are concepts and names as resulting from the text analysis described in the previous chapters.

2. It should provide support for Query Analysis. Query analysis is characterised by less syntactic structure than running text, and by a kind of closed vocabulary: Only terms occurring in the index need to be analysed; this is just a subset of the total vocabulary, restricted to a special domain and document collection. Query Analysis could use a shortcut of text analysis by using static data instead of full-fledged components. Query Analysis is time critical, so a direct link between textforms and concepts as represented in the index would be preferable. It would include normalisation and lemmatisation for single words, compounds and multiwords.

3. It should provide support for **Query Expansion**. Query Expansion is a key element in retrieval. Expansion would follow several action lines:

- **Textform expansion** collects all textforms which represent a given index term. This requires a link from the concept to its occurrences as provided by text analysis. Such an approach is superior to an inflection generation component as it only delivers text forms which really occur in the document collection, which by experience is only a fraction of all possible inflections of a term.
- **Concept expansion** provides links to other concepts, in terms of synonyms, broader and narrower terms, part-of relations etc.; such links can be found in synonym dictionaries or in collections like WordNet. In case ontological information is available links from the index terms to ontology nodes would help in retrieval.
- **Relevance feedback data** are provided by links from index terms to textforms and from there to document hits. These hits are the data based on which data-driven components can operate.

It depends on the concrete systems if such expansions are done by human selection, semi-automatically, or fully automatic.[25] The resource used for expansion is identical in all these cases, user involvement and the acceptance of noise in the search would be relevant decision parameters.

4. It should of course also support **Query Translation**. The requirements for the translation resource can also described in more detail:

- It should contain **translations**, i.e. link terms in different languages. It should cover all languages represented in the document collection, i.e. its index terms must match the document collection which is supposed to be searched. Relevant foreign language terms that are in the corpus but not in the resource will lead to poor search results.
- If a specific document collection must be supported, the index terms will be domain-specific, and so must be the translation resource. In particular the MT

[25] Cf. (Azad and Deepak 2019; Chandra and Dwivedi 2019). Interactive term selection for query expansion is offered in REALIST (Thurmair et al. 1986).

system used for document retranslation must be able to provide domain adaptation possibilities, like adding/removing dictionary modules and specialised glossaries.
- It should be **bi-directional** as it must translate from query to document language in search, and from document into query language in result presentation.
- It should be able to add **annotations** to the transfers, like full/partial equivalent, transfer conditions etc.
- The resource must be **consistent**. If query translation uses translations which differ in query and document translation, or show different coverage, then the users will become confused, and the overall system acceptance will drop.

Technically, as not all system components need all information items, the preferred approach would perhaps be to have one central repository where the whole resource is stored and maintained, and compile the necessary information items into the dictionaries of the respective components (like MT lexicons, extraction gazetteers, expansion ontologies etc.), without losing consistency. The LtConceptNet is designed to support all these requirements.

8.2 The Dynamic Multilingual LtConceptNet

This section describes how all the results of the different analysis components are collected, integrated in a common knowledge base, and made accessible for as many applications as possible. The following questions need to be answered:

- How can the text analysis results be collected, streamlined and handed over to the following component? How will the Document Model have to be adapted?
- How does the resulting resource called Multilingual Concept Net look like? How is it designed, which information will it contain? How will it communicate with backend systems, like IR systems?
- How can the analysis results of new documents be integrated into the Concept Net?
- How can it be used e.g. in Transparent Retrieval, and what is its added value? How does it support search, with query expansion, query translation, and document retranslation?

The design of the LtConceptNet reacts on these requirements. It is a network offering modular resources, supports both monolingual and multilingual applications, provides special support for Query Analysis and Translation. It interacts with the Document Model for newly analysed documents, it supports the compilation of resources e.g. into MT dictionaries, and offers a single point of management/administration.

The LtConceptNet idea has been realised in TINA in two flavours and in two domains, one being Law Enforcement (Aventinus: (Thurmair 1997)), the other one being Assistive Technology (EASTIN: Andrich 2011).[26]

8.2.1 LtConceptNet: Design and Implementation

The LtConceptNet is the main contribution of the TINA text analysis to multilingual search applications. It is not just oriented towards single documents but collects the knowledge of a whole collection into which newly analysed documents are integrated. It is intended to support several multilingual applications:

The first application is to support **classical IR systems**. Such systems build their indices based on document terms providing the term, its frequency, and a list of hits in the document or in parts of it. The LtConceptNet needs to collect this information and hand it over to backend systems like ElasticSearch[27] for efficient storage and search, but also other backend systems should be supported.

The target system environment are not huge web-based systems but medium size commercial systems which contain both proprietary document data and web documents: They should have common support. As such systems cover many different domains and languages the LtConceptNet should be domain-independent in its basic implementation but domain-adaptive to improve search results; for this reason the whole application should offer as much domain-independent knowledge as possible but on top also tools for efficient domain adaptation.

A second application is **abstracting**. If users want to obtain an overview of a document content, the LtConceptNet can help to identify the key concepts and support a summarisation of the document context.

A third aplication is **question answering**. Such systems often follow a retrieval—reader architecture by which first relevant documents are collected which undergo fine analysis in a second step. a ConceptNet would help in improving the retrieval component of such systems.

Question Answering is also supported by providing Named Entities. They often function as anchor points in questions, helping to answer '*who*', '*when*', '*where*' requests.

Meanwhile other techniques for question answering have been successfully implemented based on QA datasets (like the WikiQA dataset, cf. (Yang et al. 2015)) and word embedding techniques.

---

[26] The term ‚Concept Net' has first been used in Thurmair (2004a). Meanwhile there is another technology stream using the term ‚ConceptNet' (cf. Liu and Singh 2004b; Havasi et al. 2007; Speer et al. 2017) with quite different objectives. To avoid misunderstanding the name has been slightly modified in TINA.

[27] www.elasticsearch.com??

8.2.1.1 Monolingual ConceptNets

There are many systems which are just monolingual, and do not want to carry the overhead of a multilingual environment. For such applications, a monolingual flavour of a LtConceptNet will do the job. However there are specific properties which distinguish LtConceptNet from other resources.

1. Conventional Indexing

Indices are (alphabetically sorted) lists of (usually) textforms, with information on frequency and position of hits. Most operations, retrieval strategies, statistical models etc.,[28] start from there. They take the index terms as given, and inherit all problems encountered during the creation of the index list.

As opposed to this, the LtConceptNet would provide improvements on such index lists:

- The LtConceptNet is based on **lemmata**, not on textforms. This is a difference to many systems which use textforms and a kind of stemming.[29] Such a procedure often deteriorates the results as has been shown above (Sect. 5.5). In turn, the LtConceptNet must provide links from lemmata to textforms as textforms are the tokens occurring in the texts.
- Beyond the conventional index files the LtConceptNet is supposed to cover **multiword terms**. Not taking multiwords into account is a significant source of errors not just in the indices but also in all forms of computation which are based in index terms and their frequency: As already explained above, a term like $_{\text{en}}$ '*hold frame*' ($_{\text{de}}$ '*Raumspant*', in shipbuilding) should not count in the frequency lists of '*hold*', and sometimes not even for '*frame*'.
- The ConceptNet would be able to differentiate between different **word senses**, based on the results of the LtSensExtract components. Sense distinctions would improve precision, they are also required for translation selection in multilingual contexts.

2. Thesaurus Links

Historically the thesaurus was the first approach to create links between the index terms. A thesaurus creates a network of concepts instead of a single list, the nodes being the terms, the edges being the thesaurus relations.

[28] Cf. (van Rijsbergen 1979; Salton and McGill 1983; Ingwersen 1992; Frakes and Baeza-Yates 1992; Croft 2000; Manning et al. 2008).

[29] E.g. Gurusamy and Kannan (2014), Chen and Gey (2004). Airio et al. (2003) state that stemmed indices give much worse results than morphologically normalised ones.

Thesauri were first used for manual indexing of documents, cf. (Foskett 1997), and provided a controlled vocabulary to achieve consistent indexing behaviour between different individuals. As a consequence, thesauri are normative, not descriptive. Together with the terms also the types of links were standardised. The first ISO standard on thesauri was created in 1974 (ISO 2788) and refined since then until its current version (ISO 25964). The most important links which are provided in thesauri are:

- **broader terms/narrower term (BT, NT)**: Hyperonyms and hyponyms, to create hierarchies of concepts
- **synonym (SYN)**: Synonyms for a given term: '*mad cow disease*', '*bovine spongiform encephalopathy*', '*BSE*' would be marked as synonyms
- **related terms (RT)** are terms which are not really synonyms but similar or relevant in this context[30]

As thesauri were intended to be normative also the links between the terms were supposed to be normative. A standard for this kind of links is SKOS.[31]

There are hundreds of thesauri in different domains, even in the same domain user-specific thesauri can be found, e.g. the Daimler thesaurus differs from the Opel thesaurus in the field of automotive, due to different product lines, task organisations etc. Thesauri reflect a very specific view to a domain: Nobody would subclassify a '*vehicle*' into '*armoured vehicle*' and '*non-armoured vehicle*', and into '*tracked vehicle*' and '*wheeled vehicle*', except an army thesaurus. As the thesauri always reflect a specific way to see the world many attempts to merge them into larger ones, or create ontologies by merging thesauri, as discussed by e.g. (Sintichakis and Constantopoulos 2002), suffer from mismatches and unsatisfactory links and lose the link to the particular usage situation.

The consequence for the LtConceptNet is a network architecture distinguishing between nodes containing the terms and the links between them, called cross-references (xref). The links are typed, and the types must at least cover the classical thesaurus relations (BT, NT, SYN, RT).

Although there are many attempts to create thesaurus links automatically[32] the results would not be stable enough to allow thesaurus links updates at runtime when new documents are processed. Thesauri and their relations would have to be produced offline, and be subject to maintenance periods and versioning.

[30] '*an authorized, preferred heading that is related in definition and scope, but not broader or narrower* ' https://www.archives.gov.

[31] SKOS, cf. https://www.w3.org/2004/02/skos/, cf. also (DuCharme 2011).

[32] E.g. (Specia and Motta 2006; Yang and Powers 2008; Lanza et al. 2020; Hodge and Austin 2002).

3. WordNets

Like thesauri, WordNets[33] also are networks consisting of nodes and links between them.

- The **nodes** consist of *synsets*, i.e. lists of concepts considered to be semantically similar, or "*that are roughly synonymous*"; a short definition of the meaning of a synset is given in a *gloss*, with frequency information for this gloss; and the node may contain *usage examples* like example sentences. The single members of synsets are lemmata, presenting both single and multiwords. The synsets are accessed by index files returning pointers to synsets for a given lemma.
- The **links** between the nodes extend the thesaurus relations by other types of relations, like *holonym*, *meronym*, *entailment*, in addition to the classical *hyponym* und *hyperonym* relations. Based on *hyponym/hyperonym* links, the nouns and verbs are organised in trees, with a small number of *beginner 'trees'* to start with.

WordNets extend theasuri in two ways: By providing synonym terms they extend the normative thesaurus terms by a descriptive component, i.e. terms which really occur in texts. And by allowing different types of links they offer more flexibility in the linking of nodes. Overall, WordNets combine lexical and theaurus information elements.

The LtConceptNet follows the WordNet structure in its layout. It also uses lemmata as basic elements, instead of textforms.[34] Moreover, like WordNets, the LtConceptNet can be organised as a monolingual resource: All WordNet relations, as well as the *xref* relations in the LtConceptNet, stay within one language. However it differs from the WordNets in several respects:

1. It keeps **concepts** as heads of the nodes, instead of synsets. Synsets incorporate a specific semantic link, '*synonym*', into the node structure. However, in most cases concepts are synonyms *only in a certain respect*, in other respects they are quite different. The words of a given synset may themselves be members of several synsets, which may complicate sense detection in texts. Therefore the LtConceptNet treats the '*synonym*' relation not as part of the node structure but as an explicit link between concepts. The focus on single concepts instead of sets of them enables the net also to add links to a dictionary containing morphosyntactic description for a concept; this is not possible for a synset.
2. Nodes in LtConceptNet do not contain glosses or example sentences. They describe semantics by using the means found in the analysis components, like senses or domain attributes. They also allow for additional feature—value annotations. Adding glosses to a LtConceptNet node would be an improvement, however.

---

[33] https://wordnet.princeton.edu/, also Fellbaum (1998).

[34] To create lemmata from textforms, WordNet offers special lemmatisation components in its toolset.

3. There is also a difference in the links as the LtConceptNet links can contain annotations which enrich the links with additional information, like constraints in usage.
4. The main difference, however, is that WordNet is a static resource, basically describing the vocabulary of a given lexicon, whereas the LtConceptNet is a dynamic resource, describing a given document collection, like an index: Every newly analysed document changes the LtConceptNet, in terms of new terms, of frequency counts, of reference sentences, and sometimes even in the links, e.g. in cases of Named Entities or head-modifier links.

Of course a core resource must be provided as a starting point, and as domain adaptation is an important feature of the TINA analysis it also affects the LtConceptNet. So LtConceptNets may grow in different directions, depending on different domains or document collections.

4. Ontologies

Like the other representations, an ontology is a set of concepts and relations between them. Most of them originated in taxonomies used to describe specific domains. Like in the effort to merge thesauri, the challenge is to harmonise ontologies which themselves mirror a certain view to the world. Taking this into account attempts to merge and harmonise them have led to the postulation that at least the top nodes of the ontologies should be standardised. This consideration resulted in proposals for the construction of *upper ontologies* like SUMO,[35] DOLCE (Borgo et al. 2023), Cyc (Lenat 1995) and others, to define interoperability at least in the top nodes.

Ontologies are closer to conventional taxonomies or thesauri in the attempt to standardise or normalise the nodes and links they use, by defining constraints and formal representation languages to describe them and to draw inferences on them. Significant research investigates ontologies in a Web service context, cf. (Buitelaar and Cimiano 2014).

The LtConceptNet would not support such formalisation attempts. It operates more on a descriptive than on a normative side, and text analysis is not in the focus of ontology construction. However there are attempts to link the WordNet resources to such upper ontologies. If however an ontology is available for a given segment of the LtConceptNet, then a link of type 'ontology' could be set.

5. The Monolingual LtConceptNet

As a result, monolingual LtConceptNets are formally similar to the representation forms just described.

[35] (Niles and Pease 2001), cf. https://www.ontologyportal.org/.

The **nodes** describe concepts, consisting of a lemma, a POS attribute, and a semantic description, with a lemma frequency count related to the respective document collection. They can be enriched by additional annotations.

The **links** are subsumed to cross-references (XREF) defining semantic relations between concepts. LtConceptNet follows the standard link types as used in thesaurus and WordNet environments: Hypernym, hyponym, synonym, related etc. Links can have annotations which define flavours of the links, e.g. *SYN -><QUASI-Synonym>*, or *REL -><valid for {domain}>*. The purpose of the XREF links is to support mainly Query Expansion.

Three more link types can be found in LtConceptNet:

- IDX type links point to hits of the concept in texts, defined by textform, onset and offset.[36] Indices are used to retrieve documents for searches as in classical IR systems.
- LEX type links point to a dictionary entry in the Basic Lemma Dictionary where phonological, morphological and syntactic properties are stored. The representation of lexical entries like in LEMON[37] is not intended in the LtConceptNet; just a pointer to such an entry is stored. Systems like LEMON describe lexical models while the LtConceptNet describes document collections.
- ONTO points to a node in an ontology, if available. The link would serve as one lexicalisation of this ontology node

The difference of these link types to XREF links is that these link types do not point to other LtConceptNet nodes but to other system resources: the document collection, a morpho-syntactic dictionary, and an ontology.

It is important to note that all these links are monolingual. This fact enables the use of a LtConcepNet in purely monolingual setups.

8.2.1.2 Multilingual ConceptNets

Multilinguality extends monolingual processing by a multilingual dimension, and poses specific questions to an environment like the LtConceptNet. It is also an issue for IR system backends as they pose constraints on the use of multilingual document collections.

1. Indexing

The most straightforward way of multilingual indexing is to ignore the language issue **and store all terms of all languages in one index**.

[36] Depending on the possibilities of the backend system it may be necessary to build an intermediate resource with pointers from lemmata to their textforms.

[37] (LEMON 2019).

Such an approach faces the known disadvantages: Terms of different languages are collapsed in the index and create false hits, as mentioned above. The more languages are supported the more severe the problem becomes. And if the term is a stop word in one of the languages, it disappears altogether.

By adding a language tag to the index term the problem could be eased, and by also tagging the documents by language such crosslingual effects could be reduced. But tagging the index term by language means creating language-specific indices.

A second drawback is that even with language-tagged index terms only multilingual retrieval (MLIR) is possible: English terms search English documents, French terms search French documents, etc. This is not sufficient if foreign language documents are of interest as well. Therefore the index terms of the different languages need to be linked, i.e. to be translated. The need to translate terms would require an extra structure describing the translations, in addition to the index itself.

Like for the index terms, the issue of multiwords aggravates: Translations of index terms very often relate to multiwords. If only single words are available, then terms like $_{es}$ '*resistencia al corte*' -> $_{en}$ '*shearing resistance*' will not be translated, in particular if the prepositions are stopped away. For this reason multiword treatment is much more important for multilingual setups than for monolingual ones.

Conventional index components only contain single words, and rely on post-coordination to identify multiwords. This strategy fails in case of multilingual indices: only pre-coordinated terms can be translated.[38] Therefore the index must contain the multiword term as a whole.

As a result, the Multilingual LtConceptNet will keep the monolingual indices of the respective languages in separate repositories, without trying to merge them.

2. Thesaurus Links

The nodes of thesauri, or taxonomies, are sometimes multilingual. This is the case in domains which are highly technically standardised. Each node in such a taxonomy can be lexicalised with multilingual terminology. An example for the assistive domain is the EASTIN portal.[39] A similar example is the Eurovoc thesaurus of the European commission, cf. Fig. 8.4.

It should be noted that like in monolingual setups such multilingual taxonomies have a normative purpose; they define which terms *should* be used in documents and in search. The equivalents are often defined by the responsible institutions,[40]

[38] Cf. (Gaus 1995) for the description of pre- and post-coordination in indexing. Cf. also Buder et al. (1990).

[39] https://www.eastin.eu.

[40] An extended description of EASTIN is given in Sect. 8.3.2.

	Eurovoc
ID	182
es	control de cambios
de	Devisenkontrolle
el	έλεγχος συναλλάγματος
en	exchange control
ro	controlul schimbului valutar
pt	controlo cambial
it	controllo dei cambi
fr	contrôle des changes
hu	árfolyam-szabályozás

	EASTIN
ISO	121204
POS	NoC
src	abledata
da	pedalforlængelse til speeder
de	Gaspedalverlängerung
en	accelerator pedal extension
et	gaasipedaali pikendus
it	estensione del pedale dell'acceleratore
lt	akceleratoriaus pedalo pailginimas

Fig. 8.4 Examples of multilingual thesauri (Eurovoc, EASTIN)

they are not extracted from document collections. As a result they are more suitable for ontology production than for a multilingual LtConceptNet; however the existing data can easily be integrated.

Multilinguality does not just have consequences on the thesaurus terms but also on the link structure: The organisation of a multilingual thesaurus as in Fig. 8.4 assumes that the presented terms are fully equivalent. This assumption only holds if domains can be structured in an isomorphic way in several languages. This is true for many domains, mainly with technical content. However, other domains are organised quite differently in different countries and languages. Examples are the educational system, the legal system, also political institutions etc. Terms of such domains hardly find a full correspondence in other languages, and also the links between them differ. As an example, it requires special handbooks to compare and sort the national ranks of military or police officers in international organisations.

Such examples indicate that the preferable solution is to keep the relations of the hierarchies of the different languages separate from each other, rather than to try to find a hybrid hierarchy that fits all. Not just the nodes but also the links between them will be treated as language-specific. For the above-mentioned examples it means that the terms would be assigned to different (monolingual) LtConceptNets. This creates a certain redundancy, however it enables a consistent representation and structure layout. So, the multilingual LtConceptNet will keep the relational structures between the single language terms separate, with the consequence that it consists of several language-dependent nets. Several options then exist how to link them.

3. Multilingual WordNets

To make WordNets multilingual, two options can be taken: One is to make the synsets multilingual, the other one is to keep the synsets language-specific and create special means to cover multilinguality.

The first option attempts to augment synsets with foreign language equivalents. Such an approach is taken in **BabelNet** (Navigli and Ponzetto 2012), using Wikipedia links and Machine Translation to create the respective equivalents and merge them into 'Babel synsets'. Babel synsets are nodes which contain terms and their multilingual equivalents. The links between them were collected from the Princeton WordNet, and from Wikipedia categories.

The second option keeps the synsets as monolingual references and merges them into a common structure. Two main lines were proposed:

- create a common link structure as a backbone for the participating synsets, their nodes linking the synsets by common IDs shared with the backbone,
- keep the link structures of the different languages separate, and create a layer of interlingual links

The first alternative was expanded in the MultiwordNet projects (cf. Pianta et al. 2002), the second one in the EuroWordNet projects (cf. Vossen 2004). Challenges for both approaches consist in

- lexical gaps (lexicalisations which exist in one language but not in others),[41]
- denotational differences (synset members are more/less specific in one language compared to another one[42]), occurring even within one language (cf. the $_{de}$ *'präfekt'* means something different in Austria and Switzerland),
- different structuring of a domain as a whole (like educational systems), leading to different links between the nodes of the participating languages.

The **MultiWordNet** project proposed an approach by which language-specific synsets are linked by a common backbone relational structure. They contain databases containing the synsets for the participating languages (like ITALIAN-DB or ENGLISH-DB), and a COMMON-DB containing the relations which hold for all language-specific synsets, those synsets being linked by common identifiers. Cases where the relations between synsets differ by language are collected in a COMMON-ADD-ON or<language>-ADD-ON, overwriting parts of the COMMON-DB links.

Such a structure, containing common and language-specific add-ons, including language-specific synsets with no common identifiers as a consequence, indicates that the idea to create a WordNet based on an 'interlingual' common backbone of hierarchical links is a challenging approach. It assumes an isomorphism of the conceptual structures of the different languages.

Such an assumption has been given up in the **EuroWordNet** projects (cf. Vossen 2004). EuroWordNet does not try to merge snysets or relations between them. It preserves the language-specific structure of the single WordNets, and in

[41] "*7.8% of the English words correspond to lexical gaps in Italian*" (Bentivogli and Pianta 2005:253).

[42] In Wikipedia such entries are sometimes marked like: „*The examples and perspective in this article may not represent a worldwide view of the subject* ". (e.g. the entry for ‚*Travel Agency* ': https://en.wikipedia.org/wiki/Travel_agency .

order to link them it creates a data structure of an Interlingual Index (ILI), with the sole purpose of connecting synsets across wordnets. It is a special type of node only to link the synsets of the languages to each other. The ILI itself is an unstructured list of records. No relations between the records are maintained.

In later versions, like the **Open Multilingual Wordnet** (e.g. Bond et al. 2016; Bond et al. 2020), wordnets are linked by the Collaborative Interlingual Index (CILI). The CILI records are extended by definitions in English, as an universally understood definition is considered to be a prerequisite for a worldwide and universal collection of concepts. The CILI nodes differ in structure from the WordNet nodes: They contain an ID (and an English gloss) as identifier, and just links to nodes in the other wordnets of the system.

While the idea to create a list of all concepts of the participating wordnets in a Global Wordnet[43] is at least ambitious, the realisation creates two type of nodes: monolingual ones, with synsets and links between them, and multilingual ones (ILI, CILI) with links between monolingual nodes.

The two approaches towards interlingual links resemble to other multilingual systems: In Machine Translation, for instance, there were two main approaches: transfer-based and interlingua-based. Transfer-based systems link two languages directly, interlingua-based systems analyse all participating languages into a common structure and generate from there.

The Open Multilingual WordNet would belong to the interlingua type of systems. It links a given synset to a interlingua type structure, and from there it links to a target language. Even if it does not take over the monolingual relations between them, the differences are indicated by the efforts to create concepts without synsets (for lexical gaps, with a feature '*lexicalized=no*'), to add (English) definitions, avoid potential duplications of concepts, add validation measures and confidence scores, and other means.

An alternative to the interlingua approach would be to follow more a transfer-based development, by creating direct links between the languages. This is what the LtConceptNet intends to do: A concept would have different links into different target languages, which even could be annotated differently.

There is one more issue to take into account: The WordNets discussed so far are multilingual, i.e. they describe lexical relations in many languages. However they are not crosslingual: They do not provide translations. The reason is the notion of a synset: Nodes in WordNets are built by synsets, not by single terms. Members of a synset share certain semantic properties, most notably synonymy. While there is a certain range of acceptance which terms could be considered as synonyms[44] it is also obvious that such terms differ in many other semantic respects, like connotations, language register, collocations and others. But all these factors come into play when a translation for a concept has to be found. As a result, all members of a synset may require different translations. This fact cannot be expressed if such terms are glued together into synsets.

---

[43] http://globalwordnet.org/.

[44] There is also the opinion that real synonyms are close to non-existent.

As a consequence, translations are used to build multilingual wordnets but it is difficult to use wordnets for translation purposes: They are not able to propose translation equivalents for the members of their synsets even if a corresponding synset exists in the target language.

The LtConceptNet tries to avoid such difficulties by using less ambitious design features:

- It keeps the different concepts in different nodes, and does not favour one semantic relation (synonymy) into a system property (i.e. synonymy is not an inherent property of the node but just another link).
- It does not try to create a interlingual kind of backbone structure but links the concepts of the languages by direct transfer links.
- It creates language-specific relational structures, independent of whether or not they are isomorphic in other languages, similar to the EuroWordNet approach.

This way, only one type of node is required throughout the network, and only one additional type of link, 'XFR' (transfer), is required for the representation of multilinguality in LtConceptNet.

4. Multilingual Ontologies

Most of the multilingual ontologies originate in multilingual taxonomies. As Gillis-Webber and Keet (2022) show, most of them use multilingual labels. Very few ontologies use linguistic models or mapping models based on a common interlingua. The multilingual labels are intended to be translations of each other, like in the case of thesaurus relations, based on a semantically rather strict domain model.

As the LtConceptNet does not intend to provide formal processing means, it would just add the translations as links to the existing concepts as described above in the processing of thesauri.

5. The Multilingual LtConceptNet Design

The extension of the LtConceptNet into the multilingual and crosslingual domain consists in adding an additional type of link, 'XFR' (transfer), to the nodes of the net. The nodes themselves will be annotated with a language attribute, the links with (source and) target language. The target of the link is a (sense-disambiguated) node of the target language. The XFR links can also have attributes, defining the scope of the translation, additional tests and actions, similar to the links in the transfer dictionaries of MT systems. Like the other links in the LtConceptNet links are directed, i.e. a source-to-target link does not imply its reverse target-to-source relation.

8.2.2 *LtConceptNet Layout*

8.2.2.1 Design Features of the LtConceptNet

The LtConceptNet collects the results of the text analysis. Its first and main property is therefore to be a dynamic resource, as opposed to WordNets. It links the knowledge found in the document analysis to an existing kernel network.

Obviously the LtConceptNet contains static and dynamic parts.

- **Static** parts are the basic lexical nodes with knowledge from monolingual dictionaries, translations resulting from the lookup of bidirectional dictionaries, and thesaurus and other crossreference relations available from conceptual hierarchies.
- These structures are **dynamically** extended by adding the results of the TINA text analysis. Everything laid down in the document model of an analysed document will be part of the LtConceptNet for further analysis and search.

This fact makes the LtConceptNet not just domain-dependent but even specific to a given document collection. Therefore it is an essential requirement that updates of the LtConceptNet run fully automatically. Domain dependency here is not a drawback but an advantage as it enriches the static backbone with information that is specific for a document collection, this way improving the search results without overgeneration.

The fact that the LtConceptNet also maintains text analysis results has of course consequences for the net layout and content. Its basic structure is like all networks.

- The **nodes** store the terms and all analysis results which can be linked to them. In particular for Named Entities they store their feature annotations, like *day* and *month* for dates, *amounts* for units and prices, *types* of communication means. As the nodes represent concepts they maintain lists of the text forms for these concepts, including normalisation. Such forms will be used to access the documents.
- The **links** contain additional link instances resulting from the text analysis, like head-modifier specifications, links between Named Entity parts, and others. In addition, links can also have attributes, like type information for Related Terms, constraints and additional tests for transfers etc. Of course the most prominent extension is the IDX link from the node to the document hits.

To compare to LtConceptNet with other existing Knowledge Representations, the following differences can be stated:

- As already stated the LtConceptNet is **dynamic**, i.e. it changes with the progress of document analysis. It therefore is **domain** and document collection specific. Its most prominent feature is the link to an index of document hits, updated every time a new document is analysed.

- The LtConceptNet is a **multilingual** resource, containing links to foreign language nodes. Instead of an interlingual approach like EuroWordNet's ILIs it is transfer-based, with direct language-indicating links between concept nodes. This way it avoids problems like universal concept structures, gaps in lexicalisation and other such challenges. Multilinguality does not only refer to the support of more than one language but also the support of crosslingual tasks, especially crosslingual search. Therefore the LtConceptNet will contain links between nodes of different languages.
- The LtConceptNet is **not a dictionary**. It does not describe the linguistic properties of a given lexical entry like in the Lexical Markup Framework (LMF, cf. Francopoulo 2013), or in web-based representations like lemon, cf. (Montiel-Ponsoda et al. 2011). Its focus is on the occurrences of terms in document collections. Many lexical properties of a term are irrelevant for this purpose, however if a term has a monolingual entry, the LtConceptNet can contain a link to such a dictionary entry. But the LtConceptNet also contains nodes for unknown terms, or other phenomena occurring in text analysis.
- The LtConceptNet is **not a WordNet**. Aside from the fact that it is dynamic it mainly differs from WordNets by the fact that it does not build synsets, instead keeps the members of a synset (i.e. the single terms) as separate nodes, linked by an explicit synonymy relation. This way it can link a single term to a list of document hits but also to a specific translation.
- As compared to **ontologies**, the LtConceptNet would be just the lexicalisation part of such a knowledge base. The ontologiy itself would have to be maintained outside of the LtConceptNet. If in a given domain an ontology is available, the LtConceptNet would be able to create links to the respective ontology node.
- Of course the terms of the LtConceptNet can be compounds or **multiwords**. Their parts will be linked by appropriate relations, in particular *head-modifier* relations. This enables the search for e.g. all modifiers of a given head word.

8.2.2.2 LtConceptNet Nodes

The nodes in a LtConceptNet are sorted alphabetically. Beside an ID, they contain the following information elements:

1. The **language**. Every node belongs to a language-specific net.
2. The **concept**. As steted above, it consists of three elements: Its lemma, a part-of-speech annotation, and a semantic description. The POS annotation is an element of the BTAG set. Usually function words would not be found in the LtConceptNet.[45] The semantic description could be a domain tag for content words, a sense identifier, or one of the semantic types for Named Entities (like *person*, *location*, *price* etc.). A special glossary showing contexts of the concepts like in EuroWordNet would be desirable but is often not available.

[45] Except if they are part of a multiword.

3. The **forms**. As the concept is a lemma in base form, a link to the textform occurrences is required. In order to avoid calling inflection or lemmatisation tools during search, the node contains all inflected forms from which it is derived, including normalisation results. As the document collection from which the inflected forms originate is known only inflected forms really occurring there need to be considered, not the full paradigm. This way, many inflected forms which are possible but not used in the document collection need not be considered. The *forms* also contains a stem in case a stemmer result can be found. At indexing time, the index (IDX) can be accessed to find document hits by using the *forms* information for the respective textform.

Each member of the forms list, i.e. each textform or stem, contains frequency information, in addition to the frequency of the concept as a whole. This way operations using frequency, like needed for clustering, need not access the index table but can collect the information directly from the LtConceptNet.

4. The **frequency**. Frequency information is important not just for indexing: It plays an important role also in WordNets.[46] It is known that frequency information can differ significantly depending on the corpus used to create it, and its domain; if this is not specified, the frequency information is of only limited value. In the LtConceptNet the node frequency refers to the document collection under analysis, and can be easily verified and used from there. The frequency tag in the node refers to the frequency of the *concept*, the one of the different *text forms* is given in the *forms* attribute.

5. The node also contains a **type** attribute. Node types are '*Common word*'(cw), '*Named entity*' (ne) and '*unknown*' (unk). The distinction between common words and Named Entities is relevant because it defines which attributes of the node would be available in the attributes section. '*Unknown*' nodes are created during analysis, if no successful analysis can be given. The unknown term should stay in the index because it may be the object of a search (e.g. a strange proper name).

6. The **entry type** of the node, values being '*single word*', '*compound*' and '*multiword*'[47] is used to define the available information for the node: Compounds and multiwords provide internal structures: For searches with just a head (like $_{de}$ '*tisch*') compound structures need to be available, to return $_{de}$ '*küchentisch*'/'*arbeitstisch*' as possible hit. Similar considerations hold for multiwords ($_{en}$ '*kitchen table*') and their parts, they would also be analysed into head-modifier relations between their parts.

7. The **attributes** are lists of feature-value pairs. Their content depends on the type attributes. A tag for domain information would be found here. In case an

[46] WordNet does not give frequencies for corpus data, however, cf. https://wordnet.princeton.edu/documentation/wndb5wn.

[47] The entrytype for bound morphemes as used in lexical analysis is not needed here; bound morphemes never are members of LtConceptNet. Thee are cases, however, where bound morphemes are used, e.g. in English-German translation when English adjectives are translated by German compound specifiers, like '*royal*' ->'*Königs-*', '*functional*' ->'*Funktions-*' etc.

ontology is involved, a feature will be set, pointing to a node in the ontology, e.g. '*ISO9999=121204*'; the ontology itself would be a knowledge base outside the LtConceptNet. For Named Entities, the features extracted during analysis are stored here: '*amount*' for units and prices, '*first name*' and '*last name*' for persons, '*subtype*' for locations and communication means, '*day*', '*month*' and '*year*' for dates, or '*legalform*' for companies, if available.
8. The **links** section contains all the links from and to the node.

A gloss attribute to define the semantics of the concept node is not foreseen, rather for pragmatic than for conceptual reasons. It would be useful to give more explanation to the node entry.

8.2.2.3 LtConceptNet Links

The links relate the node to different kinds of objects.

1. **IDX**: **Index** links point from the text forms (and stem) of the concept of the node to the documents. Each textform points to a resource where document-ID, onset and offset of all matches of this textform are given, together with frequency information for this textform. The index is not a node but a special structure, like a HashMap.

The reason for this approach is that the LtConceptNet may have to cooperate with different system backups, which have different ways to deal with multilingual information. Examples like Autonomy and Elasticsearch have been mentioned above. In any case, there is additional effort required to link the LtConceptNet to a backend system, and it is advisable to keep all possibly required data in a common structure for the time being. Therefore the LtConceptNet also maintains the index part of the system.

Due to the unknown backend structure the LtConceptNet shows a bit of redundancy here. The concepts maintains a pointer to the document IDs in which the concept occurs, independent of its textform or the entrytype (also multiwords point to a document list). This is the use of the IDX pointer. In addition, the concept also contains the *forms* attribute, collecting all the textforms of the given concept. As a fallback the textforms would take pointers to document IDs as well, including frequency, onset and offset. As a textform can originate from several concepts (homographs, including multiword parts) the frequency information and the document ID would differ from the LtConceptNet nodes, and searching with textforms would lose precision for this reason. Whether or not this can be avoided depends on the backend system.

2. **LEX**: The node may point to a **lexical entry** in a (monolingual) dictionary. Again this is a knowledge base outside the LtConceptNet, containing linguistic information about the concept, like morphosyntactic features. It will be described in Sect. 9.2 below.

3. **ONTO**: In case there is an **ontology** available for the domain of the LtConceptNet, a link from the node to the respective node of the ontology would be given here. The link would point to a node in the ontology, e.g. '*ISO9999*

->121,204'; the ontology itself would be a again a knowledge base outside the LtConceptNet domain.

4. **XREF**: **Crossreferences**: This section of links is intended to link the nodes of the LtConceptNet to other such nodes. It will be used in Query Expansion. There would be six types of such links:

- SYN links to synonyms; they would be merged into one synset in WordNet
- BT (broader term, hyperonym) and
- NT (narrower term, hyponym) would give the hierarchical position of the node in a conceptual hierarchy
- RT (related term) would collect all other kinds of relationships, like PartOf, MemberOf, and the reverse HasPart, HasMember. The respective kind of relationship is given in the attribute of the RT link.
- HEAD is a pointer to all concepts which the given node has as heads (i.e. which it modifies). This link is subject to updates during indexing.
- MOD is a pointer to all modifiers of the concept node, updated every time a new document is indexed.[48]

HEAD and MOD links are created during document analysis, from compound, multiword and noun phrase analysis, as described in Sect. 7.3. Therefore they are kept aside of the other link types of XREF.

Additional RT links can also be created from the document analysis for Named Entities, like *HasFirstName/HasLastName* for persons, *LocatedIn/Locates* for locations, *PartialDate/ContainsDate* for dates, or *EmplyedBy/Employs* for links between persons and companies. There could be more depending on the depth of analysis.

Except for HEAD and MOD and some of the RT links the crossreference links are static, and inherited from the external resources used to build the LtConceptNet.

All these links can be used at query time for Query Expansion. The fact that they are specific for a given document collection prevents them from overgeneration.

5. **XFR**: **Transfer**: This set of links is created between nodes of LtConceptNets of different languages. They give the target language and the node in the LtConceptNet of this language.

All links of the ConceptNet can have **attributes**, implemented as feature-value pairs. They provide more information to the links, and add constraints or subdivisions. Attributes to XREF like RT (Related Term) give kinds of links like '*PartOf*' or '*isMember*'. Attributes to SYN could be nuances like '*QuasiSynonym*'. Attribues to XFR links could be tests or other constraints of transfer selection like '*colloquial speech*' or '*US English*'.

[48] In CoNLL-U there is the DEPREL relation which gives further details as to what kind of modification/dependency is meant. Here there is just the global relation of modification.

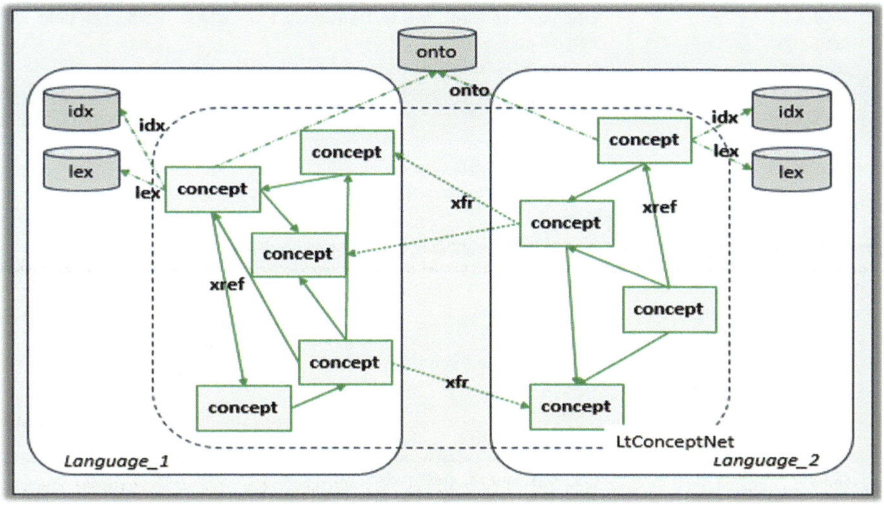

Fig. 8.5 LtConceptNet visualisation

The whole picture is visualised in Fig. 8.5. Figure 8.6 gives the complete overview of the LtConceptNet.

8.2.3 *Updating the LtConceptNet*

As the LtConceptNet is a dynamic structure it has to be updated every time a new document enters its conceptual space. The new document stores all its information in its Document Model. The integration therefore consists in importing the information items of the Document Model into the LtConceptNet. It has to consider the following configurations:

- A concept contains a proper analysis result, or it represents just an UNKNOWN (i.e. just instantiated).
- A node for a concept does exist or not, and itself is of type 'unknown' or not

Depending on the respective configuration the update procedure must execute different tasks.

When a new document is provided all entries of the Document Model are inspected. Only content words will be taken into the LtConceptNet.[49] These are nouns, verbs, adjectives, and some types of adverbs, esp. verbal modifiers.[50]

[49] Except for function words used in multiword expressions.

[50] Adverbs like ,therefore' are not used, also STAG sentential modifiers like '*probably*' are not of interest.

```
Node              ::=  language, concept, forms, frequency, type, etype, attributes, links
language          ::=  <an iso code>
concept           ::=  lemma, POS, sem
  lemma           ::=  <string>
  POS             ::=  <a BTAG>
  Sem             ::=  <string>
forms             ::=  textform+ stem?
  textform        ::=  <string>, txtfrq, idx-link+     // including stems
  txtfrq          ::=  integer                         // frequency of textform or stem
  idx-link        ::=  <pointer>
frequency         ::=  <integer>
type              ::=  co | ne | unk                   // common, name, unknown
etype             ::=  sw | cp | mw                    // singleword, compound, multiword
attributes        ::=  fvpair*
  fvpair          ::=  feature, (value | value ';')+
  feature         ::=  <string>
  value           ::=  <string>
  links           ::=  ->
```

```
links             ::=  idx, lex?, onto?, xref*, xfr*
idx               ::=  targetidx, attributes  // could also give <lemma,pos,sem> of target
lex               ::=  targetlex, attributes
onto              ::=  targetonto, attributes
xref              ::=  target, xreftype, xrefattributes
xreftype          ::=  syn | bt | nt | rt | head | mod
xrefattributes ::=     partof | haspart | memberof | hasmembr | hasfirstname |
haslastname              | locatedin | locates | partialdateof | containsdate
xfr               ::=  target, xfrtype, xfrattributes
xfrtype           ::=  <a languagecode>
target            ::=  <a ConceptNet ID>
targetlex         ::=  <a dictionary ID>
targetonto        ::=  <a, onto ID>
xfratributes      ::=  fvpair*
fvpair            ::=  feature, (value | value ';')+
feature           ::=  <string>
value             ::=  <string>
```

Fig. 8.6 LtConceptNet definition

Every such entry will be searched in the LtConceptNet. Search criteria are the language, and the features constituting a concept: Lemma, part-of-speech (BTAG), and semantics. While such comparisons are straightforward for Named Entities, the semantic information may be difficult to compare for content words; it could be sense indicators, domain tags, or no *sem* value at all. Sometimes already existing links in the LtConceptNet could help to identify the right node, like ontology links.

If the entry can be assigned to an existing LtConceptNet node, this node will be updated. Otherwise a new LtConceptNet node will be created.

8.2.3.1 Updating the Nodes

A ConeptNet node can be empty (type *unknown*) or already instantiated (type *common* or *name*). An empty node may have been created by a previous document analysis containing the node as UNKNOWN, or by links to this node, for instance if a transfer link was created from a node of some other language with this node as target, but the node did not exist yet. In the other cases it is already instantiated by previous LtConceptNet updates.

In case the node type is ***unknown***, or it does not exist yet, it must be instantiated. This operation includes the following actions:

- The **language** of the ConceptNode is taken from the language attribute of the Document Model
- The **concept** description is taken from the respective concept of the Document Model: the **lemma** and the **part of speech** (BTag) and **sem** information are copied. In case the Document Model presents an unknown lemma, the lemma and POS are set according to the Defaulter results. For the **sem** feature, the topic of the document as determined by the Topic Identifier can be used, if it is not just General Vocabulary (GV), Common Technical (CTV), or Common Social Vocabulary (CSV).
- The **type** feature is set depending on the information in the Document Model: common word, name, or unknown
- The **entrytype** feature (single word, compound, multiword) is also taken from the Document Model, the default being single word.

For **all** LtConceptNet nodes, be they newly created or already existing, the update includes the following node operations:

- The **forms** feature must be updated. New text forms of lemmata, or new stems must be added to the forms list. And the respective (textform) frequency information must be updated.
- The (concept) **frequency** feature of the LtConceptNet node must be incremented
- The **attributes** feature needs to be checked, and attributes which do not already exist must be unified with the LtConceptNet list.

8.2.3.2 Updating the Links

The update of the LtConceptNet links depends on the respective configuration.

IDX links are updated in any case as the Document Model returns new document hits. Updating the IDX implies an update in the node frequency, and in the node forms: If the textform or the stem is not already available, they are added to the forms list, and their frequency count is incremented. If the LtConceptNet node is new, a new index entry must be created, otherwise the existing index entry is enriched by the new document hits.

LEX links will be created if the Document Model term can provide it, and the LtConceptNet does not already provide it (from previous entries).

ONTO links will be validated. In cases where the ontology provides lexicalisation information all terms specified there will be taken over into the LtConceptNet, with links to the respective ontology nodes. This may create empty LtConceptNet nodes. At runtime such nodes would be instantiated but no new links would be created.

XREF links will also be inherited from the LtConceptNet, new documents would not be able to enrich such links, except for HEAD, MOD and some RT links:

- For content words which are compounds or multiwords (i.e. *etype*=*co* or *mw*) HEAD and MOD links are created by searching/creating the nodes in the LtConceptNet, and creating the *head/modifier* links between them.
- For named entities the attributes of the DocumentModel term are taken over to the LtConceptNet into the RT links. It is a design decision whether or not links like *HasLastName* or *LocatedIn* are added to the XREF links.

An example of an import of a multiword term or name is given in Fig. 8.7.

The update of **XFR** links would involve the following steps:
The term is looked up in the bilingual dictionary.

- If it is not found (1:0 case), nothing can be done. This holds e.g. for unknown terms.
- If it is found with just one translation (1:1 case), the target side is inspected: If the target node does not exist yet, an empty node will be created there, with just the language attribute instantiated: It may be completed by later documents. Otherwise it is just linked with the current source node.
- If it has several translations (1:n case), a disambiguation step must be included to find the right link: Disambiguation first checks the word sense, then the

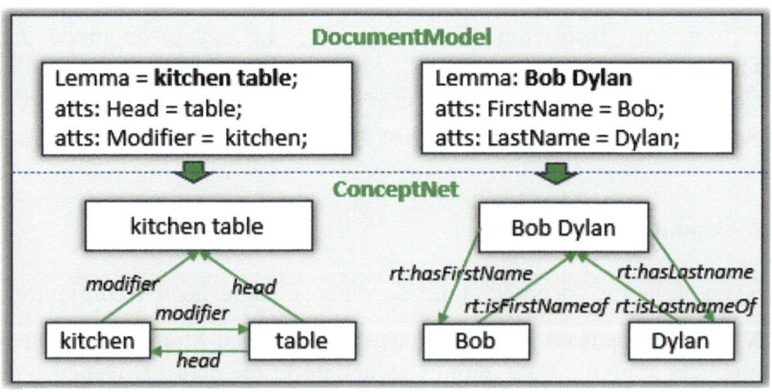

Fig. 8.7 Updating the LtConceptNet with multiwords and names

	SL node		TL node		Links	
	new	update	new	update	new	update
SL node exists	--	forms, frq, atts	--	--	--	IDX, XREF
SL node empty	forms, frq, atts	--	--	--	LEX, ONTO	IDX, XREF
SL node missing	lg, concept, type, etype, forms, frq, atts	--	--	--	LEX, ONTO, IDX, XREF	--
=>						
TL node exists	--	--	--	--	XFR	--
TL node empty	--	--	--	--	XFR	--
TL node missing	--	--	lg, concept, type,etype	--	XFR	--

Fig. 8.8 LtConceptNet update schema

domain information, and finally takes the most frequent target node.[51] The reading which comes out best is treated as described in the 1:1 case. In cases which cannot be disambiguated several XFR links must be created.

All update requirements for the different configurations are presented in Fig. 8.8.

8.3 Transparent Information Retrieval Using the LtConceptNet

It has been explained that the LtConceptNet is a dynamic structure. The previous section described how the information elements laid down in the document model of a given input document could be integrated. This section describes how the LtConceptNet supports the different phases of Transparent Information Retrieval in Text Analysis and Search.

Text Analysis as described in the previous chapters uses lexical resources which may undergo updating procedures triggered by the LtConceptNet. This may relate to Named Entities, which are not yet represented in the Named Entity Repository, or to multiwords which are found in the text. Of course unknown terms and head-modifier relations would also be looked at for dictionary maintenance.

However, the LtConceptNet is not just intended to support text analysis but also to support querying.

8.3.1 Support for Query Processing

8.3.1.1 Query Analysis

Query Analysis in the intended application requires two global parameters: language and domain.

[51] It has been mentioned above (in Sect. 7.2) that this heuristic is still better than other disambiguation attempts.

- Although a **language** identifier is available in TINA (cf. Sect. 3.2 above), it is too risky to use it for short and possible ambiguous pieces of text: A query for '*information*' does not reveal which language is intended to be searched. A wrong selection of the query language results in dramatically low search results. Therefore the better option is to use a global parameter (and have it confirmed/updated by the end users).
- A similar consideration holds for the **domain** information. It provides significant additional information for term disambiguation and word sense selection, and should also be provided as a global parameter, in particular as it does not just hold for the query but also for the document retranslation in Transparent Information Retrieval.

The LtConceptNet represents all terms represented in a given document collection. As such it provides lexical material for the query analysis.

Query analysis requires a special parsing component. This component does not need to be as explicit as a full parse; basically it just needs to identify the basic concepts mentioned in the query. Which concepts are available for search is determined in the LtConceptNet as it represents all terms in the document collection.

The fact that the terms in the LtConceptNet are a limited set can be used for the design of the Query Analysis component: It allows to build static lexical analysers and static decomposers for the analysis, and a simplified shallow multiword recogniser based on a simple comparison of the lemmata of its elements would suffice to identify multiword candidate terms. Such a resource for query analysis can be updated periodically according to changes in the LtConceptNet.

However there is always the possibility that users come forward with query terms which are out of vocabulary of a given document collection. If no hint can be found in the query term, like a spelling correction, a normalisation option, a defaulter action leading to a known term, a translateration possibility or an indicator of a Named Entity, then the users would have to be informed that the term is not searchable. Out-of-vocabulary terms are a major source of failure in retrieval, cf. (Azad and Deepack 2019).

8.3.1.2 Query Translation

Providing translations for terms is one of the most prominent tasks of the LtConceptNet; without it, query translation is not feasible.

The LtConceptNet is searched for a XFR link for a given term. Attributes of the link are tested before the target term is returned. The following constellations can be found:

- There is **no target term** for a given source. In this case, query term is expanded by examining the XREF links of the source term, to find synonyms or hierarchically related terms. If translations are discovered, e.g. a synonym, then the target terms of these expanded terms are returned. However in most cases user interaction should be initiated.

- There is more than one target term, i.e. the query terms is ambiguous. Picking the wrong word sense results in inadequate translations, and in insufficient search results as a consequence. The LtConceptNet can support query disambiguation in three ways:
  - Consulting the Word Sense Repository (WSR, cf. Sect. 9.2) may help in disambiguation. Neighbouring query terms could be identified there and help to select the right sense.
  - Disambiguation based on domain information. Such information would be available as a global parameter and guide the disambiguation process
  - In some cases it may help to compare the target term frequencies (available in the IDX links of the LtConceptNet) and select the most frequent target term.

  However it may be safer to ask for user confirmation—if the users are able to understand the differences in the target language terms.
- It there is just one translation there is still the possibility that the target term is incorrect, e.g. if lexical gaps must be taken into account. However in most cases the translation may be picked.

The LtConceptNet can also be used for validation purposes: If source language synonyms are available, then all of them can be translated and the translations can be compared; there should be interconnected target terms resulting from synonym source terms.

8.3.1.3 Query Expansion

Query Expansion with LtConceptNet is possible along two lines:

- **IDX expansion**, i.e. expansion of an index term into different inflected forms, normalisation possibilities, spellings and the like. To produce these forms, which are de facto the strings for which the backend retrieval system will search, the *forms* feature of the LtConceptNet nodes can be used. They contain all textforms that were found in the document collection. As mentioned this is superior to inflection generators which generate unnecessary terms and produce retrieval overhead this way.

  IDX expansion in Transparent Information Retrieval would only be required on the document language side, to receive the strings to use for the search.
  In case of Query Expansion based on relevance feedback or Term Association the required information to produce them would be available in the IDX links, and could also help in query language expansion, to have more terms available for translation. Users may understand the query-expanded terms better than the originally translated ones.

- **XREF expansion**, i.e. expansion of terms into semantically related terms, as documented in the XREF links of a given query term in the LtConceptNet. The expansion would return the term, and the type of link, plus possibly an attribute value like '*quasi-synonym*'.

The two types of expansion should be distinguished because they are used differently in query processing: XREF expansion would be applied both for query language and document language expansion: Expansion before translation (query language expansion) could be useful to support transfer selection as just mentioned, while document language expansion would produce the real terms for the search. In turn, IDX expansion is only required on the document language side.

8.3.1.4 Search

The search implies interaction with the backend retrieval system. The interface depends on the backend system used, and affects the way the query terms are handed over and integrated into the query language provided by the backend, e.g. ElasticSerch Query Language (ES|QL).[52]

The backend returns the document hits and ranks them. Often the backend systems offer different options for ranking,[53] most of them offer at least the Okapi BM25 method.[54]

The IDX link in the LtConceptNet provides information required by the most common ranking methods, e.g. term frequency and document frequency, which is needed to compute for instance tf·idf scores.

The backend systems support only monolingual ranking. Although procedures for multilingual ranking have been proposed (cf. Sect. 8.1.4 above), the solution for Transparent Information Retrieval is simply to rank documents from different languages after having them re-translated into the query language, and then try monolingual ranking for the result set.

8.3.1.5 Document Retranslation

Document retranslation into the query language will use machine translation techniques. Two options can be imagined:

- **Keyword Translation** does not intend a full text translation but just keyword translation. The keywords of a document are identified and translated. Such an approach is useful in cases where only information translation is intended, to decide whether a document in relevant for a given search profile or not; in case

[52] https://www.elastic.co/guide/en/elasticsearch/reference/current/esql.html.

[53] https://www.elastic.co/guide/en/elasticsearch/reference/current/index-modules-similarity.html.

[54] Cf. https://en.wikipedia.org/wiki/Okapi_BM25.

it is it will be sent to complete translation. In many cases the majority of documents can be eliminated before expensive translation resources are utilised, e.g. in OSINT contexts or other internet searches.

Keyword translation is also an option for languages and domains where no MT system with sufficient quality is available. For this type of retranslation the XFR links of the LtConceptNet can be exploited.

- **Full Translation** of the hits in document language into the query language is of course also an option. For users it is important to have a consistent view of the search system as a whole; therefore the same terminology should be used for Query Translation & Expansion and for document retranslation. Such a requirement implies that the MT system can be adapted to the document collection and domain in question. This could be possible by allowing the MT system to load additional glossaries or vocabulary modules. For data driven systems such adaptation is more complicated than for lexically-driven systems.[55]

8.3.1.6 Monolingual Ranking

As all document hits in a Transparent Information Retrieval system are available in the query language in the end, no specific ranking procedures like in CLIR applications need to be invented, and standard ranking tools can be applied. The LtConceptNet provides in its IDX link all information to run e.g. an Okapi Bm25 ranking.

8.3.2 Two Examples

The following two examples show applications of the TINA technology. The first example is a multilingual text analysis whereby texts are evaluated against a relevance profile, and keywords found in the texts are retranslated. So users can decide whether a text in an unknown language is relevant or not. The second example is a Transparent Retrieval example in the domain of Assistive Technology, with a natural language query, translation into the document language, search, and retranslation into the query language.

8.3.2.1 Information Translation for Drug Enforcement from some Middle East Languages

This application intended to support information translation from languages with low resources. The purpose was to identify foreign language documents belonging

[55] Cf. (Pecina 2016; Thurmair 2015).

to a given domain, in this case it was drug enforcement and drug trafficking, and to give an idea of their content. Document languages were Persian, Pashto and Turkish, target languages were German, English, and Spanish, with German as a pivot. Aside from key terms, a bit morphological information was required in the translations: number for nouns, and tense for verbs. Also Named Entities should be found.

Aside from Turkish, both Persian and Pashto show varieties in language: West (Kandahari) and East (Peschawari) Pashto show differences, and so do West (Tehrani) and East (Kabuli) Persian, known as Farsi (official language in Iran) and Dari (official language in Afghanistan). While there are many similarities in these language varieties which justify a common text analysis component, significant differences exist in Persian vocabulary.[56] A challenge for Pashto is that its writing is not fully standardised, so many spelling variants have to be normalised. Another challenge is that many documents in Afghanistan contain both Pashto and Dari terms.

The approach, developed in a project called AVENTINUS, consisted in text analysis of the foreign languages, identification of key terms and Named Entities, and translations of them.[57]

1. Resources

First, a **master word list** for each language was built. Its entries were collected on the foreign language side, by consulting existing dictionaries, and by term extraction from corpora. A certain percentage of the dictionaries was supposed to cover special domains, like biology, technical, political, medicine; such terms were also collected from special dictionaries or corpora. Moreover, proper names were added to the dictionaries: place and person names, trigger words for Named Entity recognition etc. Multiword entries were made an essential part of the master list by dictionary creators, so they had to be covered by the system builders as well.

Figure 8.9 gives size and composition of the three dictionaries.

To create the **Basic Lemma Dictionary (BLD)** the entries of the master list were linguistically annotated.

First the entry-defining annotations were added: **Lemmata** give the canonical spelling of a concept, which required to cope with spelling variants, esp. in Pashto. **Part of speech** information follows the Basic Tagset (BTAG), with some extensions to the Standard Tagset (STAG) in the function word categories. The **semantic** description was defined by a reading number[58] if required. The **entry type** value defined the further annotation strategy:

For single words, the morphological description consists of the following annotations:

[56] A contrastive dictionary of Farsi and Dari can be found in Newid and Thurmair (2023).

[57] The Aventinus system is described in Thurmair (1997a). Turkish was added in a later project.

[58] Reading number is also used for entries which are vowelised in different ways, like کلپ = pelk (eyelid) or palak (socket).

	Turkish > De	Persian > De	Pashto > De
Size			
No. Entries	89.200	82.300	60.800
No. Source Lemmata	62.000	59.100	42.500
No. translations per lemma	1,43	1,39	1,43
Composition			
General words	58.000	33.700	27.800
Domain specific terms	14.000	38.600	23.900
Proper Names	14.000	9.900	9.100
Entry Type			
Single words	58.500	34.300	37.500
Multiwords	30.700	38.700	22.500
Percentage of Multiwords	34.5%	47.0%	37.0%

Fig. 8.9 Dictionaries in AVENTINUS

- *Allomorphs* and their morphological information: Words and affixes can have several allomorphs e.g. in case of broken plurals or other irregular forms.
- *Inflection class*: For Turkish, only vowel harmony information is given. For Persian, about 11 noun and two verb classes are assumed[59]; for Pashto, a classification of nine main noun classes, four main adjective classes, and two main verb classes is coded.[60] Details are given in Sect. 5.5 above.

Other features coded are e.g. gender in Pashto, Ezafe in Persian, sentence mood in Turkish; binding for affixes, and the like.

The dictionary also contains affixes (several hundred per language), which have a much more detailed description, to support morphological analysis.

For multiword analysis, the lexicon must contain information which allows their identification in the analysis. As explained in Sect. 6.1 the dictionaries use only minimal annotation:

- *Lemma*. The lemma is the whole multiword; its parts are allomorphs.
- *Head position*, needed as the head can inflect.
- *Part of speech sequence*, i.e. the *mwposlist*, of the multiword parts.
- *Lemma sequence* of the multiword, i.e. the *mwlemmalist* of their parts.

Basically, the POS sequence will be used to formulate phrase structure rule covering the sequence; the lemma sequence is used to identify all and only the multiword parts; and the head is needed for inflection analysis and determination of the POS of the multiword.

---

[59] Described in Newid and Mumm (2007).

[60] Basically following Lorenz (1979).

Beyond morphology, only few annotations are coded in the lexica. As the system does not intend to do full parsing, only *syntactic function* (subcategorisation) is coded; *syntactic relations* are not coded systematically. Some *semantic* features are coded, mainly for named entities. *Dialect* is coded to distinguish between the Dari and Farsi origin of words. *Domain information* is coded to be able to assign words and translations to specific domains; the classification is a subset of the domain list given in Sect. 3.3 above. Each entry has also information on its *source*, and possibly a *comment*. Some of these annotations are used for human lookup (e.g. subject area, dialect, source), less for machine treatment. A small dictionary lookup system was created, to help with the different scripts.

The Basic Lemma Dictionaries on the target side (i.e. German, English, Spanish), in addition to the definition features lemma, POS, reading, and entry type, contain only a subset of annotations, required to generate plural forms for nouns and tenses for verbs.

The **bilingual dictionaries** (fa>de, ps>de, tr>de) contain transfers for entries described by pairs of<lemma, POS>, and support transfer selection by providing tests using domain tags and a special Boolean feature '*preferred_translation*' set by the coders, to overcome the drawback that statistics for translation selection were not available.

The **Named Entity Gazetteer** (NEG) contains entries for persons, places (many locations in Afghanistan, Iran and Turkiye), organisations and Named Entity indicators for units and dates (like month names).

The dictionaries were **validated**[61] in the following ways: Spell checking for source and target lemmata was applied; scripts were written to ensure completeness and correctness of the annotations; and human translators validated the correctness of the transfers. For the Turkish dictionary, an error rate of 1.7% was found; the other lexica are in a similar range.

As opposed to other languages, no **Full Word Dictionary** (FWD) was created, instead a morpheme list was provided, containing all lemmata, all allomorphs, all affixes, and all function words of the analysed languages.

2. Text Analysis

The text analysis follows the line described in the previous chapters:

Language Identification was tried based on word lists.[62] The quality of the component was reduced by the fact that many mixed-language document were found in the input.

A **Normalisation** component was implemented to cope with different spelling of the same word, and character encoding variations, cf. (Shamsfard et al. 2010).

---

[61] Criteria for dictionary validation are collected by the ELRA validation manual (Fersøe 2004).

[62] There were not sufficient corpus data available at that time.

The **Text Analysis** proper was implemented as a sequence of LtDeep components (cf. Sect. 5.5 above), based on the fact that whitespace-based tokenisation is not an option: While agglutination in Turkish combines several tokens into one string, in Arabic-script-based languages both agglutination (preposition plus noun) and whitespace insertion can be observed. This is described in Sects. 5.4 and 5.5 above. As a result, tokenisation is done in combination with morphosyntactic analysis.

Segmentation used the lemma list for each language to provide the morphemes, and a finite state automaton to control their sequence in the input strings.

The **analysis rules** followed the extended x-bar scheme as presented above (Sect. 5.5).

- The X-0 rules produce complete single words from morpheme sequences, by combining affixes and inflections with the word stems. This also holds for multiword parts.
- Multiword analysis operates on the X1 level. Multiwords form between 30 and 50% of the dictionary entries. The phrase structure part of the X1 rules models the POS sequences of the lexical entry of the multiword (mwposlist), and tries to unify their lemma features (mwlemmalist). The rules also cover cases of non-adjacent multiword parts, especially in verb analysis.
- Validation for the multiword recognition component was done for Persian. In the Persian lexicon, about 700 multiword structures were used by human coders. However, with a rule set of about 150 rules, 91.1% of all structures could be covered; Pashto shows similar behaviour. Turkish has much less variance in multiwords, and therefore less analysis rules.
- The constituent analysis of the system identifies XP structures. Special phenomena to be mentioned are NP analysis in Persian (ezafe and definiteness (را)), or circumposition analysis in Pashto (بې له ... نه, without). On this level also Named Entities are analysed using special named entity rules e.g. for dates.
- The resulting XP structures are collected in a flat recursive sentence analysis as described in Sect. 5, Fig. 5.29. The best-scoring interpretation is sent to transfer.

The **transfer** component does dictionary lookup. In cases of ambiguity the system chooses either the entry of the selected domain, or the entry marked as '*preferred*' in the dictionary. If necessary, the target terms are inflected for number or tense (Fig. 8.10).

The system outputs either a foreign language text with translations interspersed, or a list of translated key terms, ranked by document frequency. Users can quickly decide whether the analysed document is relevant for them or not, and decide on the appropriate follow-up.[63]

---

[63] It is interesting to note that in the tests, the end users preferred a keyterm translation to a full document translation, as the document content would be faster to understand with keywords, compared to full text possibly containing translation errors.

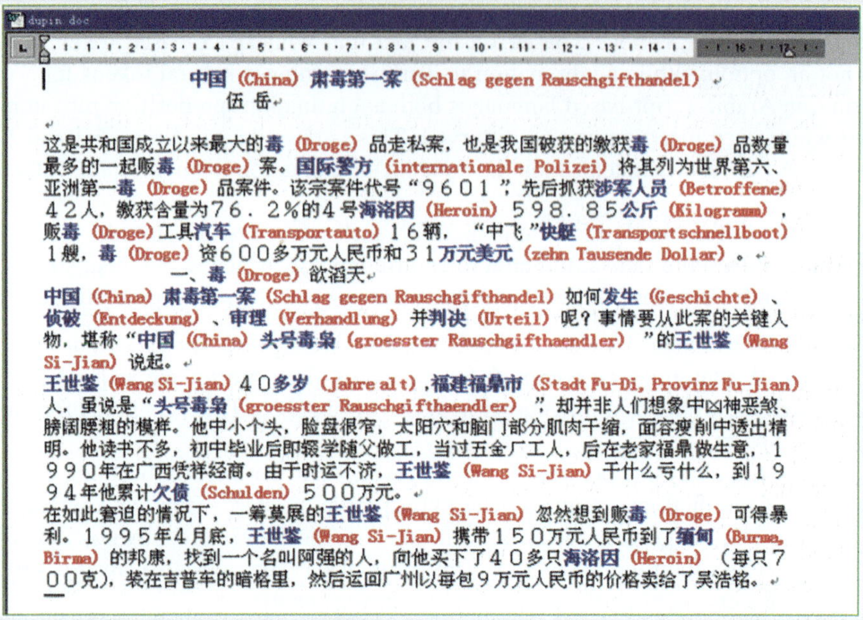
dupin doc

中国 (China) 肃毒第一案 (Schlag gegen Rauschgifthandel)
伍 岳

这是共和国成立以来最大的毒 (Droge) 品走私案，也是我国破获的缴获毒 (Droge) 品数量最多的一起贩毒 (Droge) 案。国际警方 (internationale Polizei) 将其列为世界第六、亚洲第一毒 (Droge) 品案件。该宗案件代号“９６０１”，先后抓获涉案人员 (Betroffene) ４２人，缴获含量为７６．２％的４号海洛因 (Heroin) ５９８．８５公斤 (Kilogramm) ，贩毒 (Droge) 工具汽车 (Transportauto) １６辆， “中飞”快艇 (Transportschnellboot) １艘，毒 (Droge) 资６００多万元人民币和３１万元美元 (zehn Tausende Dollar) 。

一、毒 (Droge) 欲滔天

中国 (China) 肃毒第一案 (Schlag gegen Rauschgifthandel) 如何发生 (Geschichte) 、侦破 (Entdeckung) 、审理 (Verhandlung) 并判决 (Urteil) 呢？事情要从此案的关键人物，堪称“中国 (China) 头号毒枭 (groesster Rauschgifthaendler) ”的王世鉴 (Wang Si-Jian) 说起。

王世鉴 (Wang Si-Jian) ４０多岁 (Jahre alt) ，福建福鼎市 (Stadt Fu-Di, Provinz Fu-Jian) 人，虽说是“头号毒枭 (groesster Rauschgifthaendler) ”，却并非人们想象中凶神恶煞、膀阔腰粗的模样。他中小个头，脸盘很窄，太阳穴和脑门部分肌肉干缩，面容瘦削中透出精明。他读书不多，初中毕业后即辍学随父做工，当过五金厂工人，后在老家福鼎做生意，１９９０年在广西凭祥经商。由于时运不济，王世鉴 (Wang Si-Jian) 干什么亏什么，到１９９４年他累计欠债 (Schulden) ５００万元。

在如此窘迫的情况下，一筹莫展的王世鉴 (Wang Si-Jian) 忽然想到贩毒 (Droge) 可得暴利。１９９５年４月底，王世鉴 (Wang Si-Jian) 携带１５０万元人民币到了缅甸 (Burma, Birma) 的邦康，找到一个名叫阿强的人，向他买下了４０多只海洛因 (Heroin) （每只７００克），装在吉普车的暗格里，然后运回广州以每包９万元人民币的价格卖给了吴浩铭。

Fig. 8.10 TIR: example of key term retranslation

3. Evaluation

The quality of the complete system crucially depends on the coverage of the lexicon used to analyse texts of the domain in question. Coverage tests have been performed, to evaluate which percentage of the text tokens are known to the lexicon.

For the tests, documents in the news domain were selected. Figure 8.11 shows the result. It can be seen that more than 90% of the input text tokens are covered by the dictionary in all languages.

The information translation component presented here does not support different queries; therefore some aspects of the LtConceptNet are not covered, like Query Expansion and Document Search. Instead it serves a given search profile and accesses essential components of a LtConceptNet, esp. support of multilinguality.

	Turkish > de	Persian > de	Pashto > de
No documents selected	60	350	177
No tokens in test corpus	10.930	401.600	84.308
Non-translated tokens	385	37.783	8.327
Coverage tokens	95.4	90.6	90,1

Fig. 8.11 Coverage evaluation results

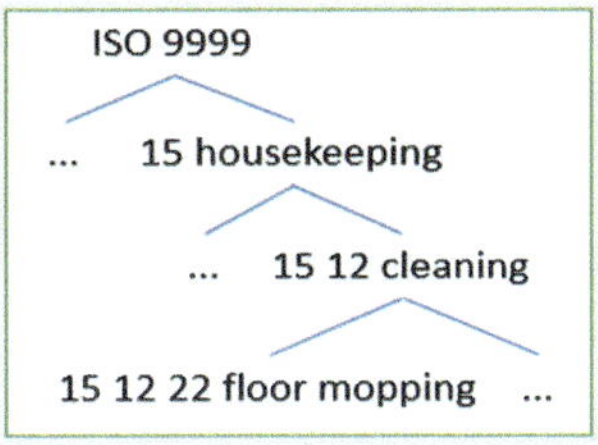

Fig. 8.12 ISO 9999 taxonomy example

8.3.2.2 Query Translation and Expansion in the EASTIN Frontend

Access to information on Assistive Technologies (AT) is a key issue in social participation and eInclusion. The UN *Convention on the Rights of Persons with Disabilities* (UN 2007) declares this as a fundamental right; all UN member states are obliged to comply with this Convention. To support people with disabilities, the member states have organised web portals which provide information about Assistive Technology products. Such portals are visited by doctors, physiotherapists and other people working in the domain.

In 2005, the major European AT information providers joined in the European Assistive Technology Information Network (EASTIN). EASTIN offers a portal (www.eastin.eu) where people can access all databases of its national members[64] simultaneously, on a European level. It is described in Andrich (2011).

The portal allows access to a database containing documents showing product descriptions. Such documents are semi-structured objects, containing formal parts (like name of manufacturer, name of product, size, price, release date) and a short (1–2 sentences) free-text description of the product, mainly explaining special features of it. Such descriptions are available in the national languages (Italian, German, etc.) and in English as a pivot. The portal is described in Gower et al. (2012) (Fig. 8.12).

The information in the Assistive Technology domain is structured along the lines of the ISO 9999 standard (*'Assistive Products for Persons with Disability—Classification and terminology'*), cf. (ISO9999, 2011).[65] The ontology[66] is not motivated by concepts but on AT product groups and their functional differences (like: '*electric vs. manual wheelchairs*'). The taxonomy has about 860 nodes, and is organised in three levels. At all levels (but mainly on the leaf nodes) there are product descriptions attached; more than 30.000 products are currently in the

[64] For EASTIN e.g. SIVA (https://portale.siva.it/), or Rehadat (https://www.rehadat.de/), cf. (Winkelmann 2011), or Hjælpemiddelbasens (https://hmi-basen.dk/), cf. (Lyhne 2011) or abledata (http://www.abledata.com/).

[65] The taxonomy is described also in Heerkens et al. (2012) and Andrich et al. (2012).

[66] Strictly speaking, the AT taxonomy is at best a lightweight ontology, as it organises concepts and (subsumption) relations between them; however it lacks a formal description and any deduction framework.

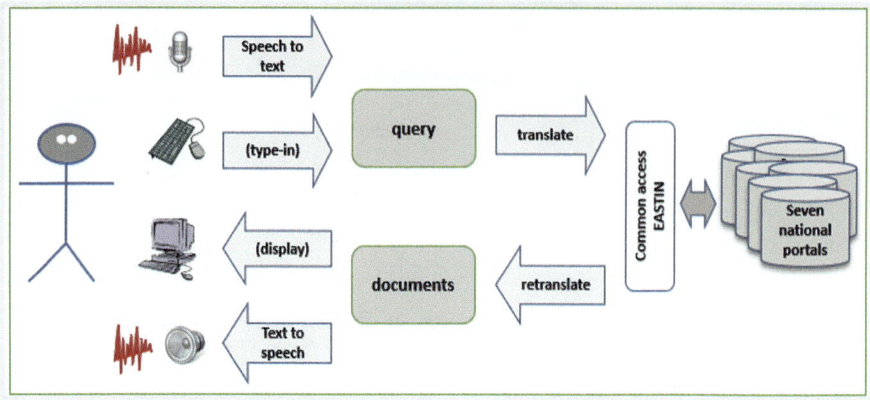

Fig. 8.13 EASTIN NL frontend

databases. Searching the taxonomy is possible by navigating in the taxonomy to find the best ISO code, e.g. '*ISO 12 06 09*' ('*walking chairs*'). Activating the code returns a set of product descriptions (Fig. 8.13).

To open the scope of this portal for additional user groups (end users), and to support people not familiar with the AT domain structure, and speaking only their native language, a natural language front-end to the EASTIN portal was built. This front-end is supposed to be *multilingual* (users should forward information requests, and receive results, in their native language), and *multimodal* (offering a speech channel, e.g. for disabled people). The retrieval process should be *transparent*, i.e. for users of the front end in e.g. Estonian language the product descriptions should be returned in Estonian.[67]

1. Resources: The EASTIN LtConceptNet

a. **Master list**

First a list of master terms describing the domain was needed. The usual way to collect a selection of domain terms is to do corpus analysis, by collecting texts of the domain, and running term extraction tools to identify domain specific terms.[68] This way was tried first, and a list of about 120.000 term candidates was produced based on English. However, an inspection showed that this list was not usable, for several reasons:

[67] The system is described in Thurmair et al. (2012).

[68] Velardi et al. (2001) present tools to support this effort. A difference may be that they construct ontologies from scratch while in ISO it is already given.

- Many of the domain specific terms did not occur in the candidate list altogether; the reason may be that the corpus data are mainly descriptions of products of the assistive domain, and do not respond to a classification interest.
- Most of the high and medium frequency candidates were not specific enough to be included into the domain term list (i.e. assign a ISO 9999 code to them)

As a result, the approach was changed, and the domain terms were collected from existing descriptors: Many AT information providers, like Abledata, Rehadat etc., offer key terms for searchers, and so does the ISO 9999 classification itself: It also offers a list of index terms. The project decided to base the domain terminology on such key terms, and to collect and merge them into a common resource. This resulted in a candidate list of about 17,000 terms. A significant part of it is multiwords.

In order to keep the master list as concise as possible, possible variants were **normalised** instead of creating term variants. Normalisation included the following operations:

- Orthography normalisation, including mapping of US to UK spelling: Only standard spelling (in English: UK spelling) and casing is represented.
- Lemmatisation of inflected forms to base forms: Only base forms are represented.
- Normalisation of hyphenations (*'bath tub'*,*'bath-tub'*,*'bathtub'* ->*'bath-tub'*).
- Normalisation of multiwords: *'wheelchair, manual'* and *'wheelchair (manual)'* ->*'manual wheelchair'*, or *'nitrogen fixation'* vs. *'fixation of nitrogen'*): Only the 'natural' word sequence is represented.
- Normalisation of the content of parentheses; parentheses serve only as a domain indicator, not as a kind of semantic descriptor; the term candidates were reformulated accordingly (not *'car hoist (person)'* but *'car hoist for persons'*).
- Attempts to use one term to describe one concept; e.g. *'backrest (bath/shower)'* were split into several entries.

Of course there were sufficiently many cases left and stayed as synonyms in the master list. The final size of this list contained about 12,700 terms.

Normalisation of the term list of course implies that at query time the query input must be normalised in the same way; otherwise no matches would be found.

b. **LtConceptNet**

The linguistic resources for EASTIN were developed as an instance of a LtConceptNet, separated from the ISO taxonomy. This is because the simple equation *'one term=one ontology node'* is not adequate as the relation between terms and links is of n:m type. (McCrae and Unger 2014) discuss the design of such an interface.

The EASTIN concept net is designed as a separate resource; it is not hardwired with the ISO 9999 taxonomy. It is based on the 12,700 terms extracted in the masterlist step. These terms were expanded manually into a LtConceptNet by adding annotations and links:

- **Node annotations**: Nodes were annotated with lemma, POS (using the Basic Tagset), and administrative information: term source, (i.e. which organisation had contributed the term in question), and comment. Abbreviations receive special annotation. For semantic description the link to the taxonomy was used.
- **IDX links**: Links from concepts to textforms (IDX) were created for the text forms found in the data.
- **XREF links**: The only explicit link type here are synonym links (SYN), and related terms (RT—'*use*'). Examples are $_{\text{de}}$ '*Sphygmomanometer; Pulsmesser*' or $_{\text{en}}$ '*IV stand; infusion stand; dripstand*'. Hierarchical links were not explicitly specified; they are incorporated in the taxonomy.
- **XFR links**: Section 8.2 above it was explained that the single languages should have either one index per language, or a common index for all languages. While the LtConceptNet is designed according to the first alternative the EASTIN resource uses the second option, i.e. one common index. Montiel-Ponsoda et al. (2009) comment: „*This model has proven to be more suitable for highly specialized domain ontologies, e.g., in engineering or technical domains.*" In the assistive technology such a decision seems to be justified, in particular as the taxonomy in itself is language-independent and consists of a numeric scheme.

  The complete English term list was translated into the EASTIN languages Danish, English, Estonian, German, Italian, Latvian, and Lithuanian. Translations were done by domain experts and domain translators. The product databases could be consulted to find the best translation, for instance in cases where a term was ambiguous, and could point to several ISO codes. The resulting term list contains translations for all 12.700 concepts into seven languages, about 90.000 terms altogether. Most terms are multiwords.

- **ONTO links**: Every term was assigned one or several links to the ISO taxonomy. This way its semantic description was defined. This link also defined how it would have to be translated into other EASTIN languages.

The complete EASTIN LtConceptNet was made available to the public via the META-SHARE infrastructure.[69]

2. Query Analysis

Query Analysis must map a query input to ISO 9999 codes, each of which points to groups of AT products. As a search term may point to several ISO codes, the challenge is to narrow down the variance of search, and guide the query to the best term of the index, and find the ISO codes from there.

[69] www.meta-share.org

Beyond the LtConceptNet described above, auxiliary language resources are available. Two considerations influenced the design of these resources: 1. Query processing is a runtime component, i.e. it is time and resource critical (maximal response time for the whole system was 2 seconds, so the query component could take only a fraction of that). 2. The EASTIN vocabulary is limited, and basically a fixed set: Not all query input words but only the words of the term list need to be recognized.

Therefore, a 'static' lemmatiser and a 'static' decomposer resource were implemented, such that in a fixed lexical resource simply inflected forms point to their lemma (lemmatiser), or word parts (decomposer; e.g. decomposing terms like '*Thorako | lumbal | orthese*', to match a query for '*lumbale Orthese*'). So, lemmatisation and decomposition consist of simple and fast list lookup.

Query analysis does tokenization, normalization, lemmatisation and decomposition. The target of this search is a list of LtConceptNet terms, which themselves point to (a list of) ISO codes.[70] A candidate list of such terms is retrieved from the input words. In case of no hit, a fall-back distance-based similarity search is tried. The link.

If several such candidate terms are found, then the next step of query analysis is ranking these terms. Ranking is based on the number of words in the query, the number of words of the candidate term, and the number of matching terms in descriptions. The terms with the highest overlap of matching terms are considered to be the best. The result is mapped to a 5-point scale, and the best ranked terms are returned, together with their ISO codes.

As the whole frontend is multilingual, this analysis workflow is built for each supported language (Fig. 8.14).

3. Search and Retranslation

Search is executed by a simultaneous access of all seven European databases linked as backends to the EASTIN portal, and a collection of the product lists returned by them, grouped under a given ISO code.

The product descriptions to which the ISO codes point in the EASTIN databases are stored in the national language, and in English. The multilingual front-end now must re-translate the product descriptions from English into the query language. For this purpose, the EASTIN front-end provided machine translation web services. The MT systems were tuned for the AT domain, using the master term list and additional corpus data for domain adaptation. The output to the queries was a list of product descriptions, each product consisting of a formal part (manufacturer, price etc.) and a translation of the textual parts.

[70] From the point of view of the ontology, this link can be seen as a lexicalisation of the ISO ontology nodes, cf. (Thurmair 2014b).

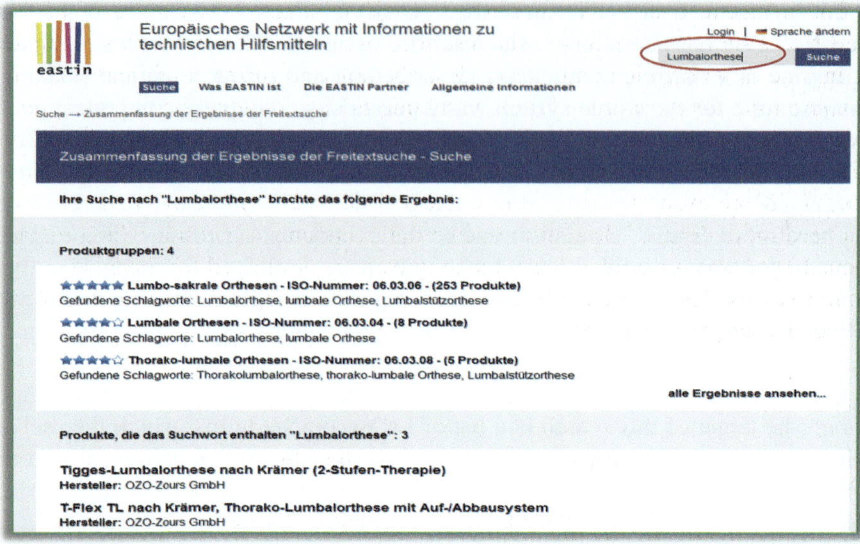

Fig. 8.14 Search in the portal for '*Lumbalorthese*': search term produces a list of ISO codes, with the terms which retrieved them underneath. Ranking is given by stars

This way, a transparent search is enabled: Query and result presentation are in the users' native language, but the searched data are in foreign language.

The system evaluation by end users (professional database searchers) showed that the language technology front-end components were considered to be a significant improvement in the accessibility of Assistive Technology by the EASTIN portal.[71]

References

Afentoulidou, V., Christofidou, A. (2021). Towards a corpus-based dictionary of neologisms. *Proc. 19th Euralex 2021, 597–606*

Aguado de Cea, G., & Montiel-Ponsoda, E. (2012). Term variants in ontologies. *Proc. 30th Conf. of Spanish Society for Applied Linguistics (AESLA), Lleida*

Aguado de Cea, G., Álvarez de Mon, I., & Montiel-Ponsoda, E. (2009). From linguistic patterns to ontology structures. Proc. Int. Conf. on Terminology and Artificial Intelligence (TIA 2009).

Airio, E., Keskustalo, H., Hedlund, T., & Pirkola, A. (2003). The Impact of Word Normalization Methods and Merging Strategies on Multilingual IR. *Proc. Intern. Conf. of the Cross-Language Evaluation Forum for European Languages (CLEF)*

Andrich R. (2011). Towards a global information network: the European Assistive Technology Information Network and the World Alliance of AT Information Providers, *In: G.J. Gelderblom et al., eds. (2011). Everyday technology for independence and care. pp. 190–197. IosPress*

[71] Evaluation details are given in Thurmair et al. (2012).

Andrich, R., Gower, V., Lyhne, Th., & Petersen, M. (2012) Taxonomy of resources. ETNA Report, ICT-PSP-(270746), ETNA project (Association for the Advancement of Assistive Technology in Europe, https://aaate.net/activities/projects/etna/)

Atkins, S. (1997). Bilingual Dictionaries: Past, Present and Future. *Proc. 6th Conf. of European Association of Lexicography (Euralex), Göteborg, Sweden, 1996*

Atreya, A., Chaudhari, S., Bhattacharyya, P., & Ramakrishnan, G. (2012). Building Multilingual Search Index using open source framework. *Proc. of the 3rd Workshop on South and Southeast Asian Natural Language Processing (SANLP), pages 201–210, COLING 2012, Mumbai, December 2012.*

Azad, H.K., Deepak, A. (2019). Query Expansion Techniques for Information Retrieval: a Survey. *Information Processing and Management: an Int. Journal Volume 56 Issue 5, Sep 2019 pp 1698–1735*

Bashir, F., Mansoor, H. (2003). Ontology Construction for Structured Textual Data. BASHIR, Faisal I.; MANSOOR, Hassan. Ontology construction for structured textual data. *2003-04-12*). http://multimedia.eecs.uic.edu/fai-sal/downloads/research/cs580-proj6.pdf, 2003.

Bel, N. (2010). Platform for Automatic, Normalized Annotation and Cost-Effective Acquisition of Language Resources for Human Language Technologies: PANACEA. *Proc. Intern. Conf. of the Spanish Society For Natural Language Processing (SEPLN), Valencia*

Bentivogli, L., Bisazza, A., Cettolo, M., & Federico, M. (2016). Neural versus Phrase-Based Machine Translation Quality: a Case Study. *Proc. 2016 Conf. on Empirical Methods in Natural Language Processing*

Bentivogli, L., & Pianta, E. (2005). Exploiting parallel texts in the creation of multilingual semantically annotated resources: the MultiSemCor Corpus. *Natural Language Engineering 1 September 2005*

Boguslavsky, I., & Dikonov, V. (2008). Universal Dictionary of Concepts. Proc. MONDILEX First Open Workshop, 2008, Moscow, Russia, 31–55

Bond, F., Vossen, P., McCrae, J., & Fellbaum, Chr. (2016). CILI: the Collaborative Interlingual Index. *In Proc. 8th Global WordNet Conference (GWC), Bucharest, Romania, pp. 50–57*

Bond, F., da Costa, L.M., Goodman, M.W., McCrae, J., & Lohk, A. (2020). Some Issues with Building a Multilingual Wordnet. *Proc. 12th Int. Conf. on Language Resources and Evaluation (LREC 2020), Marseille, pp. 3189–3197*

Borgo, St., Ferrario, R., Gangemi, A., Guarino, N., Masolo, Cl., Porello, D., Sanfilippo E., & Vieu, L. (2023). DOLCE: A Descriptive Ontology for Linguistic and Cognitive Engineering. *Applied Ontology 0 (0) 1 1 IOS Press.* arXiv:2308.01597

Buder, M., Rehfeld, W., & Seeger, Th., eds., (1990). Grundlagen der praktischen Information und Dokumentation. K.G. Saur.

Bugara, A. (2020). Designing an optimal multi-language search engine with Elasticsearch. All things Elastic series. Four different approaches to design a multi-language Elasticsearch index Jul 13, 2020 https://codarium.substack.com/p/designing-an-optimal-multi-language

Buitelaar, P., & Cimiano, Ph. (2014). Towards the Multilingual Semantic Web: Principles, Methods and Applications. Springer

Buitelaar, P., Cimiano, P., Magnini, B. (2005). Ontology Learning from Text: An Overview. In: Buitelaar, P., Cimiano, P., Magnini, B., eds. Ontology Learning from Text: Methods, Evaluation and Applications. Frontiers in Artificial Intelligence and Applications. Vol 123. Amsterdam: IOS Press; 2005: pp. 3–12.

Camacho-Collados, J., & Pilehvar, M.T. (2018). From Word to Sense Embeddings: A Survey on Vector Representations of Meaning. *J. Artif. Intell. Res. (2018).*

Chandra, G., & Dwivedi, S.K. (2019). Query Expansion for Effective Retrieval Results of Hindi–English Cross-Lingual IR *Applied Artificial Intelligence, 33,7, pp. 567-593*

Chen, A., & Gey, F. (2004). Multilingual Information Retrieval Using Machine Translation, Relevance Feedback and Word Decompounding. *Information Retrieval Journal 7(1/2):149-182*

Chrizman, N., & Itai, A. (2014). How to construct a multi-lingual domain ontology. Proc. 9th Int. Conf. on Language Resources and Evaluation (LREC 2014), , Reykjavik, Iceland, pp. 4345–4350

Croft, B., ed. (2000). Advances in Information Retrieval. *Kluwer Academic Publishers*
DuCharme, B. (2011). Introduction to SKOS semanticweb inc., October 6, 2011.
Duden (2024). Duden Das Synonymwörterbuch. *8th Edition. 2024,* https://www.duden.de/synonyme
Efthimiadis, E.N. (1996). Query expansion. Annual Review of Information Science and Technology, 31, 121-187
Elasticsearch Elasticsearch https://www.elastic.co/guide/en/elasticsearch/reference/current/index-modules-similarity.html
Espinoza, M., Gómez Pérez, A., & Mena, E. (2008). Enriching an Ontology with Multilingual Information. Proc. The Semantic Web: Research and Applications, 5th European Semantic Web Conf., ESWC 2008, Tenerife, Canary Islands, Spain
Fellbaum, Chr., ed., (1998). WordNet, An Electronic Lexical Database. *MIT Press*
Ferret, O., Fluhr, Chr., Rousseau-Hans, Fr., & Simoni, J.L., (2002). Building domain-specific lexical hierarchies from corpora. Proc. Int. Conf. on Language Resources and Evaluation (LREC 2002) 2002, Gran Canaria
Fersøe, H., (2004). Validation Manual for Lexica. *ELRA*
Foskett, D.J., (1997). Thesaurus. *in: Sparck Jones, K., & Willett, P., eds., 1997: Readings in Information Retrieval. Morgan Kauffmann. pp. 111–134*
Frakes, W., & Baeza-Yates, R. (1992) Information Retrieval, Data Structures and Algorithms. *Prentice Hall*
Francopoulo, G., ed. (2013). LMF Lexical Markup Framework, *ISTE/Wiley 2013*
Franco-Salvador, M., Rosso, P., and Navigli, R. (2014). A knowledge-based representation for cross-language document retrieval and categorization. *Proc. 14th Conf. of the European Chapter of the Association for Computational Linguistics (EACL), Gothenburg, Sweden, pp. 414–423.*
Galuščáková, P., Oard, D.W., Barrow, J., Nair, S., Shing, H.C., Zotkina, E., Eskander, R., & Zhang, R. (2020). MATERIALizing Cross-Language Information Retrieval: A Snapshot. *Proc. Cross-Language Search and Summarization of Text and Speech Workshop, at 12th Int. Conf. on Language Resources and Evaluation (LREC 2020), Marseille, pages 14–21*
Galuščáková, P., Oard, D.W., Nair, S. (2021). Cross-language Information Retrieval. https://doi.org/10.48550/arXiv.2111.05988
Gaus, W. (1995). Dokumentations- und Ordnungslehre, Theorie und Praxis des Information Retrieval. *Springer (2nd ed.)*
Gilarranz, J., Gonzalo, J., & Verdejo, F. (1997). An Approach to Conceptual Text Retrieval Using the EuroWordNet Multilingual Semantic Database. Papers from the 1997 AAAI Spring Symposium
Gillis-Webber, F., Keet, C.M. (2022). A Review of Multilingualism in and for Ontologies. arXiv:2210.02807v1
Giunchiglia, F., Marchese, M., & Zaihrayeu, I. (2006). Encoding classifications into lightweight ontologies. University of Trento Technical Report # DIT-06–016
Gormley, Cl., & Tong, Z. (2015). Elasticsearch: The Definitive Guide: A Distributed Real-Time Search and Analytics Engine. O'Reilly
Gower V., Andrich, R., Agnoletto, A., Winkelmann, P., Lyhne, Th., Rozis, R., & Thurmair, Gr. (2012). The European Assistive Technology Information Portal (EASTIN): Improving Usability Through Language Technologies. Proc. Int. Conf. on Computers Helping People with Special Needs (ICCHP), Linz
Gurusamy, V., & Kannan, S. (2014). Preprocessing Techniques for Text Mining. https://www.researchgate.net/publication/273127322_Preprocessing_Techniques_for_Text_Mining
Hassel, A. (2018). Anpassung des Soundex Algorithmus für die deutsche Sprache. https://www.groupxs.com/archives/anpassung-des-soundex-algorithmus-fur-die-deutsche-sprache/132.html
Havasi, C., Speer, R., & Alonso, J. (2007). ConceptNet 3: A flexible, multilingual semantic network for common sense knowledge. *MIT Media Lab* https://www.media.mit.edu/publications/conceptnet-3-a-flexible-multilingual-semantic-network-for-common-sense-knowledge/

Heerkens, Y.F., Bougie, T., de Kleijn-de Vrankrijker, M.W. (2012). Classification and terminology of assistive products. In: Stone, J.H., Blouin, M., eds., Int. Encyclopedia of Rehabilitation. http://cirrie.buffalo.edu/encyclopedia/en/article/265/

Hodge, V., & Austin, J. (2002). Hierarchical word clustering — automatic thesaurus generation. *Neurocomputing Vol. 48, Issues 1–4, October 2002, Pages 819–846*

IATE xxx IATE https://iate.europa.eu/assets/IATE_Handbook_public.pdf

Ingwersen, P. (1992). Information Retrieval Interaction. *Taylor Graham Publishing*

Jacobs, J.R. (1982). Finding words that sound alike. The SOUNDEX algorithm. Byte, 7, 473-474.

Jiang, Z., El-Jaroudi, A., Hartmann, W., Karakos, D., & Zhao, L. (2020). Cross-lingual Information Retrieval with BERT. *Proc. Cross-Language Search and Summarization of Text and Speech Workshop, pp. 26–31, Int. Conf. on Language Resources and Evaluation (LREC 2020), Marseille, 11–16 2020*

Jovic, A., Prcela, M., & Gamberger, D. (2007). Ontologies in Medical Knowledge Representation. 29th Proc. Int.Conf. on Information technology Interfaces.

Kang, S.J., & Lee, J.H. (2001). Semi-automatic practical ontology construction by using a thesaurus, computational dictionaries, and large corpora. https://doi.org/10.3115/1118220.1118226.

Khan, L., Luo, F., & Yen, I. (2002). Automatic ontology derivation from documents. IEEE Transactions on Knowledge and Data Engineering (TKDE)

Khozani, S.M.H., & Bayat, H. (2011). Specialization of Keyword Extraction Approach to Persian Texts. Proc. Int. Conf. of Soft Computing and Pattern Recognition (SoCPaR 2011), Dalian, China

Knight, K., & Graehl, J. (1998). Machine transliteration. *Computational Linguistics, 24(4):599–612.*

Lanza, Cl., Hazem, A., & Daille, B. (2020). Towards Automatic Thesaurus Construction and Enrichment. *Proc. 6th Int. Workshop on Computational Terminology (COMPUTERM 2020), pp. 62–71, at 12th Int. Conf. on Language Resources and Evaluation (LREC 2020), Marseille, 2020*

Lavelli, A., Sebastiani, F., & Zanoli, R. (2004). Distributional term representations: an experimental comparison. *Proc. 13th ACM Int. Conf. on Information and Knowledge Management. pp. 615–624.*

LEMON (2019). LEMON: The OntoLex Lemon Lexicography Module. https://www.w3.org/2019/09/lexicog/

Lenat, D.B. (1995). CYC: a large-scale investment in knowledge infrastructure. *Communications of the ACM, Volume 38, Issue 11. pp. 33-38*

Li, X., Strassel, St., Ji, H., Griffitt, K., & Ellis, J. (2012). Linguistic Resources for Entity Linking Evaluation: from Monolingual to Cross-lingual. Proc. 8th Int. Conf. on Language Resources and Evaluation (LREC'12), Istanbul, Turkey, pp. 3098–3105.

Liu, H., & Singh P. (2004b). Commonsense reasoning in and over natural language, *Proc. of the 8th Int. Conf. on Knowledge-Based Intelligent Information and Engineering Systems (KES-2004).*

Lorenz, M. (1979). Lehrbuch des Pashto (Afghanisch). *VEB Verlag Enzyklopädie Leipzig*

Lourdusamy, R., Florrence, J.M. (2016). Methods, approaches, principles, guidelines and applications on multilingual ontologies: a survey. ICTACT journal on soft computing, October 2016, volume: 07, issue: 01

Ludwig, O. (2005). Wörterbuch der Kfz-Technik. VDE Verlag

Lyhne Th. (2011). The Danish National Database on Assistive Technology, In: Gelderblom G.J., et al., eds.: Everyday technology for independence and care. pp. 205–213. IosPress

Madsen, B.N., Thomsen, H.E., Halskov, J., & Lassen, T., (2010). Automatic Ontology Construction for a National Term Bank. Proc. TKE

Manning, C.D., Raghavan, P., & Schütze, H. (2008). Introduction to Information Retrieval. *Cambridge University Press*

McCrae,J.P., & Unger, Chr. (2014). Design Patterns for Engineering the Ontology-Lexicon Interface. *In: Buitelaar&Cimiano, eds. 2014. Towards the Multilingual Semantic Web: Principles, Methods and Applications. pp. 15–30. Springer*

Merhav, Y., & Ash, St. (2018). Design Challenges in Named Entity Transliteration. *Proc. 27th Int. Conf. on Computational Linguistics (COLING), Santa Fe, New Mexico, USA, pp. 630–640*

Mikolov, T., Sutskever, I., Chen, K., Corrado, Gr., Dean, J. (2013b). Distributed Representations of Words and Phrases and their Compositionality. *Advances in Neural Information Processing Systems 26 (NIPS 2013).* arXiv:1310.4546

Montazeralghaem, A., Rahimi, R., & Allan, J. (2019). Term Discrimination Value for Cross-Language Information Retrieval. *Proc. 2019 ACM SIGIR International Conference on Theory of Information Retrieval (ICTIR), Santa Clara, CA, USA, pp. 137–140*

Montiel-Ponsoda, E., Aguado de Cea, G., Gómez-Pérez, A., & Peters, W. (2009). Enriching Ontologies with Multilingual Information. *Natural Language Engineering 1,1, p.1-27*

Montiel-Ponsoda, E., Aguado de Cea, G., & McCrae, J. (2011). Representing term variants in lemon . *Proc. 9th Int. Conf. on Terminology and AI (TIA),*

Moreno, A., & Pérez, Ch. (2000). Reusing the Microcosmos Ontology for Concept-based Multilingual Terminology Databases. Proc. 2nd Int. Conf. on Language Resources and Evaluation (LREC 2000), Athens, Greece

Navigli, R., & Ponzetto, S.P. (2012). BabelNet: the Automatic Construction, Evaluation and Application of a Wide-Coverage Multilingual Semantic Network. *Artificial Intelligence Vol. 193, 2012, pp. 217-250*

Newid, M.A., Mumm, P.A. (2007). Persisches Lesebuch. *Wiesbaden (Reichert)*

Newid, M.A., Thurmair, Gr. (2023). Wörterbuch Farsi – Dari, Dari – Farsi. *Buske*

Nie, J.N. (2010). Cross-Language Information Retrieval. *Synthesis Lectures on Human Language Technologies. Springer*

Niles, I., & Pease, A. (2001). Towards a Standard Upper Ontology. *Proc. of the 2nd Int. Conf. on Formal Ontology in Information Systems (FOIS-2001pp.2–9, Ogunquit, Maine, USA,*

Niles, I., & Pease, A. (2004). Linking Lexicons and Ontologies: Mapping WordNet to the Suggested Upper Merged Ontology. *Proc. Internat. Conf. on Information and Knowledge Engineering 2003, Las Vegas, Nevada, USA*

(Oxford 2014) The Oxford Dictionary of Synonyms and Antonyms *Oxford 2014*

Pazienza, M.T., Stellato, A., eds. (2012). Semi-Automatic Ontology Development: Processes and Resources. (IGI Global)

Pazienza, M.T., Stellato, A., & Turbati, A. (2010). A Suite of Semantic Web Tools Supporting Development of Multilingual Ontologies. In: Armano, G., de Gemmis, M., Semeraro, G., Vargiu, E. (eds) Intelligent Information Access. Studies in Computational Intelligence, vol 301. Springer, Berlin, Heidelberg.

Pecina, P. (2016). Adaptation of Machine Translation to Specific Domains and Applications. *Prague. Univ. Habil. Thesis*

Pecina, P., Toral, A., Papavassiliou, V., Prokopidis, P., & van Genabith, J. (2012). Domain Adaptation of Statistical Machine Translation using Web-Crawled Resources: A Case Study. *In Proc 16th Conf. of the European Association for Machine Translation (EACL), pp. 145–152, Trento, Italy.*

Peters, C., Braschler, M., & Clough, P. (2012). Multilingual Information Retrieval, From Research to Practice. *Springer*

Pianta, E., Bentivogli, L., & Girardi, Chr. (2002). MultiWordNet - Developing an aligned multi-lingual database. *Proc. 1st Int. WordNet Conf., 2002, Mysore, India, pag. 293–302*

Pirkola, A. (1998). The effects of query structure and dictionary setups in dictionary-based cross-language information retrieval. *Proc. 21st Int. ACM SIGIR conf. on Research and development in information retrieval, pp. 55–63, 1998*

Popović, M., & Ney, H. (2011). Towards automatic error analysis of machine translation output. *Computational Linguistics, 37(4), pp.657–688.*

Roy R.S. (2013). Analyzing linguistic structure of web search queries. *WWW '13 Companion: Proc. 22nd Int. Conf. on World Wide Web. pp. 395–400.* https://doi.org/10.1145/2487788.2487949

Rubino, C. (2020). The Effect of Linguistic Parameters in Cross-Language Information Retrieval Performance Evidence from IARPA's MATERIAL Program. *Proc. of the Cross-Language Search and Summarization of Text and Speech Workshop, pages 1–6, Int. Conf. on Language Resources and Evaluation (LREC 2020), Marseille, 11–16 May 2020*

Ruder, S., Vulić, I., & Søgaard, A. (2019). A Survey of Cross-lingual Word Embedding Models. *Journal of Artificial Intelligence Research 65 (2019) 569-631*

Saleh, S., & Pecina, P. (2016). Reranking Hypotheses of Machine-Translated Queries for Cross-Lingual Information Retrieval. Proc. 7th Int. Conf. of the Cross-Language Evaluation Forum for European Languages (CLEF), Évora, Portugal, pp. 54–66, Springer International Publishing, 2016

Salton, G., & McGill, M.J. (1983). Introduction to Modern Information Retrieval. *New York, NY (McGraw-Hill)*

Savoy, J. (2003). Report on CLEF-2003 Multilingual Tracks. *In: Peters, C., Gonzalo, J., Braschler, M., Kluck, M., eds. (2003). Comparative Evaluation of Multilingual Information Access Systems. CLEF 2003. Springer*

Schwarz, Chr. (1990b). Automatic Syntactic Analysis of Free Text. *Journal of the American Society for Information Science, 41, 408–417, 1990. Series Number: 6*

Sennrich, R., Haddow, B., & Birch, A. (2016). Neural Machine Translation of Rare Words with Subword Units. *Proc. 54th Annual Meeting of the Association for Computational Linguistics (ACL), Volume 1, pp. 1715–1725, Berlin, Germany.*

Seppälä, S., Hicks, A., & Ruttenberg, A. (2016). Semi-Automatic Mapping of WordNet to Basic Formal Ontology. *Proc. 8th Global WordNet Conference (GWC), pages 373–381, Bucharest, Romania.*

Shamsfard, M., Jafari, H.S., Ilbeygi, M. (2010). STeP-1: A Set of Fundamental Tools for Persian Text Processing. *Proc. 7th Int. Conf. on Language Resources and Evaluation (LREC'10), Valletta, Malta.*

Sintichakis, M., & Constantopoulos, P. (2002). A Method for Monolingual Thesauri Merging. *ACM SIGIR Forum (ACM Special Interest Group on Information Retrieval), Volume 31, Issue SI, pp. 129–138*

SKOS SKOS The W3C Recommendation SKOS (Simple Knowledge Organization System) https://www.w3.org/2004/02/skos/

Specia, L., Motta, E. (2006). A hybrid approach for extracting semantic relations from texts. *Proc. of the 2nd Workshop on Ontology Learning and Population: Bridging the Gap between Text and Knowledge. 2006.*

Speer, R., Chin, J., Havasi, C. (2017). ConceptNet 5.5: An Open Multilingual Graph of General Knowledge. *AAAI 31 (2017) 4444–4451*

Spink, A., Wolfram, D., Jansen, M., & Saracevic, T. (2001). Searching the Web: The Public and Their Queries. *Journal of the American Society for Information Science and Technology 52(3):226 - 234*

Tariq, M., Manumaisupat, P., Al-Sayed, R., & Ahmad, K. (2003). Experiments in Ontology Construction from Specialist Texts. Proc. EuroLan, Bucharest

Thurmair, Gr. (1984). Linguistic Problems in Multilingual Morphological Decomposition. *Proc. 10th Int. Conf. on Computational Linguistics and 22nd Annual Meeting of the Association for Computational Linguistics, pp. 174–177, Stanford, California, USA.*

Thurmair, Gr. (1997a). Multilingual Information Processing. The AVENTINUS System. *In: Proc. National Annual Training Conference (FBINAA), Berlin.*

Thurmair, Gr. (2004a). Multilingual Content Processing. Proc. 4th Int. Conf. on Language Resources and Evaluation (LREC'04), Lisbon, Portugal. (invited talk)

Thurmair, Gr. (2004b). Sprachtechnologie in einem Informationssystem. *In: Eibl, M., Wolff, Chr., Womser-Hacker, Chr., eds. (2004). Designing Information Systems: Festschrift fuer Juergen Krause, pp. 127–143, UVK, 2004.*

Thurmair, Gr. (2014a). Conceptual Transfer: Using local classifiers for transfer selection. *Proc. 9th Int. Conf. on Language Resources and Evaluation (LREC 2014), , Reykjavik, Iceland*

Thurmair, Gr. (2014b). Lexicalizing a Multilingual Ontology for Searching in the Assistive Technology Domain. *in: Buitelaar, P. & Ciminiano, Ph., eds: Towards the Multilingual Semantic Web. Springer*

Thurmair, Gr. (2015). Evaluation of the domain adaptation of MT systems in ACCURAT. *In Proc. of the 18th Annual Conf. of the European Association for Machine Translation, pages 206–209, Antalya, Turkey*

Thurmair, Gr., Niedermair, G., Schwarz, Chr., & Wessel, A. (1986). Retrieval Aids by Linguistics and Statistics (REALIST): Systemkonzeption. *in: Schwarz, Chr., Thurmair, Gr. eds. (1986). Informationslinguistische Texterschließung. Hildesheim (Olms), pp. 251–265*

Thurmair, Gr., Agnoletto, A., Gower, V., Rozis, R. (2012). EASTIN-CL: A multilingual front-end to a database of Assistive technology products. *Proc. 16th Annual Conf. of the European Association for Machine Translation (EAMT), Trento, Italy,*

Trojahn, C., Fu, B., Zamazal, O., & Ritze, D. (2014). State-of-the-art in Multilingual and Cross-Lingual Ontology. In: Buitelaar&Cimiano, eds. 2014, 119-136

van Rijsbergen, C.J. (1979). Information Retrieval. *Butterworth-Heinemann, 1979*

Velardi, P., Missikoff, M., & Basili, R. (2001). Identification of relevant terms to support the construction of Domain Ontologies. *Proc. ACL 2001 Workshop on Human Language Technology and Knowledge Management (HLTKM)*

Verdejo, F., Gonzalo, J., Peñas, J., López, F., & Fernández, D. (2000). Evaluating wordnets in Cross-Language Information Retrieval: the ITEM search engine. Proc. 2nd Int. Conf. on Language Resources and Evaluation (LREC 2000), Athens, Greece

Vinokourov, A., Cristianini, N., & Shawe-Taylor, J. (2002). Inferring a semantic representation of text via cross-language correlation analysis. *Advances in Neural Information Processing Systems. Vol. 15*

Vossen, P. (2004). EuroWordNet: a multilingual database of autonomous and language-specfic wordnets connected via an Inter-Lingual-Index. *International Journal of Lexicography 17,2*

Vrochidis, St., Kompatsiaris, I., Casamayor, G. et al., (2015). MULTISENSOR: Development of multimedia content integration technologies for journalism, media monitoring and international exporting decision support. *IEEE Int. Conf. on Multimedia & Expo Workshops (ICMEW)*

Winkelmann P. (2011). REHADAT: The German information system on assistive devices, In: Gelderblom, G.J., et al., eds.: Everyday technology for independence and care. pp. 205–213. IosPress

Xiong, D., Meng, F., & Liu, Q. (2016). Topic-based term translation models for statistical machine translation. *Artificial Intelligence Vol. 232, 2016, 54-75*

Xu, J., Weischedel, R., & Nguyen, Ch. (2001). Evaluating a Probabilistic Model for Cross-lingual Information Retrieval *Proc. 24th annual Int. ACM SIGIR Conf. on Research and development in information retrieval September 2001, pp. 105–110*

Yang, D., Powers, D.M. (2008). Automatic Thesaurus Construction. *Conferences in Research and Practice in Information Technology Series. 74.*

Yang, Y., Yih, W.T., & Meek, Chr. (2015). WIKIQA: A Challenge Dataset for Open-Domain Question Answering. *Proc. 2015 Conf. on Empirical Methods in Natural Language Processing, pp. 2013–2018, Lisbon, Portugal*

Yu, P., & Allan, J. (2020). A Study of Neural Matching Models for Cross-lingual IR. *Proc. 43rd Int. ACM SIGIR Conf. on Research and Development in Information Retrieval (SIGIR '20), Virtual Event, China. ACM, New York, NY, USA*

Zhang, X.F., Huang, H.Y., Zhang, K.L. (2009). Cross-language information retrieval based on weight computation of query keywords translation. *Proc. IEEE Int. Conf. on Intelligent Computing and Intelligent Systems 2009*

Chapter 9
Resources

The complete TINA system depends heavily on the availability of linguistic resources. Reusability and creation of resources, above all of dictionary resources, needs to be planned carefully, in particular as many of them require human intervention. This chapter investigates which resources are needed for which component, what their interdependence is, how they could be created efficiently, and how legacy data, if available, could be integrated.

It is a mistake to believe that all terms required in the different components are available from corpus extraction tools. Issues of copyright, confidentiality etc. prevent many terms from showing up in corpus data, especially in specialised domains. Resource production should take care of this fact and enable a combination of dictionary/wordlist information and corpus-extracted material.

It is true that the creation of resources is a labour-intensive task. Therefore it is of utmost importance to plan the resource building strategically and reuse as much as possible. But also in the domain of resource creation significant progress has been made in providing computational support towards more efficiency. And it pays off: Once a resource is collected the task is done, and only minor maintenance effort needs to be planned for it.

9.1 Resource Design

9.1.1 Resource Requirements

Figure 9.1 shows which TINA components require which resources.[1]

[1] Small auxiliary lists are not shown.

G. Thurmair, *Knowledge-Driven Multilingual Text Analysis and Transparent Information Retrieval*, Cognitive Technologies,
https://doi.org/10.1007/978-3-031-91741-7_9

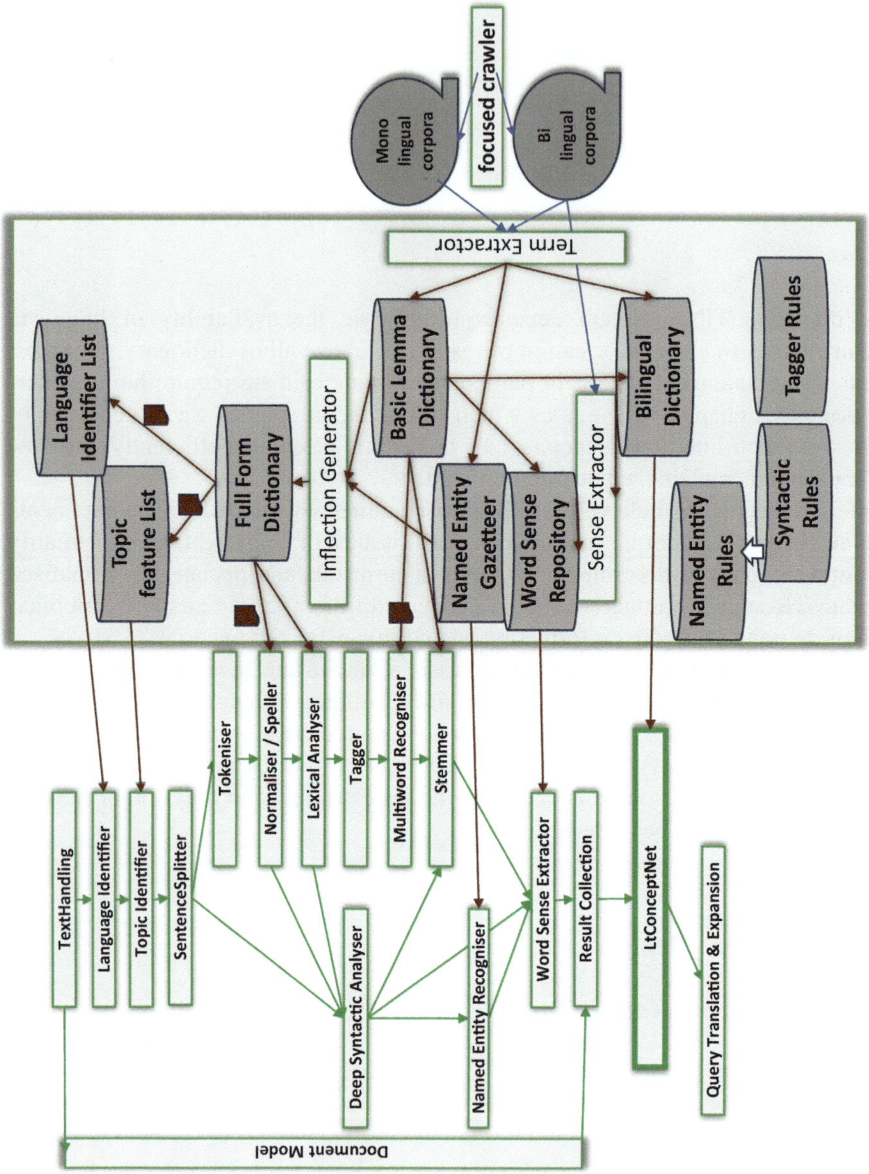

Fig. 9.1 TINA resources: overview

As for resources based on words and/or dictionary entries, the following observations hold:

- **LtLanguageIdentifier** needs, aside from ngrams, *word lists*, i.e. textforms, annotated with a language
- **LtTopicIdentifier** accesses the *topic feature list*. It provides textforms, annotated with a topic and a frequency score, built form topic corpus data
- **LtSSplit** needs *word lists* supporting sentence boundary detection, like start words, abbreviations etc.
- **LtTokeniser**, aside from the Chinese Dictionary, uses *token lists* like unit names.
- **LtNormaliser** needs correction lists of the type '*incorrect form ->normalised form*', to be used in text analysis and query processing
- As for the **Lexical Analyser**, the type of resources depends on the language. While the analysers for the (western) European languages are based on *textform*-based full word dictionaries (FWD), the analysers for Chinese, Turkish, Arabic, Farsi, and Pashto use *lemma*-based dictionaries (Basic lemma dictionaries, BLD). Lemma dictionaries will be transferred into textform dictionaries by means of flexers, cf. below).
- **Multiword recognisers** use *lemma* dictionaries; they contain also multiword entries whereas full word dictionaries only offer textforms i.e. single words.
- The **LtSensExtract** word sense extractor accesses a special Word Sense Repository, created from corpus data with the help of Bilingual Dictionaries.
- The Named Entity recogniser **LtNamer** accesses dictionary resources called Named Entity Gazetteers, containing proper names, both single and multiwords.
- Finally, the **LtConceptNet** and the translation resources for the Transparent Information Retrieval access bilingual dictionaries for translation and retranslation, and monolingual cross-reference resources for query expansion.

As far as rules are concerned, there are three main types of resources:

- Rules like foreseen in LtDeep are used for all morphological analysers which use non-whitespace-based lexical analysis, and all LtNamer components using rule-based Named Entity analysis. They use the formalism described in LtDeep.
- Finite State Analyser tables are built for German decomposers, for Farsi and Pashto analysers, and above all for Turkish lexical analysis.
- The LtTagger uses rules in a special tagging format which analyses alternative POS nodes and removes implausible analysis options.

All this is described in more detail in the respective chapters above.

9.1.2 Design

The task to organise lexical material in a consistent way has always been a challenge for computational linguists.

1. Most approaches use a top down approach, by describing which information items need to be presented, and how they should be organised in a lexical model. They intend to cover all lexical phenomena in a systematic and concise way, to serve as a reference in the description of lexical entries.

Significant efforts in the description of (monolingual and multilingual) entries have culminated in projects like EAGLES,[2] MILE,[3] LIRICS[4] and others. Their recommendations have been used as guidelines for the production of resources in many languages.

The Lexical Markup Framework (LMF[5]) collects all annotations in a data model, and compilers from and into the model were provided (Francopoulo 2013). In the context of web research, models like *lemon*, cf. Montiel-Ponsoda et al. (2011), or *Lexinfo*, cf. Cimiano et al. (2010), have been proposed, using Web based technology. There were also example implementations, i.e. encodings of lexical entries, however more for demonstration purposes and less for practical use in language analysis systems.

2. A lexical model was also developed in the context of TINA, influenced by the MULTILEX project, cf. Sérasset (2001), and an external interchange format called OLIF was defined for exchange of dictionaries in industrial environments, presented in Lieske et al. (2001). However these developments started from a different perspective: Instead of defining a model for *all* aspects and annotations in lexical description, the interest in TINA turned into finding the *minimal* annotations required to run a given component: Only these annotations needed to be coded. This was mainly to reduce the development effort.

Next, the requirements for the development of resources for the single components were compared and evaluated for reusability, and (only) the missing annotations for the new component needed to be added. In total, a lexical model grew up in a bottom-up approach, containing annotations as used by the single components, such that each component should be equipped with resources tailored to its particular requirements, without the burden of needing a huge dictionary resource. This is even more important, if components should run in stand-alone mode. So, the TINA lexical model defines the content of a central reusable resource based on the resource requirements of the single components, and compilers to extract the necessary resources for a given component.

---

[2] EAGLES (1998), EAGLES (1999).

[3] MILE (2001).

[4] LIRICS (2006).

[5] https://www.lexicalmarkupframework.org/

It can be seen that resources like Full Word dictionaries (FWD) as compiled from a Basic Lemma Dictionaries (BLD) would be helpful in many applications like spell checking as a special resource. Also the Named Entity Gazetteer (NEG) is quite useful for components like LtNamer, and would enable the system to run the Named Entity Recogniser as a stand-alone tool.[6]

Experience has also shown that the option to run stand-alone components requires a modular management of resources, as opposed to a centralised data model, and justifies the development of compilers from a central resource into such applications.

However such bottom-up efforts should be guided by a conceptual/virtual model of a resource representation, and takes into account how the different information items interact and can be developed efficiently. Such a conceptual model is outlined below.

3. The difference of the TINA lexical model to a retrieval system like LtConceptNet, despite similarities in the organisation of XREF, XFR etc., can be described as follows:

- LtConceptNet is not a dictionary but a model of linking concepts in a multilingual setup to improve search results.
- LtConceptNet is a dynamic resource, focussing on a given document collection whereas the lexical resources described in this chapter are supposed to support also general domains and serve at least as a starting point for domain adaptation.
- LtConceptNet is just one application whereas the resources need to support also other applications like Information Translation, Named Entity Recognition, text anonymisation, word decomposition and others, but without the overhead of a whole concept net.

4. In order to produce and maintain the TINA resources, in would help to observe principles of reusability, modularity, and consistency, and develop tools for validation and quality assurance, as e.g. proposed in the ELRA recommendations of Fersøe (2004).

9.2 The TINA Lexical Model and Its Derivations

It has been mentioned that one of the design principles of the resources is that the single components should be supported just by the resources specifically required by them, both in terms of language and content. Overhead providing unnecessary resources should be avoided. However, in order to keep a concise view and allow for reusability, it would be helpful to follow a kind of lexical model in the organisation of all resources, from which the resources of the single components can be derived.

[6] Or in contexts like LtAnonymizer mentioned above.

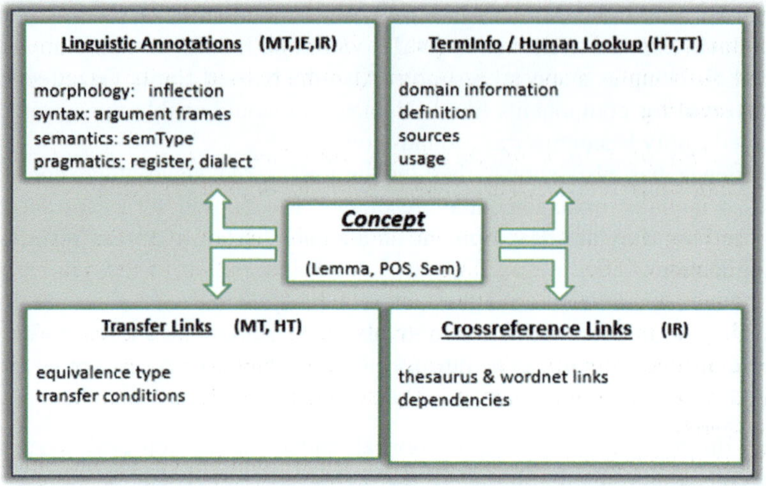

Fig. 9.2 TINA lexical model

Such a lexical model, presented e.g. in Fig. 9.2,[7] would be centred around a concept, and would divide the material along several conceptual lines:

- Information items as used in **human lookup** and/or terminological description: mainly definition, and domain categorisation. Additional information could be added to support the requirements of a terminological entry.[8] However, not just (special-domain) terms but also general vocabulary concepts want to be looked up by human users. Definitions could be used for semantic disambiguation, similar to the WordNet entries
- Information items required for **linguistic analysis**, like inflection, syntactic frames, semantic information, including Named Entity description.

In addition, **links to other concepts** would have to be established, as required by translation and retrieval applications:

- Crossreference links (XREF) to concepts of the same language.
- Transfer links (XFR) related to concepts of other languages, depending on translation conditions.

The different linguistic applications (Machine Translation, Information Extraction, Information Retrieval, Human Translation etc.) focus on different segments of this model; however the model itself allows to identify synergies and sharing of resources.

[7] Following Thurmair (1997) and Knops and Thurmair (1993).

[8] www.cotsoes.org; de: KUEDES Empfehlungen zur Terminologiearbeit.

This model serves as a reference point for the development of resources for the different components of the TINA system. For instance, full word dictionaries can be created from the morphological description of a lexical entry.

9.2.1 Basic Lemma Dictionaries (BLD)

This dictionary focuses on the linguistic descriptions of concepts. It collects all information items describing a concept: its internal structure, its morphosyntactic behaviour, its semantic properties, domain relations and usage conditions. The focus is on computational tractability; therefore all information contained in the BLD is presented, in the form of feature-value pairs. It also serves as a source for the derivation of other dictionary resources, like Full Word Dictionaries.

The Basic Lemma Dictionary contains only linguistic descriptions; i.e. it describes just a part of a complete dictionary entry: Contextual references, terminological information, diachronic aspects etc. are not covered in the BLD. Differences in their linguistic description will constitute different concepts, be it in syntactic frames, be it in semantic features.

An open issue is if non-open word classes should be part of the BLD dictionary. Function words are often kept in a separate resource, and only open class words are part of the BLD. This line is also followed in the TINA resources. One of the reasons is that the annotations of function words need to be given on a textform basis, not on a lemma basis, and also that function words often need more detailed and more extended annotations than content words. In later development steps the function words are merged with the inflected BLD entries to produce the Full Word Dictionary (FWD). This leaves the BLD entries nearly exclusively for open word classes.

The linguistic description in a BLD entry contains at least the following annotations; more can be given due to different languages or different applications:

9.2.1.1 Concept Description

As everywhere in TINA, concepts are described as triplets of<lemma, part-of-speech, semantics>. However such descriptions need consideration even on this level as can quickly be seen when trying to merge dictionaries.

1. As for the **lemmata**, it can be stated that efforts to merge dictionaries on the basis of their lemmata face surprising difficulties. The main differences in defining the canonical form of an entry can be described as follows:

- the lemma must not contain **additional annotations**, like '*rely on*', '*date sb*', '*squadra (f)*', '*hold (weather)*'. Such annotations are useful as they describe syntactic or semantic contexts but must be given in features, not in the lemma.

- **casing** should be normalised in the canonical descriptions: German nouns stay capitalised,[9] acronyms like 'DOG' stay uppercased or mixed,[10] Proper Nouns usually stay capitalised.
- verbs with **reflexives**, i.e. verbs that have the same semantic agent and patient, use clitic elements in the infinitive form in many languages. Such forms keep the reflexive indicator in the lemma: $_{\text{it}}$ '*lavarsi*', $_{\text{de}}$ '*s. waschen*', $_{\text{fr}}$ '*se laver*', $_{\text{tr}}$ '*yıkanmak*', $_{\text{es}}$ '*debilitarse*'.
- **hyphens** are used differently in different dictionaries, like $_{\text{en}}$ '*data-base*' vs. '*database*', $_{\text{de}}$ '*Brot-Aufstrich*' vs. '*Brotaufstrich*', and need to be normalised.

All these phenomena are defined differently in many resources and need to be taken into account when dictionaries are to be merged.

2. For the **part-of-speech** annotation, the Basic Tagset (BTAG) is used as described in Sect. 2.3.2. Independent of the task of mapping POS tags, differences in the dictionaries can be observed in the lexicalisation of participles. In language use they tend to behave like adjectives, e.g. $_{\text{en}}$ '*married*', $_{\text{de}}$ '*verheiratet*', $_{\text{it}}$ '*sposato*'. As this is an ongoing tendency in language development differences in dictionary comparisons can be expected in this area.
3. Information on **semantics** reach from reading numbers to domain markers to sense indicators denoting contexts. Semantic information is only required, if several senses of the lemma need to be distinguished; in many cases (e.g. technical terms like $_{\text{de}}$ '*Eisenverhüttung*', $_{\text{en}}$ '*iron smelting*') no additional semantic annotation beyond the domain tag would be expected.
4. The **entry type** is an annotation derived from the lemma and basically describes if a lemma is a single word, compound or a multiword entry. Also bound morphemes are marked as such here (as opposed to the ConceptNet). The feature is technically important as it triggers different lexical descriptions, e.g. if a multiword entry is validated, it requires the presence of *mwlemmalist* and *mwposlist* features to be well-formed. Figure 9.3 shows the BLD concept and its features.

concept ::=	**lemma, POS, sem, etype, ling, term, xfr, xref**	
	lemma	<string>
	POS	<a BTag>
	sem	[[semdescr]]
	etype	sw,cp,mw2,mw3,mw4

Fig. 9.3 Basic Lemma dictionary: concept definition

[9] Although they are lowercased when occurring as parts of compounds.
[10] Cf. the JRC names, cf. Steinberger et al. (2011).

```
morph ::=         form, flex, morphonem, morpos
    form          ::=  abbrev | acron | regular | inflected
    flex          ::=  flexclass,allodesc+
        flexclass  ::=  integer | <string>
        allodescr  ::=  allomorph, fvp+
        allomorph  ::=  <string>
        fvp        ::=  feature, value
    morphonem ::=      onset, offset, vowharm
        onset      ::=  vok, (kons | (const | conswk) )
        offset     ::=  vok, (kons | (consst | conswk) )
        vowharm    ::=  (a|e), (i|ı|ü|u)
    morpos        ::=  morad | morno | morvb
        morad      ::=  compform
            compform  ::=  clitic | split
        morno      ::=  gender, number
            gender    ::=  masc | fem | neut
            number    ::=  sg | pl | sgt | plt | du
        morvb      ::=  pfaux, separaffix
            Pfaux      ::=  haben| sein
            separaffix ::=  <string>

sem ::=           semdescr, polarity, supportverb
    semdescr      ::=  semfea | domain | readingno | wsd
        semfea     ::=  <feature>
        domain     ::=  <topiclist>
        readingno  ::=  <integer>
        wsd        ::=  wsdentry+
            wsdentry  ::=  <string>
    polarity      ::=  pos | neg
    supportverb   ::=  t | f

syn ::= synad | synav | synno | synna | sybvb
    synad ::=          subcat | args | position | compar
        subcat     ::=  attr | pred | name | adv | nst
        args       ::=  role | status | filler | feaval
        position   ::=  pre | post
        compar     ::=  t | f
    synav ::=          subcat, modtype, pos
        subcat     ::=  nomod | vbmod | clmod | pnmod | avmod | rel |
                         neg | qu
        modtype    ::=  no | vb | cl | av | pn
        pos        ::=  pre | post
    synno ::=          subcat, args
        subcat     ::=  lass | day | month | nat | title | unit
        args       ::=  role | status | filler | feaval
    synna ::=          subcat
        subcat     ::=  comp | inst | loc | fname | lname | pers
    synvb ::=          subcat, args | trtype, vform
        subcat     ::=  full | aux | mod
        args       ::=  frame+
                        frame ::= role | status | filler | feaval
        trtype     ::=  t | i | dt
        vform      fin | inf |par | grd | cvb

pragm ::=         domain, dialect, register
    domain   ::=  <the topic list>
    dialect  ::=  (fa|dr) | (uk|us)
    register ::=  fig, obsol, slang, colloq, pejor
```

Fig. 9.4 Basic Lemma dictionary: linguistic annotations

```
morph ::=       form, flex, morphonem, morpos
      form      ::=        abbrev | acron | regular | inflected
      flex      ::=        flexclass, allodescr+
                flexclass  ::=        integer | <string>
                allodescr  ::=        allomorph, fvp+
                allomorph  ::=        <string>
                fvp        ::=        feature, value
      morphonem ::=        onset, offset, vowharm
                onset      ::=        vok, (kons | (const | conswk) )
                offset     ::=        vok, (kons | (consst | conswk) )
                vowharm    ::=        (a|e), (i|ı|ü|u)
      morpos    ::=        morad | morno | morvb
                morad      ::=        compform
                           compform   ::=        clitic | split
                morno      ::=        gender, number
                           gender     ::=        masc | fem | neut
                           number     ::=        sg | pl | sgt | plt | du
                morvb      ::=        pfaux, separaffix
                           pfaux      ::=        haben| sein
                           separaffix ::=        <string>
```

Fig. 9.5 Morphology

9.2.1.2 Linguistic Annotations

Each of the concepts needs linguistic annotations. They are given in the form of feature value pairs. The focus is again on annotations which are needed by one of the analysis components.

Each feature belongs to a given **type**. The types follow the usual linguistic distinction, morphology, syntax, semantics, and pragmatics. Each of these feature types are described in a special section of the linguistic annotations. Figure 9.4 gives the overview.

1. Morphology

Morphological annotations are divided into features for form, inflectional behaviour, morphophonemics, and additional POS-specific information. The '*form*' feature distinguishes abbreviations (ending with period), acronyms and regular forms. The value '*inflected*' is needed to prevent dictionary entries which are already inflected from being used by LtFlexer for renewed inflection generation; examples of such already inflected entries can be found in every dictionary.

The inflection annotations ('*flex*') first give a flexclass; this can be either the number of the inflection class, or for better readability an example word describing the inflection parameter, something like '*inflects_like*'. Many words use allomorphs for inflection, like umlaut ($_{\text{de}}$ '*Wagen—Wägen*'), broken plural

($_{\text{fa}}$ اجر vs. اجور, $_{\text{en}}$ '*reward*'). Such allomorphs are given in the *allodescr* feature, together with a specification in which feature constellation the allomorph should be used; this is a feature-value pair collection.

Morphophonemic information refers to onset and offset (often vocalic—consonant); in Turkish, a differentiation between strong and weak consonants is needed, and also information on vowel harmony.

Finally there are POS-specific annotations: Degree formation for adjectives (clitic ($_{\text{en}}$ '*stronger*') vs split ($_{\text{en}}$ '*more content*')), gender and number information for nouns, including singulare/plurale tantum, and dual for Arabic, perfect formation in German ($_{\text{de}}$ '*haben*' vs. '*sein*'), separation of prefixes ($_{\text{en}}$ '*off*' for '*write off*', $_{\text{de}}$ '*auf*' for '*aufstehen*').

Of course the annotations depend on the languages covered by the system; the above examples relate to the language set described in TINA.

It should be noted that the dictionary provides lemmata, not inflected forms; therefore many features describing inflected forms, like case, tense, person etc. would not be given here; they are created in the process of generation of inflected forms, and would show up in the Full Word Dictionaries (FWD, cf. below) (Fig. 9.5).

2. Syntax

The syntactic annotations differ depending on the part-of-speech of the entry. All of them show a syntactic subcategorisation, following the specifications in the tagset (STAG, cf. Sect. 2.3.2).

For nouns and adjectives, syntactic arguments should be given; they are simpler in structure than full verb frames: They have just one argument, they all take the role of a modifier ($MOD), they are always optional. A set of feature-value pairs describes their behaviour: $_{\text{en}}$ '*proud*' with '*ap=of;*' (*proud of*); $_{\text{de}}$ '*sicher*' with '*ca=gen;*' ('*seiner Sache sicher*'); $_{\text{es}}$ '*rico*' with '*ap=en;*' ('*rico en vitaminas*'), etc.

Verbs specify a list of their frames. Frames can have probabilities, to be used in syntactic analysis for frame assignment. Each frame consists of a set of arguments. Arguments specify the roles they describe (objects like $DOBJ, $IOBJ, complements like $SOMP, $OCOMP for '*he teaches him Spanish*'), their status (obligatory, optional), the POS to which they can be assigned, and feature-value pairs describing the constraints the argument must satisfy (case, preposition, form etc.)[11]

Some of the TINA dictionaries follow a simpler approach for framing and just specify the transitivity type, the values being intransitive, transitive, or ditransitive. While subcategorisations are required by the tagger the framing information is only needed in case of full syntactic analysis which in the context of the TINA components is not represented, so just a rough annotation may be justified.

In addition to subcategorisation, adjectives need annotations for position (before or after the head: $_{\text{fr}}$ '*grand*' vs. '*inconnu*') and for comparison behaviour

[11] The dictionary follows the proposal of Gebruers (1988) for verb valency descriptions.

```
syn    ::=    synad | synav | synno | synna | sybvb
       synad    ::=       subcat | args | position | compar
                subcat    ::=  attr | pred | name | adv | nst
                args      ::=  role | status | filler | feaval
                position  ::=  pre | post
                compar    ::=  t | f
       synav    ::=       subcat, modtype, pos
                subcat    ::=  nomod | vbmod | clmod | pnmod | avmod | rel | neg | qu
                modtype   ::=  no | vb | cl | av | pn
                pos       ::=  pre | post
       synno    ::=       subcat, args
                subcat    ::=  lass | day | month | nat | title | unit
                args      ::=  role | status | filler | feaval
       synna    ::=       subcat
                subcat    ::=  comp | inst | loc | fname | lname | pers
       synvb    ::=       subcat, args | trtype, vform
                subcat    ::=  full | aux | mod
                args      ::=  frame+
                          frame ::= role | status | filler | feaval
                trtype    ::=  t | i | dt
                vform     fin | inf |par | grd | cvb
```

Fig. 9.6 Syntax

('comparison possible Y/N' with a Boolean value; the respective forms are built in morphology). For verbs, a feature 'verbform' is foreseen (values are '*finite, infinite, participle, gerund, converb*'): verbs in such forms can very often be found in dictionaries, in addition to the base lemma form. For numbers, the numeric value is annotated, for comparisons e.g. in the recognition of dates. Adverbs are annotated for the constituents they can modify (nouns ($_{en}$ '*only dogs*'), determiners ('$_{en}$ *nearly all*'), numerals ($_{en}$ '*around 30*'), verbs ($_{en}$ '*yesterday*'), or clauses ($_{en}$ '*allegedly*') (Fig. 9.6).

3. Semantics

The **semantic description** uses several annotation options, described by several features: The value of the feature '*readingno*' is an integer, simply counting the readings; the value of the feature '*domain*' is one or several domains as described in Sect. 3.3; the value of '*sem*' is a semantic feature like '*human*', '*animate*', '*abstract*' etc. as described in Chap. 5; the value of '*subsem*' as used for Named Entities is a subdivision of the STAGs NoPL (places), NoPC (companies), NoPP (persons) as described in the Named Entity Gazetteer (cf. Sect. 6.2). It seems that different semantic categorisations are adequate for different system components.

In addition, the semantic section of the dictionary marks also support verbs and polarity (positive—negative) (Fig. 9.7).

```
sem    ::=    semdescr, polarity, supportverb
              semdescr     ::= semfea | domain | readingno | wsd
                               semfea     ::= <feature>
                               domain     ::= <topiclist>
                               readingno  ::= <integer>
                               wsd        ::= wsdentry+
                                              wsdentry  ::= <string>
              polarity     ::= pos | neg
              supportverb  ::= t | f
```

Fig. 9.7 Semantics

```
pragm ::=        domain, dialect, register
        domain  ::=    <the topic list>
        dialect ::=    (fa|dr) | (uk|us)
        register ::=   fig, obsol, slang, colloq, pejor
```

Fig. 9.8 Pragmatics

4. Pragmatics

Pragmatic features collect three annotations: **Domain** annotations follow the list used for topic identification (cf. Sect. 3.3). **Dialects** are particularly important for English (to annotate UK vs. UK English) and Persian (Farsi vs. Dari entries) and help in language identification and in spell checking. **Register** information, like colloquial, pejorative, slang, obsolete etc., is stored, if available, it is more useful for human lookup than for machine treatment (Fig. 9.8).

9.2.1.3 Multiword Annotations

Multiwords need the same annotations as singlewords, in terms of morphological, syntactic and semantic descriptions. Morphology and POS tag can often be taken from their head constituent, however there are many multiwords which do not follow this pattern, esp. adverbs like ${}_{en}$ '*in advance*', '*for sure*' etc.: Their POS tag cannot be inherited from their constituents. Non-inheritance is also the most frequent case in the syntactic and semantic annotations, multiwords have their own verb frames and argument structures.

For multiword recognition the dictionaries provide three specific annotations. First, the position of the multiword head is given. Heuristics like '*the head is the last noun before the preposition*' help in many cases,[12] however in noun-noun combinations the head position is not always obvious. Position information is preferred to lemma information as the lemma identification may be difficult in multiwords like '*higher and higher*'.

[12] Like in the Term Association component explained in Sect. 7.3.

multiwordinfo	::=	headpos, mwposlist, mwlemmalist		
	headpos	::=	<integer>	
	mwposlist	::=	<BTag>+	
	mwlemmalist	::=	<string>+	

Fig. 9.9 Multiword annotations

Second, the multiword stores a list of the lemmata forming it, each one consisting of a lemma and a POS tag. So there is a *mwlemmalist*, i.e. a list of lemmata, comma-separated, and a *mwposlist*, i.e. a list of BTAGs, also comma-separated, to describe the parts of the multiword.

The multiword features can be generated by term extraction procedures with only little effort, cf. Sect. 9.3.2 below. Multiword treatment is required in nearly every TINA component (Fig. 9.9).

9.2.1.4 Administrative Information

Administrative information is relevant for dictionary maintenance. '*author*' and '*creationdate*' refer to the origin of an entry; '*lasteditor*' and '*lasteditdate*' refer to the last access.

Module is important in cases where the dictionary consists of different modules e.g. for different domains: The system must be able to load and unload such specific domain or customer data: Such entries can be assigned to special modules. At runtime it can be stated which modules of the dictionary should be used and in which sequence. Often a common 'general' dictionary is used in several customer projects, and the different customer dictionaries are loaded and unloaded in addition. To enable such flexibility the entries must be kept separate of each other, by means of a module annotation (Fig. 9.10).

Version number is relevant in cases where dictionary are supposed to be updated e.g. in a maintenance environment. Ad-hoc changes in a dictionary tend to produce confusion after a while. Dictionary entries behave like software: They

admin ::=		creationdate, lasteditdate, author, lasteditor, module, version, source, comment
	creationdate ::=	<date>
	lasteditdate ::=	<date>
	author ::=	<string>
	lasteditor ::=	<string>
	module ::=	<a dictionary module>
	version ::=	<version-#>
	source ::=	<string>+
	comment ::=	<string>

Fig. 9.10 Administrative information

Language	ar	de	en	es	fa	fr	it	ps	pt	ru	zh
Lemmata (k)	49.4	254.5	83.5	68.9	68.0	45.6	114.7	47.5	84.1	163.2	864.9

Fig. 9.11 BLD dictionaries size

should undergo version control, and have releases like other software components, in particular if changes of the dictionary structure are involved.

The **source** of an entry helps in dictionary maintenance, e.g. if a dictionary has been merged from several sources, and often **comments** are a helpful information element.

The size of the Basic Lemma Dictionaries in the different languages is given in Fig. 9.11.[13] They differ in explicitness, however, from just the concept information to detailed annotations.

9.2.2 Full Word Dictionaries (FWD)

These dictionaries are either created from Basic Lemma Dictionaries (BLD) by using LtFlexer (cf. Sect. 9.3 below), or they result from corpus analysis or legacy resources. They also include the entries for function words, which in general are not part of the BLD, because they need to be annotated on a textform and XTAG level, not at a lemma level. All entries are manually validated in several validation runs.

The process of the Full Word Dictionary creation is depicted in Fig. 9.12.

The entries of full word dictionaries (FWD) consist of four information elements:

1. **Textforms** are the basic information items. Textforms are inflected single words, and they are in normalised form by default: standard casing, standard accent positions etc. In reality however compromises must be found, e.g. in $_{\text{en}}$ *-isation/-ization* alternations where occasionally both forms are given. Textforms of lemmata with separable prefixes as well as multiword elements contain all parts as textforms, and the combined form as lemma:

    bumps-> bump_off
    off-> bump_off

    gear-> gear_shaft
    shafts-> gear_shaft

[13] The Chinese resources are given in Fig. 5.2 above. Here just the entries of the PKU tagset are given. The Farsi resources are given in a standard spelling, as used in a speech recognition dictionary.

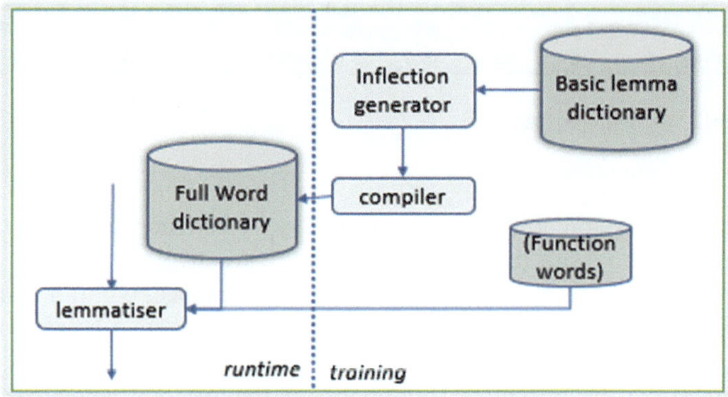

Fig. 9.12 Full word dictionary creation

2. They also contain **linguistic annotations**, depending on the tagset used:

- one of the basic parts of speech (BTAG);
- basic tags plus distributional information of syntactic subcategores as in standard tagsets (STAG); the German decomposer dictionary uses a special tagset (DTAG, cf. 5.2.4)
- extended annotations (XTAG) storing also morphological information.

To avoid redundancies, the linguistic annotations are separated into annotations which are full-form specific (like number, case, person) and others which are lemma-specific, i.e. identical for all textforms (like gender, verb arguments, semantics). Only the full-form specific annotations are relevant for the Full Word Dictionary, the others should be found in the Basic Lemma Dictionary.
3. They contain references to the **lemma** to which they belong. Such references can be given by providing the lemma directly, or more efficiently by providing a lemmatisation instruction based on *cut&append* information, as described in Sect. 5.2.2 above.
4. In cases where this is required (e.g. for LtTopicIdentifier) the full word dictionary also contains **frequency** information (e.g. in form of a score). However such annotations are only meaningful with reference to a given corpus (Fig. 9.13).

FWD	**::= FWDentry+**					
	FWDentry	::= textform, POS, lem				
		textform	::= <string>			
		POS	::= BTag | STag | XTag			
		lem	::= cut, append			
				cut	::= <integer>	
				append	::= <string>	

Fig. 9.13 Full word dictionary annotations

Language	ar	de	en	es	fa	fr	it	pt	ru	zh
Textforms (K)	161	841	143	330	110	201	461	454	1574	--

Fig. 9.14 Full word dictionaries size

Differences in linguistic annotations or in lemma constitute double entries, sharing the textform but showing differences in the other annotations. This way a given textform can instantiate several lemmata in lexical analysis as shown in Sect. 7.1.

Full word Dictionaries are available in TINA for the languages listed in Fig. 9.14. The size of the Full Word Dictionaries is also given in Fig. 9.14

9.2.3 *Named Entity Gazetteer (NEG)*

The design and content of a Named Entity Gazetteer (NEG) is seldom discussed. Most NEGs are used in special domains or even projects. However the NEG in TINA is supposed to support general purpose texts, similar to a general word dictionary.

It is obvious that simply accumulating names is not a useful option. There is no point in making the name lexicon as big as possible: The system will be confronted with handling problems, and will give false alarms for spurious occurrences of names being homographs with normal words. Just for place names the geonames server[14] provides more than 80 k names just for Afghanistan. DbPedia reports 900 Mio triplets in its collection.[15] On the other hand, it is also no option to deliver an empty NE gazetteer to a customer system installation. So a 'general-purpose' NE gazetteer would be required. But to offer such a 'general-purpose' gazetteer requires a bit of consideration, esp. in questions of maintenance.

9.2.3.1 Design

1. A rule of thumb would claim that a gazetteer of names should contain all names which a 'well educated' person could know: '*Rome*', '*Napoleon*', '*Microsoft*'. If more specific names occur in a text, like '*Kleinheubach*', then the context would often contain indicators what kind of name this may be.
2. The NEG collection should follow certain principles. This is necessary to give indications when it is extended, modified or validated in later versions, e.g. asking '*Does the gazetteer contain all abbreviations of units/all month names/…?*'

[14] https://www.geonames.org/

[15] https://www.dbpedia.org/resources/

3. Some names are language-independent, like '*Venda do Pinheiro*', others are language-specific, like $_{\text{en}}$ '*Rome*'—$_{\text{it}}$ '*Roma*'. Therefore the gazetteers must be language-specific, i.e. each language needs its own NEG (NEG_en, NEG_fr etc.).
4. Names in different languages which have the same reference and therefore are translations of each other (like $_{\text{de}}$ '*München*'—$_{\text{en}}$ '*Munich*'—$_{\text{it}}$ '*Monaco*'—$_{\text{tr}}$ '*Münih*') must be collected in special translation resources to support multilinguality, similar to transfer dictionaries in common word areas. The difference would be that the resources for name translations can be non-directed and multilingual, as they share a common referent.
5. The language-specific NEGs contain many entries which may be less relevant for other languages: More local person names, names of smaller cities, local company names and the like. Therefore the names will be split into general and language-specific parts. The latter contain the more language-specific names.
6. The NEGs contain not only names but also name indicators, like '[person] *lives_in* [place]', '[company] *sells* [company]', '*shares_of* [company]' '*Prof.* [person]', '*president_of* [country/company]'. Such name indicators are language-dependent and must be provided by the NEG of the respective languages.
7. Most efficiency can be observed if NEG's are domain or even customer-adapted. If a NEG of a motor company contains all place names of its production sites (like '*Chattanooga*'), its coverage and usefulness will increase. The consequence for the gazetteer architecture is that the NEG must be modular, with the option to load and unload domain specific name modules, like in the dictionary sections.

As a result, there will be one NE gazetteer per language. NEGs of a given language are a combination of several resources, both language-dependent and language-independent, enriched by domain-specific Named Entities. Cases where names have translation links need to be considered as well in a multilingual system. As NE gazetteers will be provided per language, general named entities will be added to each of them (Fig. 9.15).

9.2.3.2 Collection

The collection of the Named Entity Gazetteer follows the following principles:

- For **persons**, the most frequent last names per language have been collected from the internet (between 500 and 1000 names). They are assigned to the all_language NEG, as e.g. '*Sanchez*' would not only occur in Spanish texts but could be found elsewhere.

  In the language-specific dictionaries, more last names of the respective language were added (for instance, additional last names for the German gazetteer taken from the German phonebook, for the French gazetteers from French resources, etc.).

- For **locations**, all countries, and the biggest cities worldwide and in Europe were selected, in addition to oceans, the most prominent rivers, mountains, lakes etc.

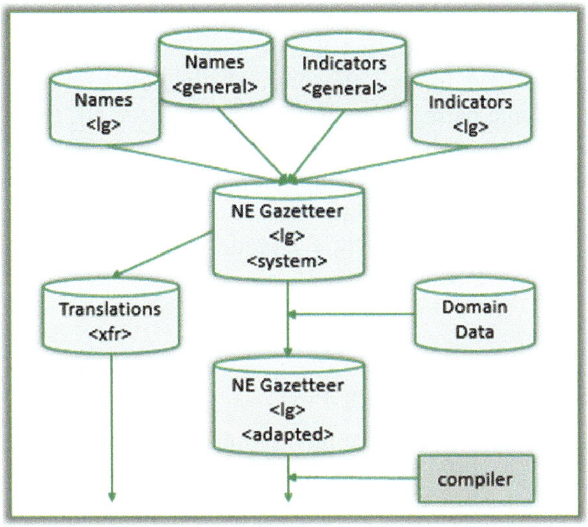

Fig. 9.15 Named entity glossary creation

In the language-specific sections, smaller regional cities and counties/region names are collected. On top, most important local rivers, mountains, lakes etc. are added (rivers in France/in Germany etc.).
As many locations have language-specific denotations a glossary of translations was built ($_{en}$ '*Ivory Coast*', $_{de}$ '*Elfenbeinküste*'). The language-specific gazetteers take their entries from there.
There are activities to standardise place names in international bodies and organisations (cf. Berman et al. 2016), however in everyday discourse the conventional names are used, and therefore must be found in the gazetteer.

- For **companies**, the biggest enterprises, listed in the stock exchanges of the European, US and Asian stock markets were collected. They were not separated by language as many of them operate internationally.

  As for the names themselves, legal forms and country subsidiaries are not taken as name parts; so '*Siemens AG*' -> '*Siemens*', '*Siemens Italy*' -> '*Siemens*'). The legal parts of companies are stored in a feature of the respective entry.

- For **institutions**, available glossaries like the JRC-Names[16] have been examined, and parts of the entries have been taken over.

The collection of the **indicator lexicon** entries is a language-specific action, and consists of the evaluation of contexts of NEs in training data: Frequent

[16] https://ec.europa.eu/jrc/en/language-technologies/jrc-names-cf. also https://islrn.org/resources/328-863-023-410-2/

contexts indicating a certain NE type are identified and added to the lexicon: Frequent verbs, frequent appositions etc. are part of the gazetteers.

The indicator lexicon follows the basic distinction between name parts and name indicators: Name parts belong to the lemma of the NE (like '*bank*' in '*Bank of Skiggs*'), while indicators do not (the lemma of '*city of Skiggs*' is just '*Skiggs*'). This difference is expressed by a feature in the gazetteer. (Rules containing indicators can distinguish between the phrase structure to identify the NE, and the span of the NE itself: The phrase structure includes indicators etc., but the NE does not). This is explained in Sect. 6.2.

9.2.3.3 Gazetteer Description

Each entry of a gazetteer consists of tab-delimited information items, organised as feature-value pairs, providing the following annotations:

- textform. The NE gazetteers do not use the lexical analysis as described in Chap. 5; many names have special morphological descriptions. As a shortcut the textforms are given, and used as access key.
- lemma. The lemmata give the canonical form of the name, like '*Ford Motor Company*', '*Pablo Picasso*', '*European Commission*', '*Pakistan*'. The multilingual glossary uses the lemmata of the participating names.
- entry type. As everywhere, single (sw), compounds (co), and multiword entries (mw2, mw3, mw4) are marked here.
- Part of Speech: One of the STAGs described in Sect. 2.3.2: *NoPL* for locations, *NoPC* for companies, *NoPI* for institutions, *NoPU* for units etc.
- Semantic subcategorisation: Feature '*semsubcat*' with values which annotate the POS tags in more detail, like '*city*' or '*country*' for place names, '*sportsclub*' for institutions.
- Features: A collection of different feature-value pairs. For instance, subcategorisation information is collected in the '*nesubtype*' feature, and the classification of the indicators is given in the '*nefunction*' feature (values being '*<ne>-part*' or '*<ne>-indicator*'). In addition, features like gender, case, position are available. Examples are '*legalform=Plc*' or '*gd=fem;*' for indicators of gender as annotations for firstnames or places.
- Source: This annotation controls the composition of the gazetteer: The origin of an entry, its membership to a domain-specific module, and the maintenance version. The source can be used to load or unload modules of the NE gazetteers.

The structure of the Named Entity Gazetteer in TINA is given in Fig. 9.16, their size is given in Fig. 9.17.

Named Entity components were only created for some languages. The languages analysed with LtDeep (ar, fa, ps, tr) contain the names in the Basic Lemma Dictionaries and don't use a special resource.

NEG ::= NEGentry+		
NEGentry ::= textform, POS, lemma, etype, semsubcat, features		
textform	::=	<string>
POS	::=	<STAG>
lemma	::=	<string>
etype	::=	sw, co, mw2, mw3, mw4
semsubcat	::=	name-indic | country | city | ...
features	::=	feaval
source	::=	<string>

Fig. 9.16 Named entity glossary structure

Language	de	en	es	fr
Names (k)	83.8	67.9	63.5	66.2

Fig. 9.17 Named entity gazetteers size

9.2.4 Word Sense Repository (WSR)

This repository stores the results of the word sense disambiguation of LtSensExtract. It is language-specific of course.[17] The Word Sense Dictionary contains the terms which form the context for the identification of a given word sense. It is a list of entries, each entry consists of three information items:

- the **entry**, in the form of lemma and POS (BTAG)
- the **sense**. This field is a label to describe the term cluster defining it. In LtSensExtract the label is a target language lemma,[18] it could however just be a sense number. The important aspect is that the context terms give an indication what the sense is.
- a comma-separated list of **context terms** which frequently cooccur with the entry, sorted by frequency.

The list of context terms is editable so users can add terms which support the respective sense. The reason is that not enough corpus data were available to find sufficiently good contexts. The process of building the Word Sense Dictionary is described in Sect. 7.2 on LtSensExtract.

The structure of the resource is given in Fig. 9.18, examples are given in Fig. 7.11 above.

The current dictionary for German covers 6.6 k lemmata with 21.9 k senses.

[17] In TINA it was only built for German, however.

[18] With the target lemma and a BTAG. The tag is only needed in translation applications. It is not relevant for sense disambiguation.

WSD	**::=**	**wsdentry+**	
	wsdentry	**::=**	**term, POS, sense, POS?, features**
		term	::= <string>
		POS	::= <BTAG>
		sense	::= <string>
		features	::= <string> , (, <string>)*

Fig. 9.18 Word sense dictionary structure

9.2.5 *Bilingual Dictionary (XFR)*

The transfer dictionaries are a collection of links from a source language entry to a target language entry, observing certain conditions. The transfers (XFR) maintain multilingual information in the TINA system.

9.2.5.1 Design

The XFR dictionaries are bilingual and directed. They are organised as a *collection* (or *package*) of entries, each one consisting of a source language concept, a target language concept, and a (possibly empty) set of conditions.

1. The XFR links are **directed**, i.e. not reversible. Non-reversibility is an obvious requirement in cases of 1:n transfers, which cause the dictionary entries to be reorganised if reversed, and reversals also require additional disambiguation information for transfer selection.

But even in 1:1 transfers care should be taken if entries are to be reversed. Experience shows that source entries, if translated, point to a slightly less specific meaning in the target language in many cases. If such entries are reverted, a more general meaning nuance will be transferred into a very specific term. An example may be $_{\text{de}}$ '*Lichtbogen* ', translated into $_{\text{en}}$ '*arc*' in many dictionaries. However when reversed an '*arc*' would receive a much too specific translation as '*Lichtbogen*', instead of the more general '*Bogen*'.

For both reasons the TINA dictionaries are not reversible. Reversal should be perceived to be a special project, including specific validation requirements.

Similar observations hold in cases where new bilingual dictionaries are to be produced by merging two existing dictionaries into a new one (e.g. *en>es* and *es>it* into *en>it*). In addition to the challenges just mentioned, the task would also be to pick the right reading in cases of polysemes. To give an example: In the Farsi Dictionary, the three terms نقش ('naɣʃ, outline, shape, contour), تصویر (tas.'vir, illustration), and تابلو (tāb.'lu, sign) can all be translated into $_{\text{de}}$ '*Bild*'. Without a reference to the Persian source, a proper translation of $_{\text{de}}$ '*Bild*' into English would not be possible.

2. XFR links are not just directed but also only **bilingual**. Multilingual termbanks are only preferable in domains, mostly technical, where clear 1:1 translations are possible. This however is just a special case, usually translations inherently are organised bilingually. Organisations of concepts in a multilingual or interlingual way, like EuroWordNet, cf. Vossen (2004), but also TBX,[19] require additional annotations to keep control of all transfers.

Multilingual entries can be used in specialised domains, like in Assistive Technology (cf. Sect. 8.3.2). In the domain of general vocabulary multilingual entries cannot be maintained and validated in a consistent way. Moreover many application areas are bilingual, only few require multilingual processing, like in the European Union, cf. Steinberger et al. (2002).

3. The nodes of a translation link are concepts as described in the Basic Lemma Dictionary (BLD). The transfer could therefore be expressed by simply using their entry ID: *SL-BLD-ID -> TL-BLD-ID*; however this would negatively affect readability. So the XFR dictionaries use the conventional view and define a transfer as '*SL-Lemma + SL-POS -> TL-Lemma + TL-POS*'. Semantic information, if available, can be used as transfer selection test.
4. The translation equivalents may be subject to a series of conditions and modifications, e.g. they may only hold in the medical domain. Such information items are described as attributes to the XFR link.
5. The dictionary also contains Named Entities. Many of them need to be translated ($_{\mathbf{en}}$ '*Egypt*', $_{\mathbf{de}}$ '*Ägypten*', $_{\mathbf{tr}}$ '*Mısır*', $_{\mathrm{fa}}$ 'مصر'), and allows the entity to participate in a multilingual scenario. The entries have the same structure as the concepts, the only difference is that the nodes of the entries do not point to the Basic Lemma Dictionary (BLD) but to the Named Entity Gazetteer (NEG). The entries themselves have the same structure in the XFR resource.

In cases where no translations exist proposals have been made to include transliterations or similarity computations (cf. Knight and Graehl (1998), Klementiev and Roth (2006)); such techniques however are not included in TINA. Only dictionary lookup is supported.

Named Entities can be ambiguous ('*Livorno*' can refer to a city but also to a province in Toscana) and will show several target entries in the bilingual dictionary as city and province name may differ in other target languages.

[19] TBX: https://www.tbxinfo.net/. In terminology, TBX has been developed as a standard. Terms there are organised in a non-directed and multilingual way. This model was not adopted as it is for terminology only and does not support general vocabulary, and it does not provide operations between members of a term information group (like tests and actions). For details cf. Thurmair (2006a).

9.2.5.2 Transfer Entries

An entry consists of a source language term and a target language term. The entries are identified by a key, consisting of source lemma and source POS. Lemmata are defined as in the Basic Lemma Dictionaries, multiwords being represented in their canonical form.[20]

The translation equivalents may be constrained by conditions, and may trigger actions on the target side. Such conditions and actions are represented as annotations to the links. Conditions for a valid transfer link are for instance:

- Syntactic constraints: they could be local (like gender: $_{de}$ '*Hut*' if *{gd=masc}* ->$_{en}$ '*hat*', if *{gd=fem}* ->$_{en}$ '*protection*') or contextual, referring e.g. to verb arguments: for instance '*bestehen {+ auf} ->insist upon*', '*bestehen {aus} ->consist of*', '*bestehen {DOBJ exam} ->pass*'
- Semantic constraints: e.g. $_{en}$ '*eat*': if {*subject=human*} –>$_{de}$ '*essen*'; if {*subject is animated*} –>$_{de}$ '*fressen*'
- pragmatic constraints, like register or dialect ('*dialect=farsi/dari*')
- domain information (e.g. $_{en}$ '*hoist*' ->$_{de}$ [Bau] '*Bauaufzug*', [energy] '*Windwerk*', ->$_{fr}$ [tech] '*plate-forme {f} élévatrice*', [gv] '*palan*'
- probability information: Probabilities are essential for data-driven applications. However, probabilities are always related to a given corpus of text, and cannot be more than an indication in general situations like simple dictionary lookup. To overcome such situations, some TINA XFR dictionaries offer a Boolean feature '*preferred translation*' for transfers which the coding people consider the most general one.
- contextual information like access to the Word Sense Repository (WSR) to identify the sense and its corresponding translation.

Annotations may also trigger certain actions on the target language side, like number and case mapping, negation-drop and the like; examples are given in Thurmair (1990).

All these constraints are laid down in the form of feature-value pairs in the annotation section of the transfer entry.

9.2.5.3 Transfer Dictionaries

The available transfer dictionaries in TINA are listed in Fig. 9.19. An important information item is how many *source language* concepts are contained in the dictionary as this indicates its coverage in text analysis. The structure of a bilingual dictionary is given in Fig. 9.19.

---

[20] As opposed to earlier single-word-based MT systems Examples were multiwords were collected on the basis of transfer tests are given above (Sect. 6.1). This leads to wrong entries e.g. if human-readable dictionaries are to be produced from such MT dictionaries.

XFR	::=	xfrentry+	
	xfrentry	::=	slconcept, tlconcept, annotations
		slconcept	::= lemma, POS
		tlconcept	::= lemma, POS
		annotations	::= feaval*

Fig. 9.19 Transfer Dictionary Structure

SL\TL	ar	de	en	es	fa	fr	it	ps	pt	ru	tr	zh
ar	--	83.1										
de		--	222.2			84.6						
en		222.0	--	221.0		165.9	169.6		152.0			
es			221.1	--								
fa		71.8			--							
fr		84.7	165.5			--						
it			169.8				--					
ps		60.7						--				
pt			152.3						--			
ru		87.9								--		
tr		91.0									--	
zh		150.2	1093.4									--

Fig. 9.20 Size of XFR dictionaries

The dictionaries were also used for human lookup, so their readability in the user experience was an issue. Because users should be able to understand why a given translation was selected in a given context transparency was an important requirement.

As for the dictionary quality, validation tests were run on many of the dictionaries: Formal tests on consistency (does each concept have a correspondence in the monolingual dictionaries?) and correctness (spelling etc.) were made, for the translations themselves. Translation experts were given samples of entries to validate their correctness. Although such a procedure depends on many questionable parameters (expertise of the experts, selection of the samples etc.) there is hardly an alternative, in particular as not all translations find their way into the corpus data which could be used for validation, as has been shown in the LtSensExtract description above (Sect. 6.1) (Fig. 9.20).

9.3 Tools for Resource Production

The resources described in the previous section do not exist independently of each other. The TINA system links them by providing tools for their efficient production and maintenance (cf. Fig. 9.21).

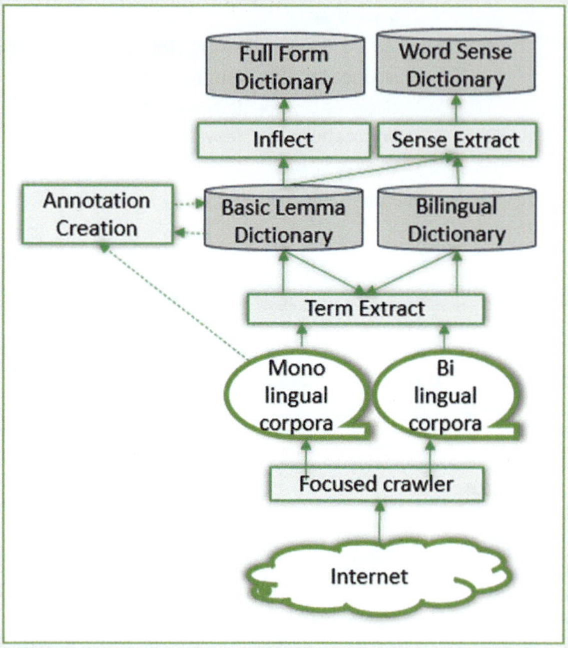

Fig. 9.21 Tools overview

- The document collections needed to build domain-specific resources can be collected from the internet with the aid of a **focused crawler**.
- The text collections are used to **extract terms** for the dictionaries, both monolingual and multilingual.
- The full word dictionaries are built from the Basic Lemma Dictionaries with the use of a **flexer** which generates inflected forms.
- The creation of the **word sense** dictionary needed for semantic disambiguation is the task of the tool called LtSensExtract.

Additional tools could of course be added e.g. to support the annotation process: Local defaulters for POS and gender, contextual defaulters e.g. for verb frame identification as described e.g. in Korhonen (2002), Ushioda et al. (1993); this would make the resource production even more efficient.

9.3.1 Focused Crawler (ILSP-FC)

One of the first tasks in linguistic analysis is to create a document collection. If such collections are not available e.g. from customer systems, then such a collection can be produced by crawling the internet. A focused crawler is a device which searches documents belonging to a given topic or domain. To do so, a standard crawler must be enriched by a topic identification component which decides whether or not a found document belongs to the topic and should be added to the document collection.

Details of a focused crawler, called ILSP-FC, designed and developed by ILSP Athens, are described in Papavassiliou et al. (2013), or in Pecina et al. (2011), Pecina et al. (2012). The ILSP-FC crawler was used in some TINA projects. Here just the outline of this component is given.

1. The crawl starts with a set of **seed terms**, describing the topic to be found, and a set of **seed URLs**, defining where the search should start. A starting size for these resources turned out to be about 50 terms and about 50 seed URLs. In multilingual contexts these sets need to be constructed for each language involved.

The URLs are used as initial elements in an agenda. The crawler processes the agenda by fetching each URL from the internet for inspection, and by adding found links to the agenda.

2. Every crawled document must be inspected in order to decide whether or not it should be added to the topic of the collection. The analysis performs a sequence of steps, each of which eliminates unwanted document candidates:

- It analyses the **character code** of the document and converts it into a defined standard; in the component described here it is UTF8.
- It removes the parts of the document which do not contain texts, like advertisements, link collections, navigation elements etc., often called '**boilerplates**', cf. Kohlschütter et al. (2010). These elements do not belong to the text and would just deteriorate the results.
- It identifies **duplicates**. This is an important step in crawling as the same document can be reached by several URLs. If the duplicates are not removed, not just the document itself but also the outgoing links are duplicated and cause a loss in crawling efficiency. Duplicates could be identified using an MD5 key[21]; however the check must not be done on the original document as it may contain different advertisements leading to different MD5 values, so boilerplate removal must be done first.
- It identifies the language of the document using a **language identifier**. Documents in 'wrong' languages are eliminated.
- The **classifier** identifies the **topic** of the document, by using the terms in the seed list. If the seed list contains only lemmata, then the document must undergo a short lemmatisation. The topic decision depends on two factors: the occurrence of topic terms, and the position of the terms in the document (like: title, text, keywords, metadata). If the resulting score is below a given

[21] An MD5 hash is created by taking a string of an any length and encoding it into a 128-bit fingerprint.

threshold, the document is discarded. Otherwise it is added to the collection for this topic, together with some metadata (URL, language, MD5 keys etc.).

3. After the decision on the document, the crawler adds the outgoing links to the agenda and continues fetching. It can be tuned to different strategies how to process the agenda (breadth first, depth first etc.). Also the termination point can be specified (like: number of pages, time elapsed). Different parameters are available for the control of the crawler.
4. A special case is the crawling for **parallel texts**. The task consists in finding parallel documents based on two document sets with same topic but different languages. Criteria can be filenames, number and type of paragraphs, images, or document structure. Details are again given in Papavassiliou et al. (2013).

To be able to use the resulting document pairs they have to be sentence-segmented, and the sentences need to be aligned, the result being a corpus of parallel (or strongly aligned) sentences.
5. An example of the **evaluation** of the crawler performance is given in Papavassiliou et al. (2013). The task was to acquire pairs of parallel documents in German and Italian for the 'Health & Safety at work' domain ($_{\text{de}}$ '*Arbeitsschutz*', $_{\text{it}}$ '*sicurezza sul lavoro*'). The system reached a precision of about 77% for both languages which is in the upper range of industrial applications.

It turned out that the precision of the topic classifier strongly depends on the quality of the seed terms; so it is worthwhile to invest a bit of effort in the definition of the seed term set to improve the crawling results. A validation step on the final document collection could also help to improve the result, e.g. by using additional topic resources.

9.3.2 Term Extraction (LtTermExtract)

A tool to extract terms from texts is a basic component in every linguistic application. In the TINA toolset it is used to feed the monolingual and bilingual dictionaries.

9.3.2.1 Analysis

The first question to answer in term extraction is what is meant by 'term': What exactly should be extracted? If existing term banks are inspected, then it can be observed that translators and terminologist extract quite different terms from a text,[22] and this observation also holds for computational term extraction tools. Obviously there is no agreement what the intended result of such an activity should be: Would proper names like '*Chicago*' be considered as terms? Or

[22] This was already observed by Hohnhold (1984) when validating the Siemens TEAM termbank.

commands like '*rmdir*' as found in UNIX manuals? Or expressions like '*sports seat*' in automotive descriptions? As long as this cannot be answered the definition of relevancy or the creation of a gold standard for extraction cannot be attempted.

At least three possible definitions seem to exist:

- In **terminology**, terms are defined as concepts which belong to or define a given subject area, e.g. automotive engineering: Language for Special Purposes (LSP), Mayer (2001). They are collected in specialised dictionaries like for automotive,[23] for data processing[24] and the like. Term extraction here would mean to produce only terms belonging to this particular domain.

  The challenge here is that the borderline between general purpose and domain specific terms cannot always be clearly drawn, in particular as many domain specific terms are derived from general vocabulary, and lead to an ambiguity in meaning, e.g. '*mouse*', '*window*', '*thread*', or '*factory*' in the data processing domain.

- An alternative view would be supported by machine dictionary and **translation** experts. Here term extraction is meant to find anything that needs to be translated, based on an existing dictionary resource. It is a kind of unknown/out-of-vocabulary search, mainly in bilingual contexts, be it in specialised domains or in general vocabulary. Traditionally the challenge here is to find unknown multiwords and their translations, using parallel or strongly aligned[25] corpora.

  Obviously the result of a term extraction run will depend on the terms which are already known and stored in the dictionary. Without such a dictionary, term extraction is supposed to return all terms occurring in the corpus, be it LSP terms or general vocabulary. Proper names would possibly also be part of the result set.

- Finally, applications like **CLIR** would consider a term any unit which could be indexed and searched, including proper names, commands and other such elements. Again in multilingual contexts, at least bilingual term extraction like in translation would be required.

The result is that term extraction has to produce different results depending on the purpose. A general definition of relevancy as needed to define measures like recall and precision is therefore hard to find, and in the end it is the human experts who decide whether a candidate is added to the list of extracted terms or not.

Taking this into account the human validation is an integral part of every term extraction component. The system needs to focus on good term proposals to be presented to the validators, collected in a list of term candidates, but also on efficient decision making and validation options for users. Once accepted terms should be prepared for integration into the existing resources.

[23] Wyhlidal, https://www.wyhlidal.com.

[24] Brinkmann and Blaha (1997) still available in antiquarian bookshops.

[25] Such alignment issues were explored in the ACCURAT project, cf. Skadiņa et al. (2019).

In the context of TINA, the relevant resources would be the Basic Lemma Dictionary (BLD) for monolingual and the Bilingual Dictionary (XFR) for bilingual term extraction. They would be created (or extended) by monolingual or bilingual term extraction from corpora. The main term extraction applications in practical use have been projects for domain adaptation[26] and maintenance of bilingual customer dictionaries.

9.3.2.2 Design

1. A component for Term Extraction consists of three main subcomponents. An initial treatment performing tasks like corpus analysis, sentence splitting, and alignment, would initiate the following processing flow (cf. Fig. 9.22):

- Creation of a list of term candidates. This list incorporates the results of the term extraction proper. Term extraction analyses the input text sentence by sentence, identifies term candidates, and applies filters to them. Such filters consist of a linguistic filter, an optional lexicon filter in cases of out-of-vocabulary words, and a frequency filter.
- Human validation. The list of term candidates is validated by human experts. Their task is to delete improper term candidates and keep the good ones. This task needs special attention as it is the most expensive step in the process.
- Term import. The remaining terms must be prepared for import into the system resources. This task includes proper lemma building, creation of annotations, and producing a proper interchange format.

a. As for the **term candidate list**, it is very important that the term list contains only entries which could be considered as proper terms by the human validators. There are term extraction systems which produce up to 80% noisy entries.[27] Noisy entries are for instance:

- Several inflected forms for a term candidate,
- variants like capitalisation/non-capitalisation, spelling variants,
- duplicate entries,
- entries where the canonical form needs to be built by the validator,
- non-existing ‘multiword’ proposals.

It takes more time to go over such term lists than to go over the original corpus altogether! As a consequence, the quality of the term list requires that all presented term candidates are properly built and are free of noise. This requirement implies linguistic processing of these candidates.

---

[26] E.g. dairy, household, and energy for MULTISENSOR, (Vrochidis et al. 2015).

[27] This is one of the results of the evaluations by Heidemann and Volk (1999), Lieske (2002).

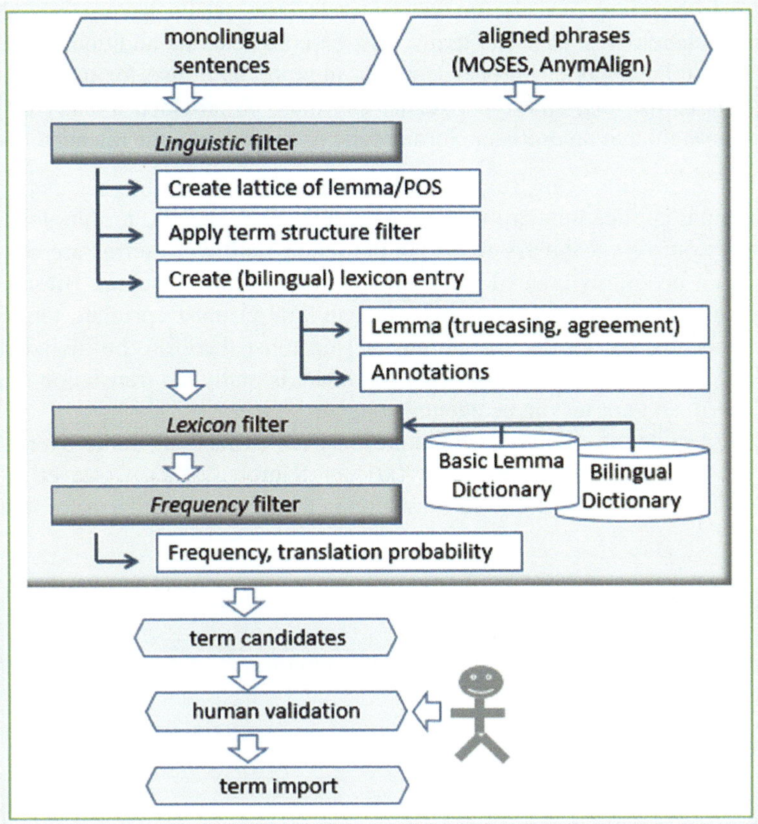

Fig. 9.22 TermExtract workflow

b. The step of **human validation** follows three criteria:

- transparency: Human need to know how the system decision was taken. Therefore they should be able to see the textforms which are subsumed under a term, as well as some example sentences from which the term candidate was created.
- simplicity: It should cost just one click to get rid of an incorrect term candidate, and no action to accept a good proposal
- efficiency: Human validators usually have a limited time budget for the extraction task. Care must be taken to use this budget in an optimal way such that a maximum of coverage is achieved in a given time period. Also, instead of presenting a candidate list for each single file it is more efficient to process a whole corpus at once and present just this resulting list.

The term candidate list must be presented accordingly, and easy options for validation must be offered.

c. **Term integration** must be prepared for all terms which were accepted by the human validation step. First the terms must be completed by additional annotations to be able to be imported. Proper lemmata must be built also for multiwords, and basic features like tag, entrytype, gender etc. must be provided. Finally the entries must be brought into an exchange format for easy import into the intended resources.

2. Term Extraction can refer to monolingual or to bilingual applications. Monolingual applications are used to collect domain-specific terminology and/or to identify out-of-vocabulary terms. In particular multiword terms are of interest, they cannot be simply created from leftovers of dictionary lookup. Bilingual term extraction (sometimes named LtBiExtract in TINA) must correlate terms of the source and the target side. An alignment step must therefore be included in the processing to ensure that a target term candidate is really the translation of a given source term, and not just an accidental cooccurrence.
3. There is quite an amount of literature on term extraction. Some overviews are given in Menezes and Richardson (2001) or Sauron (2002). Costa et al. (2016) review commercial offers for term extraction in the light of usefulness for translators.

9.3.2.3 Workflow

LtTermExtract supports most of the TINA languages.

1. Analysis

1. LtTermExtract uses two kinds of **input**:

- For monolingual term extraction, corpora containing documents in sentence-split form are used, preferably in LtText format.
- For bilingual term extraction, the input are phrase tables. This implies that parallel (or strongly aligned) sentences are required as data source. The preprocessing uses *MOSES*[28] to build phrase tables from them; other tools like *AnymAlign*, cf. Lardilleux et al. (2012), although easier to use, produced less quality results in the tests. The result are bilingual phrases with frequency information and a certain translation probability.

LtTermExtract takes either complete (monolingual) or partial (bilingual) sentences as input. Partial sentences are the result of alignment efforts as given in Phrase Tables.

2. The Linguistic Filter starts by building a **word lattice**, calling the analysis components LtTokeniser and LtLemmatiser described above for tokenisation and lexical analysis. They return a lattice of<lemma, POS>pairs for each token in the input.

[28] http://www2.statmt.org/moses/, Koehn (2010).

Term-de	::=	*AdP? NoC (NoC | NP | PP)?*
AdP	::=	*Ad | VbP*
NP	::=	*Dt (AdP)? NoC*
PP	::=	*(Ap Dt? AdP? NoC) | (ApPD AdP? NoC)*

Fig. 9.23 Multiword patterns for TermExtract (de)

Next, a term pattern matcher is applied operating in a sliding window manner. It consists of tag sequences[29] which identify potential term candidates, both single words and multiwords. Single words are assumed to be nouns (No), verbs (Vb) and adjectives/participles (Ad, AdP); all other tags are discarded.

Multiword structures follow linguistic patterns; this was explained in the section on multiwords above (Sect. 6.1). The same patterns as in multiword analysis are used to identify multiword term candidates. The multiword patterns are collected in lists; each entry contains the pattern and the position of the head. An example is given in in Fig. 9.23. These patterns are of course language-specific, like the whole linguistic filter.

In bilingual extraction, the proper alignment of the term candidates can be assumed to have been achieved in the phrase tables. Based on these data the linguistic filter must be used on both source and target side, using the resources of the respective language (lemmatisers, multiword lists etc.). Of course the patterns of source and target language may be different: A source language single word may correspond to a target language multiword, and vice versa. If no pattern can be found on either side, then the candidate is discarded.

3. If a pattern matches the input, then a proper **term candidate** must be built. This implies normalisation of the term candidate by creating canonical forms for it: **Lemma** creation includes truecasing ($_{\text{en}}$ '**security of European union*'), producing the canonical form of the head ($_{\text{en}}$ '**financial perspectives*'), proper inflection of adjectives in agreement to their head, which requires to know the gender of the head ($_{\text{fr}}$ '**acte liés*', $_{\text{de}}$ '**resistenten Gen*', $_{\text{it}}$ '*cardiopatia coronarica*', $_{\text{es}}$ '*cuestión política*'), and post-nominal parts ($_{\text{es}}$ '**sustancia peligrosa en el trabajos*'). Normalisation is essential as the same lemma may occur in several forms in the corpus, e.g. in several inflected forms or different casing; otherwise the system creates noise, and its acceptance drops.
4. To create a proper entry from a term candidate, additional **annotations** are required. Of course a POS tag is needed. It comes from lexical analysis, and for multiwords the POS tag of the head is inherited. In case gender information is required a gender defaulter as described below can be used. Multiword feature annotations specify the head position (sometimes given in the multiword pattern description), the *mwlemmalist* as built by collecting all lemmata from the candidate nodes, and the *mwposlist* by collecting all BTAG POS elements. Bilingual term candidates can inherit translation probability and frequency from the phrase table input.

[29] Based on the BTAG tagset, with an extension for (inflected) participles.

5. The created term candidates are collected in a **candidate list**, based on a comparison of the lemmata: They are either added to the list, or their frequencies are added up, if already known. This is required to avoid duplicates, and reduce noise in the output. The resulting list presents the results of linguistic filtering.
6. The term candidate list may be **filtered** according to different criteria:

- The use of a **stoplist** is recommended, to free the list from ballast
- If only in-domain terms should be extracted, then a filter can be applied which eliminates general-vocabulary terms
- If just unknown words are to be found, a **lexicon filter** containing an existing dictionary will be applied, both on the monolingual (for the Basic Lemma Dictionary (BLD)) and the bilingual side (for the Bilingual Dictionary (XFR)). It would return a list of unknown terms.

7. At the end of the analysis, a **frequency filter** is applied to the candidate list. It removes all entries which occur less frequently than defined in a threshold. The threshold definition depends on many parameters, one being the size of the Phrase Table, another one the size of the corpus. The resulting list is prepared for human validation.
8. Of course there are many factors which influence the generation of the term list:

- The treatment of **proper names**: They could be left in the term list: Some proper names may be relevant for a given domain, e.g. names of pharmaceuticals in the MED domain, or names of cars or production sites in the AUTOMOTIVE domain, names of politicians in NEWS etc. Improper names can be eliminated by human validators.
- Sometimes **non-terms** pass the term filters which happen to have the 'right' tag structure, like $_{\text{en}}$ '**strategy for example*', $_{\text{it}}$ '**formazione a favore*', $_{\text{de}}$ '**Flüchtlings-Fonds für den Zeitraum*'. Such errors create noise in the output, and must be removed by the human validators
- A more difficult case is the treatment of **inclusions**. These are cases where a multiword pattern includes another one, like $_{\text{en}}$ '*power plant*' vs '*nuclear power plant*', or $_{\text{en}}$ '*Lufthansa Frequent Flyer program*' vs. '*Frequent Flyer program*'. Such phenomena are discussed in both (Frantzi et al. 2000) and (Daille et al. 1996); results are mixed. One option would be to include both terms in the candidate list. In LtTermExtract, the decision is to check the frequency: If the included term is more frequent than the bigger one, i.e. there are also other occurrences of the included term than in the longer one, then both terms are kept. If they always have the same frequency, then only the longer term is kept. However, this heuristics is not safe in all cases.

An example of the final term candidate list is given in Fig. 9.24.[30] This is what human evaluators would receive to work on.

[30]The example does not show sentences as the phrase table alignment does not provide sentence information.

Term it	POS it	val.	Term de	POS	Frq
educazione sessuale	No		Sexualerziehung	No	37
festa nazionale	No		Nationalfeiertag	No	37
genetica umana	No		Humangenetik	No	37
momento epocale	No		historischer Moment	No	37
richiesta di supporto	No		Antrag auf Unterstützung	No	37
rifugiato afgano	No		afghanischer Flüchtling	No	37
volume di rifiuti	No		Abfallmenge	No	37
parta israeliana	No		israelische Seite	No	36
sicurezza di stabilità	No	x	Sicherheit Stabilität	No	36
aliquota centrale	No		Leitsatz	No	35
arresto in massa	No		Massenverhaftung	No	35
industria bovina	No		Rindfleischindustrie	No	35
inquisizione	No		Inquisition	No	35
latino	Ad		Lateinisch	Ad	35
preselezione del vettore	No		Betreibervorauswahl	No	35
regolamentazione ragionevole	No		vernünftige Regelung	No	35
regolamento ragionevole	No		vernünftige Regelung	No	35
appoggio critico	No		kritische Unterstützung	No	34
fiscalità sul risparmio	No		Besteuerung von Zinserträgen	No	34

Fig. 9.24 Example of term candidate list, with a mark of a non-term

2. Human Validation

The term candidate list is presented to human experts for validation. To make this task as easy as possible the list must be as clean as possible: No duplicates, no different textforms, no improper POS structures etc. In addition LtTermExtract should present what is needed to decide on a given term. Therefore each entry of the candidate list shows the following items:

For monolingual term extraction:

- the lemma of the candidate
- all textforms from which the lemma is generated
- some example sentences where these textforms were found
- the part-of-speech of the lemma
- its frequency.

For bilingual term extraction, no example sentences can be made available by the phrase tables, so only source and target terms are presented with lemma and POS information.

These items are presented in a table format. An example is given in Fig. 9.24.[31]

[31] It shows only a minimal setup: Source lemma and POS, target lemma and POS, frequency, and a column to mark the candidate as 'to delete'.

The candidate lists are sorted downwards to present the most frequent candidates first. Validators can then work down the list as far as time constraints permit, and be sure that the best coverage can be achieved in the given time period. They mark the entries which they do not want to stay in the list, this is a one-click task. In the end all marked entries are deleted.

3. Term Import

The remaining ‘good’ candidates are prepared for import into the system resources. There are special formats for term import, like TBX,[32] OLIF,[33] or XLIFF.[34] Most of them are based on tab-delimited records; and this is the format generated by LtTermExtract.

The import of a set of new entries is the easier the more annotations these entries can provide. To complete the definition of a concept, in addition to lemma and POS, a semantic annotation can often be provided in the form of a domain annotation, which the corpus collection could contribute. Beyond these obligatory features, the term import could also try to provide additional relevant features. Two such features were investigated: gender and inflection class. Both annotations are based on the local defaulting mechanism described in Sect. 5.2.5 above.

The same approach to use discrimination trees can also be applied to the defaulting of features. Instead of providing pairs of <textform, POS> as input to the defaulter, pairs of <textform, gender> or <textform, lemma-instruction> would be learned by the training component. The training data would again be restricted to the open word classes.

1. To learn **gender** annotations, the gender-bearing POS tags (adjectives, common nouns, inflected verb participles) are used. Gender annotations were created for the Roman languages (Spanish, French, Italian, Portuguese).

There is a difference between POS annotation and gender annotation: While POS defaulting can return alternatives as the following taggers can do the disambiguation, this does not hold for gender (and the other feature) annotations. Only one value can be returned.

Figure 9.25 shows results for gender defaulting. It can be seen that about half of the ending strings can be disambiguated. For the other half, either a random value can be taken, or the relative probability of the two gender values can be used. Overall about 75% correct assignments can be assumed. However, human validation would be required to create a clean resource.

Inside the system, gender information is relevant when proper lemmata for multiword need to be created: $_{es}$ ‘*animal infectada’. If the head noun of such a multiword term is unknown, a gender defaulter may still be a useful tool.

---

[32] TBX (2021).

[33] OLIF (2002) http://www.olif.net/

[34] XLIFF (2008).

language	es	fr	it	pt
# lines for training (k)	149.1	71.9	227.6	125.5
# lines gender-defaulted (k)	22.5	20.9	53.4	32.5
# lines disambiguated (k)	10.6	12.2	27.9	15.8
in %	47.1	58.4	52.2	48.6

Fig. 9.25 Gender defaulter

2. Learning **inflectional classes** would be a very helpful activity, as inflection paradigms usable in the creation of full text dictionaries could be achieved. While intuition can determine valid proposal for certain endings (like $_{\text{de}}$ *-ation: feminine, pural = -ationen*) it is unclear which coverage could be achieved. It may be more advisable to default only the safe entries and leave the unclear cases to human expertise than to default all entries and leave the human evaluators with the task to validate each single entry.

9.3.2.4 Evaluation

LtTermExtract has been used in many projects, in most cases to produce domain or customer specific terminology for import into MT dictionaries in translation projects. Evaluation was a special topic in the projects PANACEA[35] and ACCURAT.[36]

1. Data

Both projects needed bilingual data, so the bilingual Version of LtTermExtract was evaluated. Figure 9.26 gives the corpora and languages of the tests: Texts on environment and labour legislation, both en-fr; texts on health & safety regulations in several languages, different sets of automotive texts (de-en).

From all corpus data sets, term candidates were extracted by the term extraction system. From these candidates, term candidates were selected randomly. These candidates were evaluated manually by two evaluators.

Overall, 99 K bilingual term candidates were extracted of which 17.2 K (17%) were manually evaluated; details are given in Fig. 9.26.

[35] PANACEA, cf. Bel et al. (2012).

[36] ACCURAT, cf. Skadiņa et al. (2019).

Corpus	Lang	Sentences (K)	Phrase Table size (K)	Glossary size (K terms)	Translation errors (%)	TermExtract errors (%)	Total errors (%)
environment	en-fr	29	0.4	2.8	5.2	1.3	7.8
labour	en-fr	21	0.8	4.5	4.9	1.2	7.3
health&safety	en-fr	52	2.9	10.7	11.3	1.3	13.9
health&safety	en-es	48	2.6	13.2	10.9	0.4	11.7
health&safety	en-it	40	2.1	9.9	9.8	2.3	14.4
health&safety	en-pt	14	0.6	4.4	12.7	0.4	13.5
automotive1	en-de	155	7.9	15.7	5.7	2.8	11.3
automotive2	en-de	1483	85	23.2	1.5	3.3	8.1
automotive3	en-de	1595	83.9	23.3	1.7	3.1	7.9

Fig. 9.26 TermExtract evaluation

2. Quality

The basic quality measure in term extraction is recall and precision.[37] Both measures are defined in terms of relevancy, i.e. which relevant terms are found in the total document set and in the retrieved term set. The fundamental problem with this approach is that it is very difficult to define what relevant terms are. As pointed out above, the concept of relevancy differs strongly depending on the intended application. Attempts have been made to define 'termhood'[38] but the problem remains. And while there is an intersubjective intuition of what a possible translation equivalent of a given term in the target language is, there is no such intuition as to which terms should considered to be relevant.

However there is some common opinion what an extraction component should *not* produce, i.e. about the errors of such a component. Cases like duplications, incorrect lemmata, noise in lemmatisation, incorrect translations etc. are commonly considered as errors. Such an error analysis has also been undertaken for LtTermExtract. The results are also given in Fig. 9.26. They start from aligned sentences, give the size of the phrase tables created, the glossaries built with LtTermExtract, and the errors detected.

Two kinds or errors are relevant for the evaluation:

- **Translation errors**, i.e. the candidates are not translations of each other. These errors are produced by the aligners, as explained above. Translation errors seem to correlate with the size of the phrase tables: Larger phrase tables show a lower translation error rate for the extracted terms. This is not particularly surprising, as more data usually lead to better performance.

---

[37] Cf. Heidemann and Volk (1999), Sauron (2002), and most others.

[38] Cf. Vu et al. (2008), Wong et al. (2007), Kit (2002).

Translation errors are produced by the phrase table alignment,[39] and are not accessible to the LtTermExtract tool; however, they contribute to the total error rate.

Translation errors of course are relevant only for the bilingual version of LtTermExtract.

- **Lemma and annotation errors**; these errors are created by the LtTermExtract tool. They are obviously language-specific; an error analysis is given below.

LtTermExtract errors vary from 0.4 to 3.3%, depending on the languages involved,[40] with an average error rate of 2.1%. Main of errors are:

- errors in **linguistic filtering**: either homograph words pass the filter ($_{en}$'*are permanent*' as '*are*' etc. has also a noun reading; similar $_{it}$ '*sono*' in '**sono piccolo*', etc.). Or patterns pass the filter which are no terms but happen to have the 'right' structure: $_{en}$ '**strategy for example*', $_{it}$ '**formazione a favore*', $_{de}$ '**Flüchtlings-fonds für den Zeitraum*', as mentioned above.
- errors in **lemma creation**: either errors in truecasing ($_{en}$ '**fujitsu*', '**flemish port*'), mostly due to lexicon gaps, or errors in agreement, ($_{de}$ '**freundlicher Wort*', $_{fr}$ '**force élevées*', $_{es}$ '**animal infectados*').

Many of these errors can be corrected by improvements of the backend components (dictionary, gender defaulters etc.), which would bring the LtTermExtract error rate down by an estimated 1%.

The LtTermExtract errors do not depend on the size of the data. They are of course language-dependent: Errors in German result from more complicated gender agreement; in Italian homograph problems, in English casing problems are the main source of errors. Variations of error rates within one language in the different test sets do not seem to be significant.

The total error rates can be given as follows:

- In monolingual settings the translation errors of course are irrelevant as no translations are produced. Here only the errors of the component itself count, with rates between 0.4% and 3.3%
- In bilingual settings the translation errors have to be taken into account as well. If a bilingual dictionary is envisaged, the error rates accumulate, so the overall error rate of the tool is two LtTermExtract errors (for source and target side) plus the translation errors; the total error rate is somewhat linear to the translation error rate. In total it is between 7.3% and 14.4%. This means that for bilingual settings 8 entries out of 100 need to be corrected by human reviewers.

This can be considered a reasonable result of a term extraction component.

---

[39] With a translation probability threshold set to 0.6 and a frequency threshold set to>1.

[40] LtTermExtract supports the languages en de fr es it pt.

3. Usability Revisited

From a usability point of view, there are several aspects of term extraction which are important:

1. Can users run the tool at all? Which hardware resources are required to run the tool? Depending on the frequency filter, the system processed between 45 K (no filter) and 170 K (0.8 filter) entries per second on a standard PC (of the year 2015). This would be fast enough for practical use.
2. Does the tool produce all candidates? If not, the texts would have to be looked at by human validators. Tests showed that lowering the probability threshold in the phrase table would increase recall, but also increase the candidate list. A compromise using a translation probability of 0.8 leaves about 15% of term candidates undetected. Details are given in Thurmair and Aleksić (2012).
3. How **good** are the results? If term extraction is made for external clients (e.g. as part of a maintenance contract) then customers could ask for a high quality level and complain if too many erroneous entries are delivered. The answer depends on the load for the human validators.
4. How much **effort** is involved in validating the term candidate list? Usability here depends on two factors, namely noise production and speed of editing.

- How much **noise** does the tool produce? I.e. how many irrelevant candidates need to be inspected and then eliminated? This again depends on two other factors, namely how many candidates are produced, how many of them are incorrect and just create noise, and how many of them can be filtered before the validation run, e.g. by a lexicon filter (or simply a stop filter for noisy words). So noise reduction is an important factor in usability comparisons.
- How **fast** can a correct term list be produced? This involves editing issues (e.g. if a term candidate is already in base form, it needs not be edited) and also easy deletion of noise candidates. A term extraction tool should provide good support for doing this.

Experience in customer translation adaptation projects using LtTermExtract has shown that candidate lists on average comprise between 5000 and 15000 entries, and can be validated in 2–4 person days with this tool.

5. How can people make use of the term extraction output? This aspect is important because term extraction is only one step in a larger processing chain, which after cleaning up the extraction results imports the 'good' terms into the available global resources.

As LtTermExtract offers a tab-delimited interface there is no big technical restriction. Of course contentwise there are the usual challenges: producing (pseudo) duplicates, adding missing annotations, running consistency and validation tools and the like.

9.3.3 *Word Sense Disambiguation (LtSensExtract)*

This component has been extensively described in the chapter on word sense disambiguation (Sect. 7.2). So only the general outlines are repeated here.

1. Input: First input is a bilingual dictionary. The translations a a given term are taken as different senses. Second input is a parallel corpus, aligned on sentence level.
2. Creation of subcorpora: Sorting on the source language side creates subcorpora for the *terms*; sorting of these subcorpora on the target side creates subcorpora for each *sense* of the respective term.
3. Validation: GIZA++is used to ensure that the terms are really translations of each other, not just coincidental pairings.
4. Identification of contextual terms in each sense subcorpus, by simple co-occurrence and frequeny analysis
5. Collection of the found terms in the Word Sense Dictionary.

As described above, the main challenge was data sparsity. But even then the results, measured in correct assignment of translations as representing senses were better than comparable translation systems. Data sparsity requires the system to be open to user adaptations and additions.

Details, figures, and test results are all given in Sect. 7.2. The structure of the Word Sense Repository (WSR) is presented in Sect. 9.2 above.

A fallback strategy, taking simply the frequency of the target language terms, still outperformed the then available rule-based systems.

9.3.4 *Inflection Generation (LtFlexer)*

An inflection component generates inflected forms for a given lemma. It knows how to produce all morphological variants of it, or it returns a proper textform depending on a specified set of feature annotation (e.g. 'third person singular past tense indicative'). As input it uses a lemma (base form) and a specific feature configuration or an inflection class. Such an inflection class is an instruction how the different forms for a lemma must be built.

Of course such an inflection device is language dependent, i.e. it must be built for each new language. LtFlexer supports German, English, Spanish, French, Italian and Portuguese. And of course it depends on the part of speech of the lemma: noun inflections differ from verb inflections. LtFlexer creates inflected forms for nouns, verbs, and adjectives; other parts of speech, in particular function words, are treated differently in the language processing environments.

The inflection generator plays an important role in the TINA environment. It links the lemma-based Basic Lemma Dictionary (BLD) to the Full Word Dictionary (FWD) as used in the lexical analysis component of many languages, as described above in Sect. 5.2.

It also completes the morphological description of the dictionary entries themselves in the Basic Lemma Dictionary (BLD) by instantiating the inflection-specific features (case, gender, number, tense, voice etc.).

And it supports tasks like the creation of multiword lemmata e.g. in the LtTermExtract component where some multiword parts need to be inflected in order to produce a proper canonical form.

The LtFlexer component consists of two basic parts:

- a training component which extracts inflection information from training data and creates inflectional class information for a given lemma
- a runtime part which creates a specific form for a feature configuration, or all inflected forms of a lemma, based on the lemma and its inflection class.

Outside of the TINA project, it was also used as a stand-alone component in the area of Machine Translation dictionary management.

9.3.4.1 LtFlexer Training

LtFlexer creates inflectional classes for content words (nouns, verbs, adjectives) of a given language. The idea is.

- to create an inflection table containing all possible inflected forms of a given part-of-speech,
- to fill all cells of this table with forms of the training data, with one line for each lemma,
- to create inflection instructions for each cell of each row in the table, by comparing the inflected form with the lemma,
- to conflate all lines with the same lemmatisation instructions into an inflection class.

The resulting class information is added as annotation to the entries in the Basic Lemma Dictionary, and can also be used to default inflection classes for hitherto unknown words (Fig. 9.27).

1. Training Data

The training data consist of lists of textforms of a given language. The textforms should be annotated with categorial information. The information used in TINA is described in the extended tagset (XTAG) in Sect. 2.3.2 above; other annotation schemata can be converted into this scheme.

The LtFlexer uses several thousand such entries for training, details are given in Fig. 9.28.

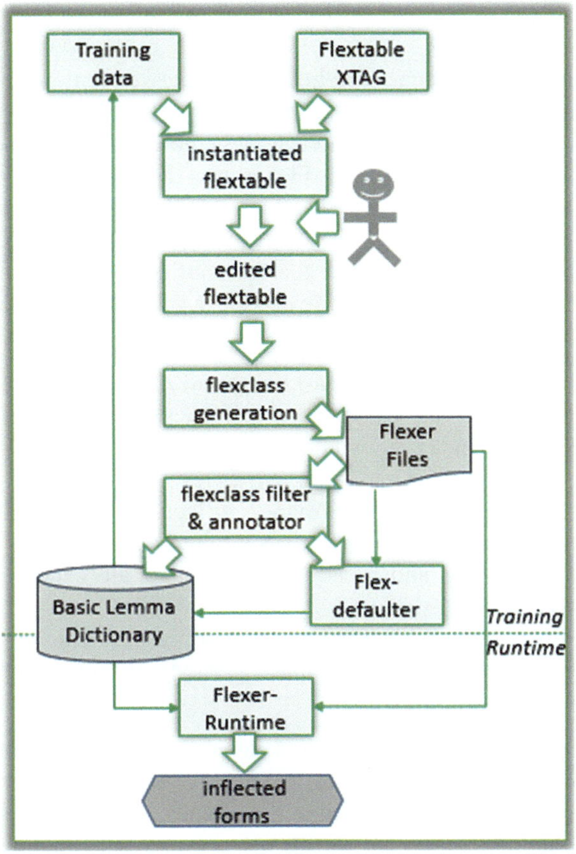

Fig. 9.27 LtFlexer structure

training data	de	en	es	fr	it	pt
nouns	104.8k	21.3k	27.5k	13.4k	54.9k	26.5k
adjektives	21.6k	6.7k	10.3k	1.2k	4.8k	2.4k
verbs	11.6k	8.9k	3.6k	3.7k	6.5k	6.8k
total wordforms	138.1k	36.9k	41.6k	18.4k	66.4k	35.7k

Fig. 9.28 LtFlexer training data

The training record should also provide the lemma for the textform; if not available, a lemmatiser could be used.

All training records are brought into a UTF8 tab-delimited format containing three elements: textform, POS (in XTAG form), lemma.

2. Inflection Table

Nearly all traditional grammars use inflectional classes to describe how inflected words are formed: Verbs ending in $_{\text{it}}$ *-are* form one class, those on *-ere* another one, those on *-ire* a third one etc. For each class the grammars present tables how a given member of this class inflects. LtFlexer uses the opposite way: It tries to conclude from inflected forms the inflection class to which they belong.

The first step in this approach is to create a table for each POS of each language. It contains a column for each feature configuration forming an inflection. This step requires an analysis which features a given language uses for inflection, and which values must be considered.

- For nouns it would be *gender* (mask/fem) and *number* (sing/pl), four fields in total for Italian, German would need six (mask/fem/neutr). Languages with morphological *case* marking would take this also into account ($_{\text{en}}$ '*house*' vs '*house's*', $_{\text{tr}}$ '*evden*' '*evi*' etc.
- Adjectives would need to provide inflections for adverb forms ($_{\text{fr}}$ '*égal*'—'*également*') and comparatives ($_{\text{en}}$ '*easy*'—'*easier*'). German adjectives can have strong and weak inflection and all gender—number—case combinations.
- Verbs would need to provide combinations of person—number—tense—mood, as well of non-finite forms: participles, gerunds, infinitives ($_{\text{en}}$ '*produced*', $_{\text{de}}$ '*abzugeben*').

As can be seen the linguistic description determines the shape of the inflection: Each inflection table for each of the languages' part-of-speech looks different. There was a significant effort in finding common definitions for those paradigms, resulting in recommendations like EAGLES or Universal Dependencies, for multilingual setups. The TINA XTAG tagset (cf. Sect. 2.3.2 above) is based on such research.

There is a difference between the inflection tables of LtFlexer and the traditional grammar descriptions: The LtFlexer only handles single word forms. Inflections involving auxiliaries ($_{\text{en}}$ '*has built*', $_{\text{de}}$ '*hat gekocht*') or particles ($_{\text{en}}$ '*more exhauseted*', $_{\text{it}}$ '*più grande*') are not taken into account as they require different treatment in computational analysis altogether. So all cells of inflection tables which would contain compositional elements are neglected in the inflection tables.

An issue here is which features constitute an inflection paradigm. In particular gender information could be questioned as gender changes for adjectives ($_{\text{fr}}$ '*bon*'—'*bonne*') but is invariant for nouns: $_{\text{it}}$ '*tornio*' is maskuline in any inflection configuration.[41] LtFlexer keeps gender as inflectional feature also for nouns, with the consequence that the cells for the other general stay empty in the instantiations of the inflectional tables.

[41] There are polysemes of course ($_{\text{de}}$ '*Mast*', $_{\text{it}}$ '*banco*'—'*banca*').

3. Instantiation

Once the inflectional tables have defined which inflected forms are morphologically possible the next step is to fill the training data into these tables. Each lemma of the training data constitutes a table row; the XTAG information defines which position in the table line the respective textform will take.

The result of this activity will be:

- There are lemmata which can provide inflected forms for *all* positions.
- There are lemmata which have *empty fields* in their inflection table line
- There are lemmata which have *more than one* inflected form on a given position.

Accordingly there are three types of records in the output: good records, deficient records, and duplicate records.

Deficient records need a decision whether or not an empty field is intentionally left empty, or if it is just a flaw in the data. Examples of fields which were intentionally left empty are: Singularia and pluralia tantum ($_{de}$ *'Obst'*—* *'Obste'*) Spanish feminine gender inflections for masculine nouns ($_{es}$ *'filme'*, English compositional comparative forms ($_{en}$ *'beautiful'*—*'*beautifuller'*), adjectives without comparison forms, and other such cases.

Duplicate cases describe inflection variants, i.e. more than one inflected form for a given position in the inflection table, like $_{en}$ *'leveling'* vs. *'levelling'* for gerund forms, $_{de}$ *'Bänke'* vs. *'Banken'*, $_{en}$ *'indexes'* vs *'indices'* or $_{en}$ *'lemmas'* vs. *'lemmata'* for nouns. There are different phenomena behind such cases, they need a more detailed inspection. It could be argued that for a generator it would be sufficient to generate just one form; however if the result is used for a Full Word Dictionary (FWD), then better coverage would be achieved, if all variants could be recognised.

4. Validation

It could be argued that the process of inflection generation could be continued without such ill-formed lines, just with less data. However,

- such lemmata could then not be inflected at all, even for forms which are available
- in many cases the deficient records outnumber the good ones, in particular in cases of systematic gaps (like in the gender example above). In particular the adjective patterns are not complete in most cases, for instance missing instantiation of comparatives in adverbial forms (for instance $_{es}$ *'visualísimamente'*)

Therefore it turned out to be a better option to inspect the file containing the deficient inflection table lines, and decide on each position how to handle it: If there is just a flaw in the training file the position is completed manually; if there is an intentional gap, then a special mark is set to prevent the position from being further processed.

For the cases with more than one word on a given position several options can be followed. Mostly there are rather systematic variations, often in the context of normalisation. One could think of an 'inverted' normalisation, creating all non-normalised forms from normalised ones. Also duplicating the inflection line for a given lemma is an option, in particular if the inflection behaviour mirrors a semantic distinction.

It turned out that the majority of the records of the inflectional tables had missing entries, and needed manual inspections. As most cases were systematic (gender with nouns, comparatives with adjectives etc.) could be quickly validated. After inspection all these records can be merged with the entries with complete inflection table lines to produce the input file for the definition of the inflection classes.

5. Generation of Inflection Classes

The inflection classes are generated in two steps:

1. In a first step all inflected forms are replaced by **inflection instructions**. Such instructions can be produced automatically, by using the cut&append technique as described in the chapter on lexical analysis (Sect. 5.2). It is based on an automatic comparison between inflected form and lemma: Just as a lemma can be created from a textform (inflected form), an inflected form can be created from a base form. The cut is computed as the first non-matching character in the comparison of the two forms, and the remaining string of the inflected form needs to be added.

The append strings are collected in a special table, and replaced by a pointer to this table in the instruction. The result is an inflection instruction consisting of two numbers, one the cut, the other one a pointer to the append string table.

It should be mentioned that the cut&append approach is much faster and requires much less computational resources than comparable approaches. Basically it requires two hash accesses and a bit of string append operations. This is a big advantage in industrial environments.

2. The second step collects all table lines containing identical inflection instructions and collects them into inflection classes. The resulting list is sorted by the number of their members, and the most populated classes come first.

The result of the generation of inflection classes consists of three files:

- one file assigns an inflection class to every lemma seen in the training file. This information will later be imported into the Basic Lema Dictionaries (BLD) which serves as the main collection of morphological annotations.
- one file describes the inflection classes, by providing the two-number inflection instructions (cut&append) for every position in the inflection table.
- one file contains all append strings required in the process.

This procedure leaves some points to discuss:

1. The **names** of the classes are just numbers. This is only a good option if the class names are not seen by users. An alternative which can often be observed in grammars is to name classes after a representative, used as value for a feature '*inflects_like*'.
2. The procedure is quite formalistic. The number of inflection classes is much higher than in ordinary grammars. This is due to the fact that it cannot abstract away. For instance, de '*Tanz—Tänze*' and '*Bach—Bäche*', both inserting umlaut and adding an 'e', fall into different classes as they have different append strings ('*-änze*' vs. '*-äche*').
3. The procedure mirrors the heuristic that most inflection changes can be found at word endings. However there are many phenomena which contradict this assumption: Changes in stems like umlaut/ablaut ($_{en}$ '*go*'—'*went*'—'*gone*'), inflection-triggered prefixations, internal inflection ($_{en}$ *editor-in-chief vs. editors-in-chief*, $_{pt}$ *aguapés-brancos* vs. *aguapé-branco*), changes in accentuation ($_{es}$ *vocálico vs vocaliquísimo*). Such phenomena can of course be handled but they produce a significant number of classes and miss important generalisations.

An option to reduce the consequences has been to annotate the inflected forms in the inflection tables, by suffixing them with hints how to process them: de '*Bache+e*' and '*Tanze+e*' instead of '*Bäche*' und '*Tänze*' would fall into the same class then, and so would '*laufen+ge*' and '*kommen+ge*' for '*gekommen*' and '*gelaufen*'. The inflection generator would need small functions to handle such inflection hints.

4. Figure 9.29 shows the number of inflection classes for the languages in question.

The following observations can be made:

1. Most of the classes result from phenomena just described: internal inflection in English and Portuguese nouns, accents in Spanish adjectives etc.
2. The procedure clearly favours the exceptions over the regular cases: A closer look on the resulting classes reveals that classes with just one member (i.e.

no. flex classes	de	en	es	fr	it	pt
nouns	115	149	61	122	242	412
adjectives	20	45	500	35	96	147
verbs	195	156	127	106	72	75
total classes	330	350	688	263	410	634

Fig. 9.29 Number of inflection classes

coverage top 10 classes	de	en	es	fr	it	pt
nouns	84,2	96,6	98,4	98,1	95,4	92,6
adjectives	99,2	96	75,5	94,1	97,8	83,4
verbs	80,3	92,3	88,8	95,9	97,3	96,8
average coverage %	87,9	94,9	87,6	96	96,8	90,9

Fig. 9.30 Coverage of the top 10 inflection classes

class	example	frq	posit.	comp.	superl.	comment
0	fern	15155	0:0	0:1	0:5	Standard, 75% of all words
1	fest	5288	0:0	0:1	0:2	e-Insertion in superlative: festeste
2	klug	257	0:0	0:3u	0:6	umlaut: klügstes
3	leise	142	0:9	0:10	0:5	e-Elision in positive: *leise-es
4	dunkel	117	2:7	2:8	0:5	e-Elision: before l in positive: *dunkeles
5	kalt	109	0:0	0:3u	0:11	combi of 1 und 2: kälteste

Fig. 9.31 Example: top inflection classes for German adjectives

idiosyncratic formation) have a share between 16% (German[42]) and 66% (Italian) of all inflection classes (Fig. 9.30). Overall 30 to 50 percent of the classes have only one member, mainly due to exceptions and idiosyncratic append strings.

Such classes are not productive, so they are not used for an inflection generator. Such classes can be collected in special exception lists, accessed if those particular words would have to be inflected. But no new incoming words would have to be formed along such patterns.

3. On the other hand, when looking at the productive classes it can be seen that an approach using the 10 top classes would produce quite acceptable results, as shown in Fig. 9.31. It shows which percentage of the training words are covered by these 10 top classes.

The table shows that only a few classes cover the lion's share of the words of the training words. They seem to represent the most productive inflection classes.

This observation is relevant as there is a missing point in the component: All words found in the training files have been assigned an inflection class, but there are other words, both in the dictionaries and in the out-of-vocabulary field which also would need an inflection instruction.

6. Dictionary Information and Defaulting

The words which have an inflection annotation in the LtFlexer files can just copy this information into the Basic Lemma Dictionary (BLD).

[42] The rather low figures for German result from the fact that there the above-mentioned heuristics for umlaut and ablaut vowel changes were implemented, which reduces the amount of single-word classes.

For the other words and the unknowns a component is required which defaults the inflection information. The defaulter learns from pairs of<lemma, inflection class>to which inflection class an unknown input word belongs. Like for the other defaulters (part-of-speech, cf. Sect. 5.2.5; gender cf. Sect. 9.3.2) heuristics on word endings are generated. For instance, Italian nouns ending in '*-sco*' insert '*h*' in plural, cf. $_{\text{it}}$ '*bosco*' vs '*boschi*', German nouns ending in '*-ung*' form their plural with '*en*': $_{\text{de}}$ '*Zeitung*' vs. '*Zeitungen*'. The inflection defaulter only uses the top inflection classes for defaulting as they cover most inflection patterns, and it is assumed that unknown words are more likely to follow one of those patterns because they are more productive.

The mode of procedure of the defaulter learning programs is described in Sect. 5.2.5. and needs not be repeated here. Its output is an inflection class for a given input word. It should be kept in mind that the inflection defaulter needs not just a word but also its part-of-speech and possibly (for nouns) its gender as input parameters. Such annotations would require to call the POS defaulter and the gender defaulter first, before the inflection defaulter can work.

Tests on the defaulter efficiency were made using words in the dictionaries which were not in the training data list. As most of the common words are already covered in the training data, the out-of-vocabulary words of the large scale lemmatiser tests (cf. Sect. 5.2.6) were used.

The results show that the defaulter in most cases returns correct inflection classes, with error rates between 4% (English) and 8% (Spanish). However it also shows that the assumption that unknown words often follow the productive patterns does not always hold: There are many foreign language words, proper nouns and other cases of irregular input. Such forms deteriorate the inflection defaulter's quality.

Once all entries in the Basic Lemma Dictionary (BLD) have obtained an inflection class annotation they can be used to generate the dictionary of full words (FWD).

9.3.4.2 LtFlexer Runtime

The runtime component of LtFlexer generates inflected forms from base forms, using the inflection class information.

1. The generation part of LtFlexer supports in two modes:

- a *precision* flexer produces the inflected form for a given feature configuration (e.g. $_{\text{tr}}$ '*ev*'+dative—plural ->'*ev-ler-e*'). Applications are systems which generate sentences and require a particular form, like in question answering or speech synthesis systems.
- a *generic* flexer produces all inflected forms of a given lemma. Applications are spell checking lists or dictionaries for retrieval components like in TINA.

Which mode is used is controlled by a parameter in the LtFlexer call.

2. The runtime version is called with the following parameters:

- the lemma to be inflected
- the inflection class to be used
- the scope, describing the output mode (generic/precision)
- the part of speech of the lemma (BTag)
- the gender (if not included in the inflection class)
- a feature configuration to be produced (for the precision flexer).

In case part-of-speech and gender are not available they will be defaulted by the LtFlexer. Even if the inflection class is not available, it can be defaulted at runtime.

3. The processing depends on the scope defined in the input:

The line in the inflection table defined by the inflection class is retrieved; it contains all inflection instructions for the given class.

- for *precision* inflection, the position describing the specified feature configuration. The inflection instruction is executed, and the resulting inflected form is returned.
- for *generic* inflection, all lemmatisation instruction of the respective table line are created. Duplicate entries are removed, and the remaining inflected forms are returned.

Examples are given in Fig. 9.32.

Postprocessing would include truecasing and creation of variants of the type en '*leveling*' vs. '*levelling*'.

4. The error rates of the generation component of LtFlexer depend on the respective lemma:

- many lemmata of the BLD dictionary have already be used in the training files. This is unavoidable as the training files are large and contain a significant portion of the vocabulary of a language.
- other lemmata have not been seen in the training and are produced by the defaulter. Of course error rates are higher there.

While the error rates for the words in the training files are less than 0.5% the rates for the unknown words mirror the defaulter results just described (between 4 and 8%). The error rates of the two data sets combine for a given language depending on the amount of lemmata already used in the training compared to hitherto

Instruction						Result	
dog	+	n	+	np	=>	dogs	(precision flexer)
slow	+	a	+	adv	=>	slowly	
dog	+	n	--		=>	dog dogs dogs' dog's	(generic flexer)
slow	+	a	--		=>	slow slower slowest slowly slowlier slowliest	

Fig. 9.32 Runtime flexer output examples

unseen ones. As the training data comprise a significant share of the total vocabulary, the total error rates would be between 3–5%.

Overall it can be seen that the generation of inflected forms produces a bit of noise, however less than in comparable components. In case of a dictionary validation, the defaulted forms would have to undergo manual inspection.

9.3.4.3 Discussion

1. The technique of creating **patterns** for inflection classes is well known in traditional grammars. To economise them they identified key forms for comparison, for instance in German strong verbs: '*gehe*'—'*ging*'—'*gegangen*': Many of the other elements of the inflection table follow this pattern.

Such a economisation, reducing the number of elements of the inflection tables, would also help in the LtFlexer training part.

2. A significant issue are **nonstandard inflection patterns** using different kinds of non-final strings, like infixes, prefixes, ablaut etc., examples are given above, others are $_{\text{it}}$ '*capogrupo*' vs. '*capigruppo*', $_{\text{en}}$ '*goose*'—'*geese*', $_{\text{fr}}$ '*beau*'—'*belle*', $_{\text{de}}$ '*geben*'—'*abgegeben*'. They can easily be handled on a cut&append base, however as they create idiosyncratic append strings the result would be as many inflection classes as there are words of this kind.

A way around this has been sketched above (in the case of umlaut), another one would be to use the allomorph information in the Basic Lemma Dictionaries and create inflection patterns for each of the allomorphs.

3. **Inflection variants** are not handled in the current component. If variants result from different concepts, the dictionary would foresee two entries: $_{\text{de}}$ '*Bank*'/'*bank*'—'*Banken*' vs '*Bank*'/'*bench*'—'*Bänke*'. However there are also variants below this level with dialectal, regional, sociological differences, the best known examples being US and UK English ('*familiarise*' vs. ‚*familiarize*'), others would be different placement of accents in European and Brazilian Portuguese.

Such variants could be handled by a postprocessing component, adding variants of a given inflected form to the forms produced by a generic flexer. It would be less relevant for a precision flexer as one produced form may often suffice, however a generic flexer would have to produce all forms in case they must be identified in some input text. It would be a kind of inverted normalisation component ('de-normaliser').

4. **Multiword inflection** is also an issue to consider. Not just the heads need to be inflected but potential adjectival modifiers as well: it '*imposta patrimoniale*'—'*imposte patrimoniali*' needs to inflect both parts. The investigation of this phenomenon, like many other issues, is left to further versions.

References

Arppe, A. (1995). Term Extraction from Unrestricted Text. *Proc. Nordic Conf. on Computational Linguistics (NODALIDA) 1995, Helsinki*

Bel, N., Papavasiliou, V., Prokopidis, P., Toral, A., & Arranz, V. (2012). Mining and Exploiting Domain-Specific Corpora in the PANACEA Platform. *Proc. 5th Workshop on Building and Using Comparable Corpora at the Eighth Int. Conf. on Language Resources and Evaluation (LREC-2012); 2012 May 23–25; Istanbul, Turkiye.*

Berman, M.L., Mostern, R., & Southall, H. (2016). Placing Names: Enriching and Integrating Gazetteers. *Indiana University Press*

Boguraev, B., Pustejovsky, J., eds. (1996). Corpus Processing for Lexical Acquisition. *MIT Press*

Bourigault, D. (1995). LEXTER, a Terminology Extraction Software for Knowledge Acquisition from Texts. *In Proc. of the 9th Knowledge Acquisition Workshops and Archives (KAW), Banff, Canada.*

Bourigault, D., Jacquemin, Chr., L'Homme, M.Cl., eds. (2001). Recent Advances in Computational Terminology. *Amsterdam (John Benjamins)*

Brinkmann, K.H., Blaha, H. (1997). Wörterbuch der Daten - und Kommunikationstechnik. Data Systems and Communications Dictionary. German-English, English-German. *Oscar Brandstetter Verlag, Wiesbaden*

Cabré Castelli, M.T., Estopá Bagot, R., & Vivaldi Palatresi, J. (2001). Automatic term detection: A review of current systems. *In: Bourigault, D., Jacquemin, Chr., L'Homme, M.-Cl. (eds.), Recent Advances in Computational Terminology. 2001*

Caseli, H., & Nunes, M. (2006). Automatic induction of bilingual resources for machine translation: the ReTraTos project. *Machine Translation 20,4, 2006*

Cimiano, P., Buitelaar, P., McCrae, J., & Sintek, M. (2010). LexInfo: A Declarative Model for the Lexicon-Ontology Interface. *in: Web Semantics 9,1 pp. 29-51*

Costa, H., Zaretskaya, A., Corpas Pastor, Gl., & Seghiri, M. (2016). Nine Terminology Extraction Tools: Are they useful for translators?. *MultiLingual 04, 2016*

Daille, B., & Morin, E. (2005). French-English Terminology Extraction from Comparable Corpora. *Proc. 2nd Int. Joint Conf. on Natural Language Processing, JeJu Island, Korea*

Daille, B., Habert, B., Jacquemin, Chr., & Royauté, J. (1996). Empirical observation of term variations and principles for their description, *in: Terminology. Int. Journal of Theoretical and Applied Issues in Specialized Communication, Volume 3, Issue 2, Jan 1996, pp. 197 - 257* https://doi.org/10.1075/term.3.2.02dai

de Marneffe, M., Manning, Ch., Nivre, J., & Zeman, D. (2021). Universal Dependencies. *In: Computational Linguistics, vol. 47, no. 2, pp. 255–308 (ISSN 1530–9312)*

Dias, G., Billot, S., Bassano, J.Cl., & Lopes, G.P. (2000). Combining Linguistics with Statistics for Multiword Term Extraction: A Fruitful Association? *Proc. RIAO 2000, 6th Int. Conf., College de France, France*

EAGLES (1998). EAGLES, 1998: Preliminary Recommendations on Semantic Encoding: Multiword Recognition and Extraction. http://www.ilc.pi.cnr.it/EAGLES96/rep2/node38.html *[EAG98]* https://www.ilc.cnr.it/EAGLES96/EAGLESLE.PDF

EAGLES (1999). EAGLES Evaluation Working Group, 1999: The EAGLES 7-step recipe. *Proc. MT Summit VII, Singapore*

Engel, G. (1994). DANtermBASE. *Proc. of the 3rd TermNet Symposium (Tama '94)*

Fersøe, H., (2004). Validation Manual for Lexica. *ELRA*

Francopoulo, G., Bel, N., George, M., Calzolari, N., Monachini, M., Pet M., & Soria, Cl. (2006a). Lexical Markup Framework (LMF) for NLP Multilingual Resources. *Proc. Multilingual Language Resources and Interoperability (MLRI'06), COLING 2006*

Francopoulo, G., ed. (2013). LMF Lexical Markup Framework, *ISTE/Wiley 2013*

Francopoulo, G., George, M., Calzolari, N., Monachini, M., Bel, N., et al. (2006b). Lexical Markup Framework (LMF). *Proc. 5th Int. Conf. on Language Resources and Evaluation – LREC 2006, Genoa, Italy*

Frantzi, K., Ananiadou, S., & Mima, H. (2000). Automatic recognition of multi-word terms: the C-value/Ncvalue method. *Int. J. Digit. Lib. 3, 2000: 115-130*

Gamallo Otero, P. (2008). Evaluating Two Different Methods for the Task of Extracting Bilingual Lexicons from Comparable Corpora. *Proc. Workshop on Comparable Corpora, at 6th Int. Conf. on Language Resources and Evaluation (LREC'08), Marrakech, Morocco Marrakech*

Gebruers, R. (1988). Valency and MT. Recent Developments in the METAL System. *Proc. 2nd Conf. Applied Natural Language Processing (ANLC), Austin, Texas, 1988, pp. 168–175.*

Goldmann, J., & Wehrli, E. (2001). FipsCo: A Syntaxbased system for Terminology Extraction. http://www.federation-nlp.uqam.ca/pdf/goldman.pdf

Heidemann, B., & Volk, M. (1999). Evaluation of Terminology Extraction Tools. *Report Univ. Zürich*

Hohnhold, I., (1984). Die Terminologie-Datenbank TEAM im Sprachendienst Siemens München *BDÜ-Mitteilungsblatt für Dolmetscher und Übersetzer 30 (1984) 3, p. 7.*

Ideue, M., Yamamoto, K., Utiyama, M., & Sumita, E. (2011). A Comparison of Unsupervised Bilingual Term Extraction methods Using Phrase Tables. *Proc. MT Summit XIII, Xiamen*

Kit, Ch. (2002). Corpus Tools for Retrieving and Deriving Termhood Evidence *Proc. 5th East Asia Forum of Terminology (Eafterm) 2002, Haikou*

Klementiev, A., & Roth, D. (2006). Named entity transliteration and discovery from multilingual comparable corpora. *In Proc. of the Annual Meeting of the North American Association of Computational Linguistics (NAACL), New York*

Knight, K., & Graehl, J. (1998). Machine transliteration. *Computational Linguistics, 24(4):599–612.*

Knops, E., & Thurmair, Gr. (1993). Design of a multifunctional lexicon. *in: Sonneveld, H., Loenig, K., eds. 1993: Terminology. Applications in interdisciplinary communication. Amsterdam (John Benjamins)*

Koehn, Ph. (2010). Statistical Machine Translation. *Cambridge University Press.*

Kohlschütter, Chr, Fankhauser, P., & Nejdl, W. (2010). Boilerplate detection using shallow text features. *Proc. 3rd ACM Int. Conf. on Web search and data mining, (WSDM '10), New York City, NY, USA, pp. 441–450*

Korhonen, A. (2002). Subcategorization Acquisition. *Univ. Cambridge Technical Report UCAM-CL-TR-530,* https://doi.org/10.48456/tr-530

Lardilleux, A., Yvon, F., & Lepage, Y. (2012). Hierarchical Sub-sentential Alignment with Anymalign. *Proc. 6th Annual Conference of the European Association for Machine Translation, pp. 279–286, Trento, Italy.*

Lieske, Chr. (2002). Pragmatische Evaluierung von Werkzeugen für die Term-Extraktion. *in: Mayer, F., Schmitz, Kl.-D., Zeumer, J., eds.: eTerminology, Köln 2002*

Lieske, Chr., McCormick, S., Thurmair, Gr. (2001). The Open Lexicon Interchange Format (OLIF) comes of age. *Proc. Machine Translation Summit VIII, Santiago de Compostela, Spain, 2001*

LIRICS (2006). LIRICS: Linguistic Infrastructure for Interoperable Resources and Systems. https://lirics.loria.fr/

Mayer, F., ed., (2001). Language for Special Purposes: Perspectives for the New Millennium. *Gunter Narr Verlag, 2001*

Menezes, A., & Richardson, St.D. (2001). A best-first alignment algorithm for automatic extraction of transfer mappings from bilingual corpora. *Proc. Workshop on Data-driven methods in machine translation – (DMMT '01) Vol. 14, pp. 1–8*

MILE (2001). Standards and Best Practice for Multilingual Computational Lexicons & MILE (the Multilingual ISLE Lexical Entry) https://www.w3.org/2001/sw/BestPractices/WNET/ISLE_D2.2-D3.2.pdf

Montiel-Ponsoda, E., Aguado de Cea, G., & McCrae, J. (2011). Representing term variants in lemon . *Proc. 9th Int. Conf. on Terminology and AI (TIA),*

OLIF (2008). OLIF Open Lexicon Interchange Format http://www.olif.net/

Pantel, P., & Lin, D. (2001). A statistical corpus-based term extractor. http://www.cs.ualberta.ca/~lindek/papers/ai01.pdf

Papavassiliou, V., Prokopidis, Pr., Thurmair, Gr. (2013). A modular open-source focused crawler for mining monolingual and bilingual corpora from the web. *Proc. 6th Workshop on Building and Using Comparable Corpora (BUCC), pp. 43–51, Sofia, Bulgaria, 2013.*

Pecina, P., Toral, A., Papavassiliou, V., Prokopidis, P., & van Genabith, J. (2012). Domain Adaptation of Statistical Machine Translation using Web-Crawled Resources: A Case Study. *In Proc 16th Conf. of the European Association for Machine Translation (EACL), pp. 145–152, Trento, Italy.*

Pecina, P., Toral, A., Way, A., Papavassiliou, V., Prokopidis, Pr., & Giagkou, M. (2011). Towards Using Web-Crawled Data for Domain Adaptation in Statistical Machine Translation. *Proc. 15th Conf. of the European Associtation for Machine Translation (EACL), pp 297–304, Leuven, Belgium.*

Preiss, J., Briscoe, T., & Korhonen, A. (2007). A System for Large-Scale Acquisition of Verbal, Nominal and Adjectival Subcategorization Frames from Corpora. *In Proc. of the 45th Annual Meeting of the Association of Computational Linguistics, pages 912–919*

Repar, A., Martinc, M., Ulčar, M., Pollak, S. (2021). Word-embedding based bilingual terminology alignment. *Proc. Conf. of Electronic Lexicography in the 21st century (eLex) 2021, virtual*

Sauron, V. (2002). Tearing out the Terms: Evaluating Terms Extractors. *Proc. Translating and the Computer 23 (ASLIB), London, UK*

Sérasset, G. (1993). Recent Trends of Electronic Dictionary Research and Development in Europe. *Technical Memorandum Electronic Dictionary Research (EDR), Tokyo, Japan.*

Skadiņa, I., Gaizauskas, R., Babych, B., Ljubešić, N., Tufiş, D., & Vasiļjevs, A., ed. (2019). Using Comparable Corpora for Under-Resourced Areas of Machine Translation. *Springer Intern. Publishing*

Steinberger, R., Pouliquen, B., & Hagman, J. (2002). Cross-lingual document similarity calculation using the multilingual thesaurus eurovoc. *Proc. 3rd Int. Conf. on Computational Linguistics and Intelligent Text Processing (CICLing), Mexico, , pp. 415–424*

Steinberger, R., Pouliquen, B., , Kabadjov, M., & van der Goot, E. (2011). JRC-Names: A freely available, highly multilingual named entity resource. *Proc. 8th Int. Conf. Recent Advances in Natural Language Processing (RANLP'2011), 104–110. Hissar, Bulgaria.*

TBX (2021). TBX - TermBase eXchange https://www.tbxinfo.net/

Thurmair, Gr. (1990). METAL: Computer Integrated Translation. *Proc. of a Workshop on Machine Translation, July 1990, UMIST*

Thurmair, Gr. (1997b). Ein multifunktionales Lexikon *Proc. GLDV 1997. in: Heyer, G., Wolff, Chr., ed. (1998). Linguistik und neue Medien, 10. Jahrestagung der GLDV, Leipzig. Deutscher Universitäts Verlag*

Thurmair, Gr. (2006). Exchange Formats: TBX, OLIF, and Beyond. *in: LDV Forum 1 July 2006*

Thurmair, Gr., & Aleksić, V. (2012). Creating Term and Lexicon Entries from Phrase Tables. *Proc. 16th Annual Conf. of the European Association for Machine Translation (EAMT), Trento, Italy,*

Ushioda, A., A. Evans, D., Gibson, T., & and Waibel, A. (1993). The Automatic Acquisition of Frequencies of Verb Subcategorization Frames from Tagged Corpora. *Acquisition of Lexical Knowledge from Text.*

Valderrábanos, A.S., Belskis, A., & Moreno, L.I. (2002). Multilingual Terinology Extraction and Validation. *Proc. Int. Conf. on Language Resources and Evaluation (LREC) 2002, Gran Canaria*

Vrochidis, St., Kompatsiaris, I., Casamayor, G. et al., (2015). MULTISENSOR: Development of multimedia content integration technologies for journalism, media monitoring and international exporting decision support. *IEEE Int. Conf. on Multimedia & Expo Workshops (ICMEW)*

Vu, Th., Aw, A.T., Zhang, M. (2008). Term Extraction Through Unithood And Termhood Unification *Proc. 3rd Int. Joint Conf. on Natural Language Processing: Volume-II, 2008, Hyderabad, India*

Wong, W., Liu, W., & Bennamoun, M. (2007). Determining Termhood for Learning Domain Ontologies using Domain Prevalence and Tendency, *Proc. 6th Australasian Data Mining Conference (AusDM'07), Gold Coast, Australia*
Wyhlidal WYHLIDAL Automotive, Engineering & Economics https://www.wyhlidal.com
XLIFF (2008). XLIFF (XML Localization Interchange File Format) http://docs.oasis-open.org/xliff/v1.2/os/xliff-core.html

GPSR Compliance

The European Union's (EU) General Product Safety Regulation (GPSR) is a set of rules that requires consumer products to be safe and our obligations to ensure this.

If you have any concerns about our products, you can contact us on ProductSafety@springernature.com

In case Publisher is established outside the EU, the EU authorized representative is:

Springer Nature Customer Service Center GmbH
Europaplatz 3
69115 Heidelberg, Germany

Batch number: 10584686

Printed by Printforce, the Netherlands